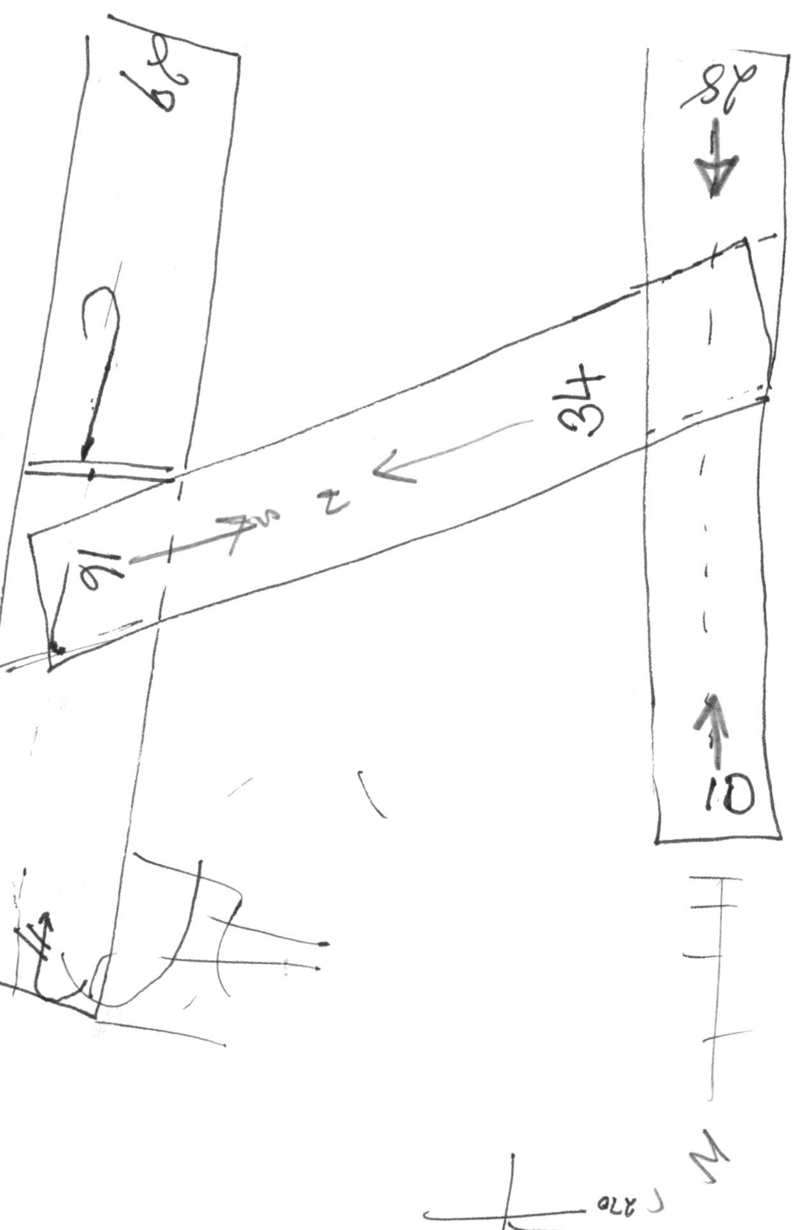
34
10

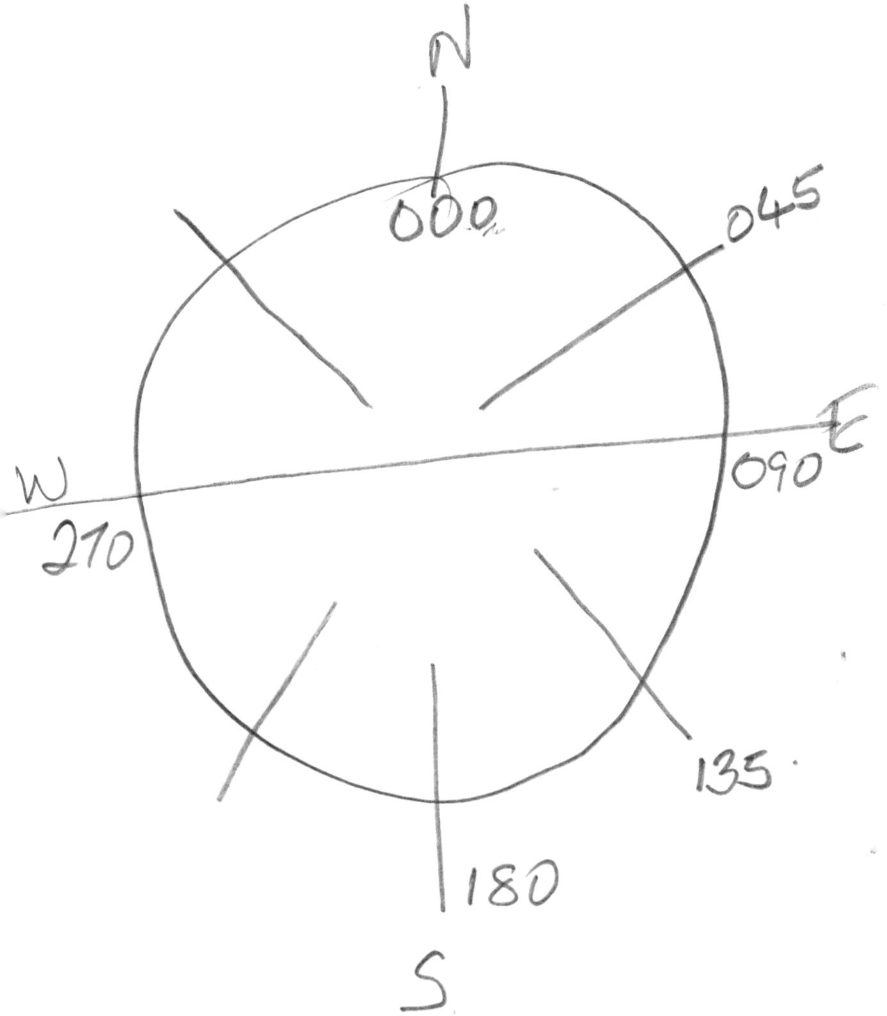

N

000

045

E

090

W

270

135

180

S

Books on Aviation by N. H. Birch and A. E. Bramson

Flight Briefing for Pilots series
Flying the VOR
A Guide to Aircraft Ownership
The Tiger Moth Story
Captains and Kings
Radio Navigation for Pilots
Flight Briefing for Microlight Pilots
Flight Emergency Procedures for Pilots

By A. E. Bramson

Be a Better Pilot
Make Better Landings
Principles of Flight (Audio-visual trainer)
The Book of Flight Tests
Master Airman, a Biography
of Air Vice Marshall Donald Bennett

By N. H. Birch

The Instrument Rating
Passenger Protection Technology
in Aircraft Accident Fires.

THE ROLE OF A SYLLABUS IN FLYING TRAINING

Most states have an aviation authority responsible for, among other matters, deciding which subjects must be understood by those training for a pilots' licence. Over the past quarter of a century the range of these subjects has grown considerably so that learning to fly, even when the student has no intention of becoming a professional pilot, is no longer as simple a matter as it was prior to the 1939–1945 war.

To ensure that all subjects and items within those subjects are taught during the course the value of a syllabus of training has in recent years assumed growing importance. It may be regarded as a shopping list of knowledge to be acquired but it has more to offer than that. A syllabus of flying training ensures:

(a) that the concept of the course is understood by the student and the instructor.
(b) that the content is known by all concerned with the course of training.
(c) that progress during the course may be monitored.
(d) that space is provided where the student may record the instructor's comments after each lesson. These are of value during revision and subsequent practice.
(e) that a summary of learning and practice is available to both student and instructor, thus ensuring that the course has been completed.

Flight Briefing for Pilots, Vols. 1, 2 and 3, have been written to conform with a Syllabus of Flying Training for the Private Pilots' Licence and the IMC Rating (currently confined to the UK) that has been recognised by the UK Civil Aviation Authority. The syllabus is published by Longman.

Throughout the text headings that cross refer to the syllabus are printed in the following type:

[Instrument Indications]

Flight Briefing for Pilots

Volume 1

An introductory manual of flying training complete with air instruction

N. H. Birch MSc, PhD MRAeS
Liveryman of the Guild of Air Pilots and Air Navigators

and A. E. Bramson FRAeS
Past Chairman of the Panel of Examiners
Liveryman of the Guild of Air Pilots and Air Navigators

Illustrated by A. E. Bramson

FIFTH EDITION

PAN is from the French
PANNĒ = BREAKDOWN

Longman Scientific & Technical
Longman Group UK Limited,
Longman House, Burnt Mill, Harlow
Essex CM20 2JE, England
and Associated Companies throughout the world.

First published in Great Britain 1961
Reprinted 1963, 1965, 1966, 1967, 1968
Second Edition 1970
Reprinted 1971, 1972, 1973
Third Edition 1974
Reprinted 1976, 1977
Fourth Edition 1978
Reprinted 1980, 1982 (twice), 1983 (twice), 1984, 1985
Fifth Edition 1986
Reprinted 1987

British Library Cataloguing in Publication Data

(Birch, N. H.
Flight briefing for pilots.–5th ed.
Vol. 1: An introductory manual of flying complete with air instruction
1. Airplanes–piloting
I. Title II. Bramson, A. E.
629.135′52 TL710

ISBN 0-582-98814-4

Set in 10/12pt AM Compedit Times
Produced by Longman Singapore (Pte) Ltd.
Printed in Singapore

Contents

The numbers of the exercises given above (and repeated at the top of each page) conform to the British Flying Training Syllabus. Australian, Canadian and South African student pilots will find it useful to relate these to their own exercise numbering which appears in Appendix 3.

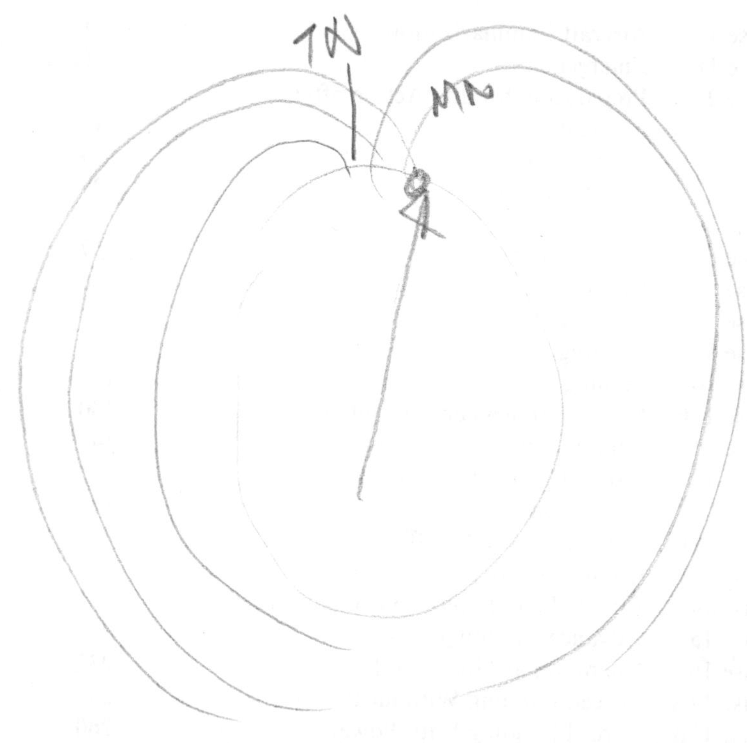

Preface

The value of formal pilot training geared to a proper syllabus was probably first demonstrated in 1916 when Lt. Smith-Barry was recalled from the Western Front during World War I and given a free hand to run No. 1 Reserve Squadron, Gosport. Here he put into operation his then very advanced ideas on flying training.

Since that time most of the Air Forces of the world have developed their flying training using the 'Gosport System' as a basis. However, although many countries produce excellent flying training manuals for their service pilots, often the civil trainee has had to make do with little more than typed notes.

At most of the smaller flying schools the main burden of training occurs during weekends and it is a long-established fact that there is a tendency on the part of hard-pressed flying instructors to take their students on air exercises before ground instruction of any kind has been given. Time and again such a practice has proved to be thoroughly bad from every point of view. It was to fill the need for a simple manual of flying training that the authors wrote Volume 1 of the *Flight Briefing for Pilots* series. The book is strictly a pilot handling manual; everything possible has been done to ensure simplicity and all background information has been confined to essentials. Those looking for complicated mathematical formulae should seek them in the various excellent books written for aerodynamicists and aircraft designers—other than a few simple graphs in the appendix this work is essentially practical in its approach to the subject.

The purpose of this book is twofold. It gives the student pilot an easy-to-read explanation of the various exercises needed to gain a Pilot's Licence. Its second function is to provide a flight practice section for each air exercise. While this step-by-step handling sequence is useful reading for the student its main value is to provide the flying instructor with a basis for his air instruction or, to use the correct term, 'patter'; this has to be synchronized with the various demonstrations in the air. Consequently Volume 1 should prove a valuable work of reference to those studying for a Flying Instructor's Rating. The syllabus and flying sequences conform to those approved by the UK Civil Aviation Authority. The book should be read with Volume 2, the ground subjects manual.

Wherever possible, the authors have tried to conform with the training techniques recommended by the UK Panel of Examiners, a body appointed by the Civil Aviation Authority and responsible for the testing of civil flying instructors in Britain. At the same time they would like to express their very real appreciation for the help and advice they have received from the Royal Air Force Central Flying School and individual members of the Panel of Examiners who have, from time to time, offered valued advice.

Volume 1 first appeared in 1961 and was joined by other manuals to form an eight-volume series. In some cases, these were published in order to replace out-of-date publications and, in consequence, the *Flight Briefing for Pilots* series was not numbered in a logical order.

The authors have now revised the series, dividing it into two, self-contained sets: Volumes 1, 2 and 3 for the private pilot and Volumes 4, 5 and 6 devoted to advanced studies for pilot improvement or professional qualifications. By 1985 the number of volumes in use had reached a total of 400,000 copies and since many of these have found a home many miles from their native land the authors have produced this completely revised edition which, it is hoped, will be more suited to international needs.

N.H.B.
A.E.B.

This book has been written to conform with a UK Civil Aviation Authority recognised syllabus of training for the Private Pilot's licence published by Longman.

IMPORTANT: *Note to Flying Instructors on how to use the Manual*

Volume 1 is a combined student/instructor flying training manual which is confined to air instruction and essential background information. Each exercise is divided into the following sections.

Background Information

This section should be read by the student before flight. It explains, in the simplest terms, aerodynamic and handling considerations for the exercise being described. Only the essential information listed in the CAA-recognized syllabus published by Longman, with which this book conforms, is included. This part of the book expands your pre-flight briefing. The detail ground subjects are in Volume 2 of this series.

Pre-flight Briefing

The pre-flight briefings will act as a guide for your 10–15 minutes 'what we are going to do and how we are going to do it' sessions before each flying lesson. They are arranged in a layout which will make good use of the blackboard. Pre-flight briefings are essential. They prepare the student for what is about to be demonstrated in the air, usually against a background of noise. Also, some students are inclined to be apprehensive in the early stages of flying training and a preview of the flying lesson has been proved without doubt to enhance the value of air instruction. During the pre-flight briefing the instructor should encourage the student to ask questions, assess the weather and comment on its suitability for the exercise in hand. The student should not hesitate to ask questions at any stage of the briefing.

Flight Practice

The final section of each exercise contains step-by-step air instruction. Although it is primarily intended to form the basis of the flying instructor's 'patter' and ensure that the demonstrations in the air follow the standard pattern, student pilots should be encouraged to read it before flight. This will prepare the student for the air exercise and later help in fault analysis.

Instructors are not intended to learn and repeat the patter word for word. Its main purpose is to present the demonstrations in a logical and concise manner while providing a basis for their own form of words.

Exercise 1

Aircraft Familiarization

Background Information

This will probably be the student's first introduction to a small aircraft and although at first impression a light trainer might appear to be somewhat frail its strength should not be underestimated. Modern light aircraft are stressed to withstand 4.5 normal loads with a 1.5 times safety factor while those capable of aerobatics are designed to accept six times normal load with a 1.5 safety factor. In effect this means that the airframe can cope with loads of up to nine times those of normal level flight.

Introduction and examination of aircraft, Internally and Externally

The Instructor will draw attention to the external features of the **Airframe,** its **Flying Controls** and the **Pitot Head** or **Pressure Head** and **Static Vent** which sample air for the airspeed indicator, altimeter and vertical speed indicator. The engine installation and method of checking oil level will be demonstrated and the dangers of turning the propeller without first confirming that the switches are off will be emphasized.

The instructor will then explain the correct method of entry/exit, pointing out areas of the wing to be avoided when placing the feet.

Explanation of cockpit layout

Before the student is able to assimilate flying instruction it is imperative that the cockpit layout – position of all controls and instruments – should be fully understood.

Most modern aircraft are equipped with a **Basic 'T' Flight Panel,** i.e. flight instruments laid out in an internationally agreed standard order of presentation. Although instrument flying is an additional skill and a qualification to be added at a later stage, modern flying training introduces instrument awareness at an early stage.

Aircraft controls, equipment and systems

Aircraft used for basic flying training are usually limited to the simplest of controls. These, and the fuel system, will be explained by the instructor along with the electric services, the **Battery Master Switch** (similar to the main switch on a domestic fuse panel) and the ignition/starter switch. The master switch in modern light aircraft is usually in two adjacent halves. This is known as a split rocker-type switch and the right half will be marked BAT since it controls all the electric power to the aeroplane. The left half will be marked ALT. It controls the alternator. Normally in flight both halves of the switch will be on. The ignition is usually controlled by a key-type switch similar to that used in a car. The electric starter may be incorporated with that switch or there may be a separate starter button.

It is standard practice to fit two separate ignition systems to aero engines (i.e. two magnetos and two sets of sparking plugs). Apart from the obvious consideration of safety, two plugs per cylinder in engines of large swept volume provide better ignition and more efficient combustion.

Although the engine is fitted with a mechanically driven fuel pump, similar in concept to that used in some cars, in furtherance of the underlying principle of safety that is a part of aviation, aircraft of low-wing design, where the fuel tanks are at a lower level than the engine, have an emergency electric fuel pump which is switched on during the take-off and landing to guard against possible failure of the engine-driven pump. It is also used for supplying fuel to the carburettor prior to starting.

Operation of the brakes will be explained during this introduction and the student will be shown the location and method of operation of all safety equipment (first-aid kit, fire extinguisher and, when water is to be crossed, life jackets).

Drills and necessity for systematic checks

All flying is based upon careful checks before important actions are taken. The average motorist will jump into a car without checking if the tyres are properly inflated. Few even take the trouble to ensure that all doors are properly closed. **Such attitudes are not good enough in aviation, an activity where even amateur pilots must have a professional approach to the task.**

Before flight, before starting the engine, pre-take-off checks and vital actions, to mention some, are remembered, in the case of simple aircraft, with the aid of mnenomics – a sequence of actions recalled with the aid of a word or combination of letters. These will be mentioned during the flying exercises that follow.

In the case of more complex aircraft **Check Lists** are used. Indeed, many flying schools adopt these even with simple training aircraft, but they must be used carefully. When check lists are to be used as a matter of policy on no account may the pilot attempt to conduct the pre-flight inspection or any other checks partly from memory and partly with the aid of a check list. That way important items can be forgotten. Systematic checks are an essential part of good airmanship; they should be regarded as 'The Pilot's Best Friend' and never taken for granted or treated lightly.

Check lists, fixed notices, aircraft/operator's manuals

Although the titles may vary (Owners Manual, Operating Handbook, Flight Manual) in essence these books are compiled by the manufacturers and, in some cases, are approved by the airworthiness authorities concerned. Suggested check lists are included in these books although many flying schools compile their own, some of them based on RAF practices. The manuals are compiled in a standard format as shown in Table 1.

Table 1

Section	*Title*	*Purpose of Section*
1	General	Description of engine and propeller type, aircraft weight and dimensions
2	Limitations	Engine limits (RPM, temperatures), airspeed limits, C of G limits, loading and other limits. List of such information affixed in view of the pilot in the cabin. These are called **Placards**
3	Emergency Procedures	Actions to be taken during an engine failure, fire, icing, loss of electrics, etc.
4	Normal Procedures	Recommended actions from pre-flight to engine shut-down. The first part of this section is in check list form; the second part amplifies each procedure
5	Performance	Full details of the aircraft's take off, climb, cruise, gliding and landing performance which may be shown in the form of tables or graphs
6	Weight and Balance	Full details of aircraft weight breakdown, methods of calculating centre of gravity and relating it to the prescribed limits for safe flight
7	Aircraft and Systems Description	Description of airframe, all systems and flying/engine controls
8	Handling, Service and Maintenance	Description of servicing schedules, ground handling, cleaning care, etc.
9	Supplements	Description of optional equipment and how it is used (radio, etc.)

These notes are of necessity written in broad terms and the student pilot is well advised to learn all he can about the aircraft which is to be his classroom during the flying course. The aircraft manual just described is an excellent source of information.

Pre-flight Briefing

Exercise 1. Aircraft familiarization

AIM — *To introduce the student to the external and internal features of the aircraft, the location and function of the controls and safety precautions while on the ground*

AIRMANSHIP — *No smoking; nearby aircraft; handling of aircraft; correct entry/exit method*

GROUND EXERCISE

External features	*Internal features*	*Check lists*
Wing, tail surface and fuselage *Undercarriage/brakes* *Engine, cowling, oil check* *Propeller and handling precautions* *Flying controls and flaps* *Pitot cover and others* *Ground handling, steering bar* *Tying down, use of chocks/covers* *Risk of smoking near aircraft*	*Location of steps/walkways* *Correct entry and exit* *Seating and harness, adjustment* *Baggage area* *Emergency equipment/first-aid box* *Use of control locks* *Flying and engine controls* *Electric switches* *Cabin heat and fresh air* *Instrument layout* *Radio installation*	*Types of checklist* *Use on the ground* *Use in the air* *Mnemonics* *Importance of using proper sequence*

Exercise 1E

Emergency action in the event of fire

Action in Event of Fire on the Ground and in the Air

Fire in a modern aircraft is today an unusual occurrence, high standards of engineering and component efficiency making the risk of fire even more remote than the possibility of engine failure. Nevertheless, however unlikely its occurrence may be fire on the ground or in the air should be looked upon as a form of lifeboat drill which is an essential part of the pilot's training.

Broadly speaking fires may be considered under two headings:

1 – Those which occur in the cabin or fuselage.
2 – Engine fires.

Cabin or fuselage fires may be caused by electrical faults. Since fuel vapour associated with the engine priming system may be present in the cockpit, under no circumstance should smoking be permitted unless the aircraft type allows.

Isolation of Fuel and Electric Systems

Under normal circumstances the origin of cabin fires may be traced by the type of smoke or fumes, electrical faults usually producing a smell of burning rubber or insulating material whereas battery shorting gives off an acrid smell. After detection by the occupants, electrical faults may be pinpointed by the smoke, and then isolated with the relevant switch although usually the circuit fuse will have 'blown' as soon as the fault developed. When the main circuit is at fault the battery can be isolated by turning off the master switch.

Non-electrical fires in the fuselage can be handled with the cabin fire extinguisher which should be removed from its bracket and directed by hand.

Most likely there will be a considerable amount of smoke/fumes as a result of the fire and steps will have to be taken to ventilate the cabin, either by opening a window/direct vision panel or by operating the fresh air vents.

Fires in the engine may be caused by faulty induction or exhaust systems, over-priming during starting, and fuel or oil leakages under pressure. Larger aircraft are equipped with a fire-warning system in conjunction with a spray ring which is arranged to discharge over all vulnerable parts of the engine.

All engines are separated from the airframe by a fireproof bulkhead which prevents the engine fire from spreading to the cabin, fuselage or wings.

On the ground the minor fire caused by over-priming before starting already mentioned can be put out by smothering the flames with a folded tarpaulin which should be held against the air intake.

In the air more serious fires caused by oil or petrol pipe fracture are complicated by the fact that while the engine is turning it will continue to pump petrol and oil. In the case of a petrol fire, turn off the petrol whether on the ground or in the air. In this way the carburettor fuel pumps and exhaust system will be cleared of petrol and only when the engine stops running should the ignition be switched off. Usually it is best to glide ahead and contain the flames behind the fire wall, but when the need arises a sideslip will keep flames away from vulnerable areas. The cabin heater vent must be closed to exclude fumes.

An oil fire can be recognized by the associated smoke. While the action of turning off the fuel will eventually stop a petrol fire in the engine, an oil fire may continue so long as the propeller rotates the engine causing it to pump oil under pressure. In the case of an oil fire, rotation should be stopped by switching off the ignition and holding up the nose so that the airflow is insufficient to turn the engine.

It will be appreciated that an engine fire can be quite minor, in which case the foregoing procedure would quickly bring it under control.

Never attempt to re-start an engine which has caught fire in the air.

Note. Exercise 1E should only be practised when accompanied by a flying instructor and then at a safe height and within gliding distance of an aerodrome.

Inspection of Safety Equipment, Stowage and Use

A real emergency is no time to discover that the location of safety equipment and how to use it are both unknown.

Before flight the pilot must confirm that a first-aid kit is in the aircraft and the fire extinguisher must be inspected to see that it is serviceable. Some exinguishers have a small pressure gauge marked

RECHARGE, 150, OVERCHARGE. The pointer should indicate 150 (or whatever pressure figure is printed on the dial). The extinguisher should carry a date of last inspection.

The pilot must also learn how to remove the extinguisher from its bracket, how to discharge it and direct the spray. A real fire is not a good time to be caught by surprise when the jet emerges in a totally unexpected arc that misses the source of the trouble.

Exercise 1E. Action in the event of fire

AIM — *To adopt the correct procedure in the event of an aircraft fire, on the ground or in the air*

AIRMANSHIP — *Precautions before practising emergency in the air:*
Height: sufficient for exercise
Airframe: appropriate to phase of flight (climb, cruise etc.)
Security: harness and hatches secure, understanding of procedure
Engine: temperatures and pressures normal. Carb. heat before closing throttle
Location: not in controlled airspace, position known, clear of cloud
Lookout: not above an airfield or other aircraft

GROUND AND AIR EXERCISE

Fire on the ground		*Fire in the air*	
Identifying source	*Action*	*Identifying source*	*Action*
Cabin fire (electric) *Shut down engine*	*Master switch OFF* *Use extinguisher (if necessary)* *Exit aircraft*	*Cabin fires*	*Handle as for ground fires but land as soon as possible*
Cabin fire (non-electric)	*Shut down engine* *Use extinguisher* *Exit aircraft*	*Engine fire (petrol)*	*Throttle CLOSED* *Fuel OFF* *Cabin heat OFF* *Ignition OFF when engine stops running* *Forced landing*
Engine fire petrol/oil	*Fuel OFF* *Extinguisher ready* *Ignition OFF when engine stops running* *Exit aircraft*	*Engine fire (oil)*	*As for petrol, then stop rotation* *Sideslip fumes away* *Forced landing*

Flight Practice

COCKPIT CHECKS

(*a*) Trim for level flight.
(*b*) No insecure items in the aircraft.
(*c*) Carburettor heat control as required.

OUTSIDE CHECKS

(*a*) Altitude: sufficient for the manoeuvre and recovery.
(*b*) Location: not over towns, other aircraft or airfields or in controlled airspace.
(*c*) Position: within gliding distance of an aerodrome.

AIR EXERCISE

(For the purpose of this demonstration imagine a fire has occurred in the engine while in flight.)

Isolation of Fuel and Electric Systems

(*a*) Throttle **Closed**.
(*b*) Petrol **Off**.
(*c*) Close the cabin heater vent.
(*d*) Wait for the engine to stop running under its own power then turn the ignition **Off**.
(*e*) If an oil fire is suspected stop the engine rotating by holding up the nose of the aircraft and reducing airspeed (simulate).
(*f*) If necessary sideslip the flames and/or smoke away from the cockpit.
(*g*) Should the fire persist operate the extinguisher.
(*h*) When a smell of burning rubber or insulation indicates an electric fault check the circuit breaker. If these are all in place and the symptoms persist turn off the battery master switch.
(*i*) Following any fire, from whatever cause, land at the first opportunity. Advise Air Traffic Control.

Exercise 2

Preparation for and Action after Flight

Background Information

Flight Authorization and Documentation

The details of a training flight are entered in the Flight Authorization Book and in the case of a student or any pilot undertaking a solo flight he will sign in the appropriate place indicating that he understands the details of the exercise his instructor requires him to practise.

Pilot's Responsibilities

It is also the pilot's responsibility to check that the aircraft is serviceable and fuelled, by reference to the Record of Serviceability and/or by confirmation with the instructor or engineer in charge. He must also ensure that the aircraft is not overloaded and that it is within its centre of gravity limits (see Airfield Performance, Weight and Balance, Chapter 8, Volume 2). The pilot must check the weather before flight and liaise with ATC.

External and Internal Checks

These commence as the pilot walks towards the aircraft. If it is standing one wing low there could be a flat tyre or a damaged/unserviceable undercarriage leg.

Before boarding the aircraft the pilot must inspect it, commencing by checking that the aircraft is suitably positioned with the tail pointing away from nearby aircraft or buildings. After starting the slipstream can cause annoyance or even damage. The ignition switches must be in the 'off' position and when applicable it is good practice to remove the ignition key as a double check. Turn on the battery master switch, note fuel contents for later comparison with a visual check of the tanks, check the anti-collision beacon/strobes. The pitot heater should be turned on for a short period so that it can be touched during the walk-around and its heating action confirmed.

With most aircraft the flaps may better be inspected if first they are partly lowered to reveal their operating links and brackets. The master switch should then be turned off and the control lock(s) removed. Now that the preliminary checks have been completed the **External Checks** can begin. Standard practice is to commence the external aircraft checks at the point of entry since the inspection will finish at this point, when the pilot can then enter the aircraft after a systematic check. Particular points for attention are listed in the aircraft manual, and will include freedom of the structure from damage, cracks and corrosion. Additionally, the security of cowlings, hatches, fuel and oil filler caps should be checked. Oil and fuel contents should always be checked visually.

When an aircraft has been standing overnight there is a risk of condensation within the fuel tanks, when moisture from the air above the fuel will collect at the bottom of the tank in the form of water. This is most likely to occur when the tanks are only partly filled. The risk is obvious – water in the carburettor, being thicker than fuel, may block the jets and stop the engine. To safeguard the carburettor and fuel lines from water and sediment contamination **Fuel Strainers** are fitted to the tanks and the lowest part of the fuel system. These must be drained into a glass or clear plastic jar before the first flight of the day, and the fuel inspected for water. Fuel is coloured (green or blue) and, being lighter, it will float on water which is, of course, colourless. The strainers should be operated *before* re-fuelling has mixed the contents of the tank and dispersed the water within the petrol. Having operated the fuel strainers, be sure that they are properly closed before entering the aircraft, otherwise the tank will lose its contents through the strainer instead of supplying the engine. Tyres should be inspected for obvious loss of pressure and **Creep.** Creep is the movement of the tyre in relation to the wheel and marks are painted on both which will be out of line when creep has occurred, with possible damage to the valve. The pitot covers (and when applicable to the aircraft type, static vent plug) should be removed and stowed in the correct place.

The condition of the propeller must be inspected and it should then be turned several times by hand before the first flight of the day. This is particularly important under cold conditions, assisting starting and providing an opportunity to check cylinder compressions. In the case of a radial or inverted engine which may be encountered at a later stage of flying, this precaution will obviate the possibility of damage from hydraulic locking caused by oil or petrol drainage into the inverted cylinders.

When the generator or alternator is driven by a belt, its tension and condition should be checked while at the front of the aircraft.

After entering the aircraft the pilot should ensure that he will be comfortable in flight and adjustments to the rudder pedals, safety harness, seat position and/or the number of seat cushions, etc., should be made. This is of paramount importance. It is essential that the doors are properly closed and a careful check must be made to ensure that the door catch(es) are engaged. A door that opens in flight can seriously disturb the airflow and some aircraft have been known to become unmanageable in these circumstances.

The pre-starting **Internal Checks** are commenced by moving all controls to ensure full and free operation in the correct sense. When the rudder pedals are connected to the nose or tailwheel this check may have to be delayed until the aircraft is moving. The instruments should be examined for any obvious signs of unserviceability. e.g. broken glass, altimeter which will not set correctly, etc.

Unless it is possible to use an external battery, all unnecessary load (radio, anti-collision beacon, etc.) should be switched off before starting so enabling the aircraft battery to provide the high current needed to turn the engine.

Starting Procedure and After Start Checks

The procedure for starting, which depends upon the type of installation, is laid down in the operating manual for the aircraft type. There are still light aircraft without mechanical starting and these must be turned by hand. This is known as **Swinging the Propeller** or **Hand Starting** and it is described in Chapter 20.

Before attempting to start any aircraft the pilot must ensure that it is parked on firm ground, free from stones which may be lifted by the propeller, with risk of damaging its blades. As previously mentioned, it must be positioned so that the tail is not pointing at other aircraft, open windows or hangar doors.

Fuel Management

The fuel system may be confined to a single tank with a simple ON/OFF selector. When two tanks are fitted it is good practice to start the engine on the tank with the lower fuel contents, taxi to the **Holding Point** (an area near the start of the take-off where aircraft may carry out their power checks) and then change to the other tank for engine checks. That way the fuel system will be checked for satisfactory fuel flow.

In Volume 2, Chapter 7, the need for a back-up electric fuel pump in some aircraft is explained. Generally these are fitted to aircraft of low-wing design where the fuel tanks are below engine level. Prior to starting the fuel line to the carburettor must be filled and when an electric pump is fitted this is switched on until fuel pressure is indicated on the **Fuel Pressure Gauge**. It may then be turned off to relieve the battery of all unnecessary load during starting. High-wing aircraft enjoy the benefit of gravity feed from the tanks which are located in the wings, consequently they are not always fitted with an electric fuel pump unless the engine is fuel injected.

When starting a car with a cold engine the choke must first be operated to provide a suitable fuel/air mixture. Some light aircraft, particularly motor-gliders, are powered by converted car engines which retain a choke for starting, but purpose built aero engines often utilize a **Primer** (described in Chapter 7, Volume 2). This takes the form of a small, plunger-type pump located on the instrument panel.

The number of priming strokes to be used when starting a cold engine depends on the outside air temperature; the lower the temperature the more priming strokes are required. Two to four strokes would be normal under average conditions although up to six may be required on a very cold day. The primer is operated as follows:

1. Release the plunger by rotating it slightly until it may be drawn back.
2. Pause a few seconds to allow the pump to fill.
3. Push the plunger fully forward to inject fuel into the engine induction system.
4. Repeat as required according to temperature conditions.
5. Rotate the plunger to lock it in the stowed position. Failure to do this can result in disturbed fuel mixture and rough running of the engine in flight.

Aircraft without primers.

Some light trainers are not fitted with a priming pump. In these cases use is made of the small accelerator pump which is built into some carburettors for the purpose of improving throttle response when the lever is moved forward to provide more power. In such aircraft priming is achieved by opening and closing the throttle several times. Warm or hot engines are not primed before starting.

Operating the electric starter.

Having completed the checks previously explained, obtained fuel pressure (when such a gauge is fitted) and primed the engine, the final act of starting is to turn on the ignition and operate the starter.

Electric starters make heavy demands on the battery and common sense demands that it should be free of unessential load before it is called upon to turn over the engine. The practice of turning on the anti-collision beacon prior to starting is therefore to be discouraged. Airline pilots do this to warn that they are about to start the engines. They have an external power source; light aircraft rarely enjoy this facility.

The throttle should be set slightly open ready for starting. People can move within striking distance of the propeller while the internal checks are in progress so before the engine is started the pilot must look from left through straight ahead to fully right to ensure that the propeller is clear. A warning should then be shouted:

'CLEAR PROP'

The ignition, usually a car-type key switch, is turned through its three positions:

RIGHT MAG – LEFT MAG – BOTH MAGS

then further rotated to operate the starter and turn the propeller. Some aircraft have a separate starter button but in either case the starter must not be operated for more than a few seconds at a time. If the engine does not fire immediately the reason may be:

Insufficient Mixture. In this case the engine will need more priming, particularly in cold weather.

Over-rich Mixture. Over-priming can cause a variety of minor problems, among them 'wetting' of the plugs, which will prevent ignition. In such cases the engine must be cleared of excessive mixture. To do this:

(*a*) Fully open the throttle.
(*b*) Operate the starter, turning over the engine until the over-rich mixture is blown out of the cylinders.
(*c*) Bring back the throttle, start the engine and adjust it to 1000 RPM.

Individual engines vary slightly and the flying instructor will advise on the amount of priming required for the aircraft in use.

After Start Checks

Immediately the engine fires and starts running the throttle should be adjusted to allow idling at 1000 RPM. The oil pressure gauge must be checked and if no pressure is indicated within 30 seconds (or as specified in the aircraft manual) the engine must be shut down without delay to prevent serious damage.

To prevent possible damage to the electric system as a result of a current surge the generator/alternator will be OFF during engine starting and this must now be brought on line by turning on the generator switch. The ammeter should then show a charge.

Some modern aircraft have an **Over-Voltage Warning Light** which illuminates in the event of a momentary high-voltage situation. The light confirms that the alternator over-voltage protection system has operated and the battery is now supplying current to the various services (instruments, lights, radio, etc.). Unless there is a serious fault causing an over-voltage situation the alternator may be brought back on line by switching off the master switch and switching it on again. The red light, usually placarded HIGH VOLTAGE should then go out.

With the engine running the instruments should be scanned to confirm correct suction for the attitude indicator and the direction indicator. Both the cylinder head temperature gauge (when fitted) and the oil temperature gauge will be showing readings as the engine warms although it is convenient to allow the engine to reach its working temperature while taxying out for take-off. The radio may be switched on at this stage and testing calls made to the tower. Also the anti-collision beacon/strobes should be switched on. 'See and be seen' is a slogan worth remembering during all phases of aircraft operation, on the ground as well as in the air.

It is usual to test the engine near the **Holding Point** prior to taxying onto the runway. By then the engine should have reached its working temperature.

Power Checks

To ensure adequate engine cooling the aircraft must be parked into wind prior to opening up the throttle. When there are two fuel tanks the selector should be moved to the fuller one *before* the run-up. On no account may the selector be changed immediately prior to moving onto the runway for take-off. If there is a fault in the fuel system the engine could stop immediately after lift-off or at some other critical phase of departure.

With the parking brake on, each magneto should be checked at idling speed to ensure proper function and to avoid unnecessary stress caused by misfiring at higher power settings. For the first flight of the day the throttle should be smoothly advanced to the fully open position, first checking oil and engine temperatures and then that full power is being developed. The aircraft manual will give the **Static Revs** that should be indicated on the engine speed indicator. Maximum power should only be applied for a few seconds before bringing back the throttle to the figure specified in the aircraft manual for magneto testing. The engine speed should be checked for RPM decrease as each magneto is tested in the following manner:

Ignition to RIGHT magneto. Check RPM decrease.
Ignition to BOTH (to clear the plugs that were switched off).
Ignition to LEFT magneto. Check RPM decrease.
Ignition back to both for normal running.

While recommended figures vary for different engine types typical readings for a light trainer would be:

Magneto check	engine at 1700 RPM
Maximum decrease/magneto	125 RPM
Maximum difference between magnetos	50 RPM

While the engine is running at magneto check speed the carburettor heat should be applied. There will be a small decrease in engine speed when it is operating correctly. By now the engine will have warmed up ready for flight and the engine temperature and pressure instruments should be giving normal readings. These are usually identified by green sectors marked on the instrument faces. These small instruments are not to be ignored and an abnormal reading on the ground should be regarded as a compelling reason not to take off until the fault has been investigated.

Finally the services (suction, electric) should be checked at the same speed before moving the throttle fully back to the idle stop to ensure that the engine continues to idle at low speed (typically 500–600 RPM). The engine should not stop when the throttle is closed.

After the first flight of the day the full power check is omitted but all other checks must be observed.

Running-down Procedure

To cater for uneven expansion/contraction rates of the different alloys and metals used in an engine it is essential that, after flight, the proper run-down procedure is followed. By adhering to this simple practice uneven cooling, leading to expensive damage and premature engine change, will be avoided.

The engine should be allowed to idle at 800–1000 RPM for several minutes, thus bringing it and the engine oil to an even temperature. Often the run-down can be enacted while taxying back to the parking ramp provided hard taxiways are used and low power is required to keep the aircraft rolling. Taxying on soft grass often demands very high power settings and on such airfields the need to idle the engine prior to shut-down is of paramount importance.

While the engine is idling at 800–1000 RPM each magneto should be tested to ensure proper function. The pilot is not looking for a specific mag drop; a dead cut would reveal a failed magneto while no RPM decrease could mean that the magneto which is supposed to be OFF is, in fact live. An aircraft with a live magneto represents a potential danger – the propeller, if turned by hand, could start the engine unexpectedly and cause a serious accident. A suspected live magneto must be reported to the maintenance engineer.

The engine is stopped by operating the **Idle Cut-off,** a valve built into the carburettor which, when operated, starves the engine of fuel. The idle cut-off comes into effect when the mixture control is moved to the fully WEAK position (i.e. pulled out fully in the case of a knob-type control or moved fully back in the case of a mixture lever).

After the engine has stopped the ignition should be turned OFF without delay and, as an added safety precaution, the ignition key is removed. All switches, radio, battery master, generator/alternator etc. are turned OFF, for prolonged parking the fuel is turned OFF, the parking brake is checked ON and then the control lock is placed in position prior to leaving the aircraft.

Parking and Security

Ideally aircraft should be kept in a hangar but when one is not available the most sheltered position is obviously preferred.

Many of the precautions to be taken are no more than a common-sense application of practices already explained in this chapter. For overnight parking the ignition, master switch and fuel selector must all be OFF. The parking brake must be ON and the control lock(s) will be in position. When the aircraft is not fitted with purpose designed locks the elevators and ailerons may effectively be safeguarded by tying back the stick or control wheel with the safety harness.

The pitot cover must be in place to protect the pressure instruments and when the aircraft will be subjected to strong sunlight a canopy cover will avoid 'crazing' of the clear plastic and protect temperature sensitive radio equipment from excessive heat which can damage its performance.

Tie-down points are located under each wing and below the tail surfaces, usually under the rear fuselage. These must be used to lash the aircraft, either to permanent tie-down points on the ground or to concrete blocks provided for the purpose. The stronger-than-forecast, overnight wind has damaged many light and some not so light aircraft in the past.

Headsets and other valuable, moveable equipment should be removed from the cabin which should then be locked.

Post-flight Documentation

At the end of the flight the Authorization sheet must be signed in the appropriate column following a training detail or a flight from the flying school and back to base. Most flying schools keep a 'snag' book for recording defects and any malfunctions should be listed therein for attention by the maintenance engineers. This applies to the airframe, engine or radio equipment.

Few pilots enjoy admitting a heavy landing but a series of undetected incidents of this kind can lead to fatigue and the collapse of, for example, the nosewheel strut. Often this has been known to occur when another pilot has carried out a perfect landing. To avoid future expensive damage, and the possibility of serious risk to other pilots, all heavy landings MUST be reported.

Exercise 2. Preparation for and Action after Flight

AIM — *To learn the correct method of preparing for flight, documentation, checks and engine starting procedure. The exercise also covers actions to be taken after flight.*

AIRMANSHIP — *Location of documents. Correct positioning of the aircraft for starting run-up and parking.*

GROUND EXERCISE

Pre flight	*Starting and after starting*	*Rundown and parking*
Aircraft documents	*Pre-start checks*	*Idle at 800 – 1000 RPM*
Tech log	*Brakes ON*	*Dead cut mag check*
Weight and balance	*Master switch ON*	*Idle cut-off*
	Switches OFF	*Check:*
	Fuel ON	*Park brake ON*
Authorization	*Throttle CLOSED*	*Ignition OFF/key removed*
Weather information	*Mixture RICH*	*Master switch OFF*
Flight Plan	*Fuel pressure*	*Generator/Alternator OFF*
Check lists	*Prime for starting*	*Radio and other switches OFF*
Aircraft position for start	*Clear for starting*	*Control lock in*
Ignition checked OFF	*Throttle set*	*When Parking*
Remove control lock(s)	*'CLEAR PROP' Call*	*Pitot cover on*
Internal checks	*Starting*	*Fuel off*
External checks	*Oil pressure check*	*Remove valuable items*
	Generator ON	*Doors locked*
	Change tank	*Tie down*
	Power checks	*Authorization sheet*
	Instrument scan	*Snag book*
		ATC (if applicable)

Ground Practice

Imagine you are preparing for a flight. These are the procedures to be followed.

Flight Authorization and Documentation

(*a*) Check the aircraft's technical log for serviceability and fuel state. Is there sufficient for the flight?
(*b*) Check that, with the load to be carried and the amount of fuel in the tanks, the aircraft is within its maximum take-off weight. Ensure that it is in balance.
(*c*) Sign the authorization sheet.
(*d*) Advise Air Traffic Control of your flight details.
(*e*) Check the weather.
(*f*) Collect maps, charts, flight plan and check lists.

External and Internal Checks

(*a*) While walking to the aircraft see that it is standing level. If not determine the reason. Check position relative to other aircraft, etc.
(*b*) Open the cabin door, check the ignition is off, turn on the battery master switch, check fuel contents indications, lower part flap, turn on the pitot heat, check the anti-collision beacon/strobes and operate the stall warning vane. Turn off the master switch. Remove the control lock(s).
(*c*) Complete the external checks, either from memory or with a check list. Touch the pitot head to ensure that it was heating while the switch was on. Compare visual state of the fuel tank(s) with the gauge indications.
(*d*) For the first flight of the day strain the fuel system using a clear jar to detect water/sediment contamination. Check that the strainers are properly closed.
(*e*) Enter the aircraft, adjust the seat and harness. Check that the doors are properly closed.
(*f*) Check free and correct movement of controls.
(*g*) Check that the instruments appear undamaged.
(*h*) Check security and state of the fire extinguisher.
(*i*) Check there is a first-aid kit.

Starting Procedure

(*a*) Check parking brake ON.
(*b*) Master switch ON.

(*c*) Switches OFF, fuel ON tank with lower contents, throttle CLOSED, mixture RICH.
(*d*) Electric fuel pump ON until fuel pressure is indicated.
(*e*) Prime the engine.
(*f*) Set the throttle for starting.
(*g*) Carefully look all around from left to right. When sure there is no danger of hitting persons or objects with the propeller shout: 'CLEAR PROP'
(*h*) Turn the ignition through all magneto positions to activate the electric starter (or use the start button when one is fitted).
(*i*) When the engine fires, release the starter, set the throttle to idle the engine at 1000 RPM.
(*j*) If the engine fails to start carry out 'too lean/too rich' procedure according to circumstances.

After start Checks

(*a*) Check oil pressure and if there is no indication within 30 seconds shut down the engine immediately.
(*b*) Turn on the generator/alternator and check that the ammeter shows a charge.
(*c*) Turn on the anti-collision beacon/strobes, switch on the radio and make a radio check call.
(*d*) Before moving off to the holding point carry out a magneto check at idle RPM to ensure both magnetos are working and that the switch is not faulty.

Power Checks

(*a*) At the holding point/run-up area head the aircraft into wind and apply the parking brake.
(*b*) Check there is an oil temperature indication, change fuel tanks and ensure there is no other aircraft directly behind.
(*c*) For the first flight of the day open the throttle fully and check that the minimum static engine speed is being achieved. Check that the brakes are holding the aircraft.
(*d*) Set the throttle to give the recommended engine speed for the magneto check.
(*e*) Turn the ignition switch to RIGHT magneto and note the drop in RPM. Return the switch to BOTH magnetos. Turn the switch to LEFT magneto and again note the drop in RPM before selecting BOTH magnetos. Ensure that the mag drop is within the limits given in the aircraft manual.

(*f*) Check: suction, electric charge, engine temperatures and pressures 'in the green', gyros erected and steady.
(*g*) Operate carburettor heat and note drop in RPM confirming serviceability. Return to COLD.
(*h*) Close the throttle and check slow running.
Note: After first flight of the day item (c) is omitted from subsequent engine checks.

Running-down Procedure

(*a*) Allow the engine to idle for a few minutes at 800–1000 RPM, either while taxying back to the parking ramp or when the aircraft is parked at the end of the flight.
(*b*) While the engine is idling check each magneto for a 'dead cut' or live magneto.
(*c*) To stop the engine operate the idle cut-off, then turn the ignition OFF and remove the key.
(*d*) Turn off the master switch, generator switch, radio and all other switches.
(*e*) Place the control lock(s) in position.
(*f*) If the aircraft is to be parked for any period other than a few hours, turn off the fuel.

Parking and Security

(*a*) Ensure that the aircraft is correctly positioned for parking.
(*b*) Put on the pitot cover and, if required, the canopy cover.
(*c*) Check all switches and fuel OFF.
(*d*) Remove items of value from the cabin.
(*e*) Lock all doors and the baggage compartment.
(*f*) *Tie down the aircraft.*

Post-flight Documentation

(*a*) Sign the authorization sheet.
(*b*) List any defects in the 'snag' book.
(*c*) Notify ATC (when applicable).

Exercise 3
Air Experience

Background Information

The eagerly awaited first flight is intended to give the student pilot a little time to become accustomed to the sensation of flying before serious instruction begins. To those without previous flying experience the apparent lack of speed often creates something of an anti-climax which rapidly gives way to the sheer pleasure of seeing the countryside from above.

On this first flight, comfort and the ability to reach the controls with ease should be checked. Visibility out of the aircraft may need attention and, unless the seat is adjustable, cushions behind and beneath may be needed. An ideal arrangement should be found and adhered to when flying a particular aircraft.

Review of Local Area Landmarks and Adjacent Controlled Airspace

Before the flight the instructor will point out local controlled airspace and other airfields, using a suitable map to highlight prominent ground features that will, during the flying course, act as valuable aids to location. Such landmarks will assist in locating the home airfield and help in avoiding controlled airspace and prohibited/danger areas. It is important not to infringe such airspace.

The Demonstration Flight

During this first flight local landmarks and their relationship to the airfield will be indicated by the flying instructor. While no serious attempt to teach is made on this first occasion, later in the flight the student will be invited to place the hands and feet lightly on the controls so that the remarkably small movements needed to produce a change in flight attitude are appreciated. This is called **Following Through**.

Throughout this book such phrases as 'move the stick' or 'push the stick' occur. In fact during most flight conditions these movements are in the nature of pressures rather than deliberate movements and to establish the necessary touch and appreciation of feel when flying a light aircraft the method of holding the control column is important. The wheel/stick should be held by the thumb and first two fingers only and not gripped in the fist, although this procedure may be necessary on heavy aircraft.

Dual instruction is based upon demonstrations flown by the instructor followed by student practice under supervision. Clearly there can be no room for misunderstanding as to who is in control. When the instructor wishes his student to repeat a demonstration he will say 'You have control' and the student must reply 'I have control'. This is known as the **Handing over** procedure.

Aircraft and Engine Instrument Indications

Study the instruments at intervals during the flight and determine the height of the aircraft (altimeter), the speed through the air (air speed indicator), and the RPM of the engine (engine speed indicator).

Above all relax and endeavour to become part of the aircraft, allowing the body to go with the bank during a turn, rather than attempting to lean away from it. The importance of keeping a constant lookout cannot be overstressed. Some students tend to allow their attention to become devoted to the cockpit with the attendant risk of collision with another aircraft or even simply becoming lost through not maintaining a lookout. At an early stage in training the student should become accustomed to using the 'clock code' for reporting the position of other aircraft, straight ahead being 12 o'clock. 90° left nine o'clock, slightly to the right one o'clock, etc. The words 'high' or 'low' may be included in the report to describe an aircraft above or below. During later flights the instructor will expect his student to find his own way back to the airfield.

The Airfield Layout from the Air

At a safe height above circuit traffic the instructor will explain the airfield layout from the air, drawing attention to the runway pattern or the disposition of the landing/take-off strips in the case of grass airfields.

The position of the signals area, control tower and windsock will be identified along with significant features (woods, lakes, villages, etc.) that will, as the flying course progresses, act as aids to lining up with a particular runway. Noise sensitive areas to be avoided will also be drawn to the attention of the student.

By the end of the flight the student will know what to expect on the next occasion and this will permit greater concentration on the lesson being taught.

Pre-flight Briefing

Exercise 3. Air Experience

AIM — *Introduction to the sensation of flying in a light aircraft and a review of the local area.*

AIRMANSHIP — *Lookout*
Suitability of weather

AIR EXERCISE

Before flight	*In the air*
Important landmarks relative to airfield *Controlled airspace/other airfields* *Restricted/Danger/Prohibited areas* *Seat and harness adjustment for best reach and comfort*	*Clock code* *Handing-over procedure* *'You have control'* *'I have control'* *Important landmarks from the air* *Recognition of areas to be avoided* *Follow-through procedures* *Holding controls correctly* *Importance of light touch* *Instrument indications* *Flight instruments* *Engine instruments* *The airfield from the air* *Runway/strip layout* *Tower/signals area/windsock* *Features on approaches* *Noise sensitive areas to be avoided*

Flight Practice

Having ensured that the student is seated comfortably, can reach the controls and is able to see out of the aircraft the instructor will allow the student to start the aircraft, assuming Exercise 2 has already been demonstrated. The student should also carry out the power check. The instructor will then take off and climb towards the area discussed during the pre-flight briefing.

Review of Local Landmarks and Adjacent Controlled Airspace

(*a*) Relax at all times and make no attempt to lean away from the direction of bank while the aircraft turns.

(*b*) Form the habit of looking in all directions. If you see another aircraft report it immediately, even if you think it has already been seen. Use the clock code.

(*At this stage the instructor will point out local landmarks and indicate areas to be avoided – controlled airspace, danger and other areas, etc. He will also draw attention to other airfields in the local area.*)

(*c*) Hold the wheel/stick lightly with the thumb and fingers. Place the feet lightly on the rudder pedals. Note that very small movements of the controls are all that is required to fly the aircraft. Try gently moving the wheel/stick from side to side and back and forth; You have control (*student to respond* 'I have control').

Aircraft and Engine Instrument Indications

(*a*) Note that the altimeter is showing that we are flying at -- feet, the airspeed indicator is reading -- knots and the engine is turning at -- RPM. The attitude indicator responds to changes in pitch and bank and the direction indicator shows our heading. If the nose is raised or lowered the vertical speed indicator will show a climb or a descent in feet per minute. The engine temperature and pressure instruments are all 'in the green' and we have -- gallons of fuel.

The Airfield from the Air

(*a*) The runway layout and taxiways may clearly be seen as we fly overhead above circuit traffic. Note the position of the control tower, signals area and windsock.

(*The instructor will complete the exercise by pointing out ground features and potential obstacles in the approach areas. Noise sensitive areas will also be indicated.*)

Exercise 4
Effects of Controls

Background information

In the air there are no buildings, roads, etc. to confirm body attitude (i.e. vertical, leaning left and so forth). However, during most of the flying exercises the instructor will relate aircraft attitude to the horizon.

Primary Effects of Controls

The three principal controls are the **Elevators, Ailerons** and **Rudder**. Their position on the aircraft is illustrated in Fig. 1.

The elevators are actuated by backward and forward movement of the control column and they control the aircraft in the **Pitching Plane** (nose up/nose down during level flight). As the aircraft pitches up and down so the angle of attack is altered, thus increasing or

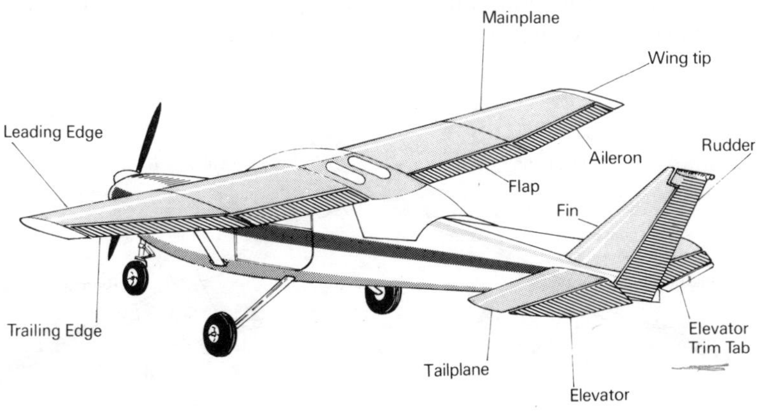

Fig. 1 General view of a typical trainer showing the flying and the control surfaces.

decreasing drag. Because of this the elevators also affect the airspeed of the aircraft. This is shown in Fig. 2.

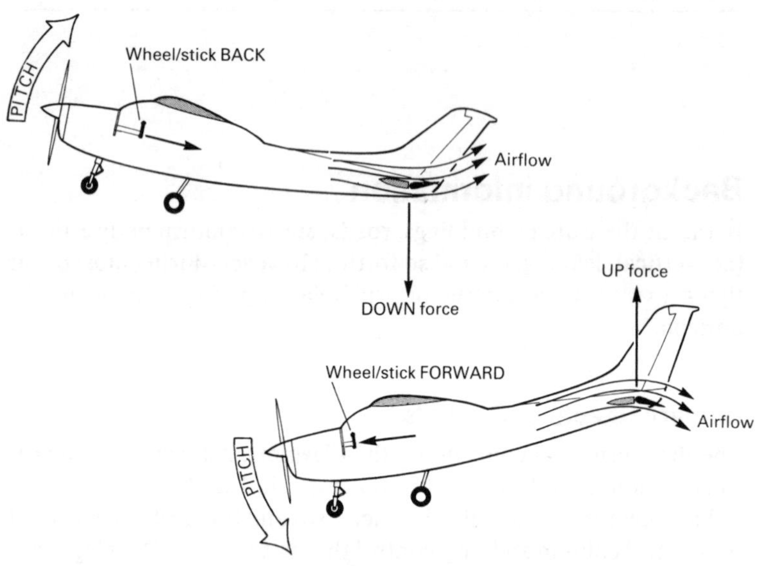

Fig. 2 The elevators control the aircraft in the pitching plane.

The ailerons are actuated by sideways movement of the wheel/stick and they control the aircraft in the **Rolling Plane** (left or right wing up or down). These movements are shown in Fig. 3.

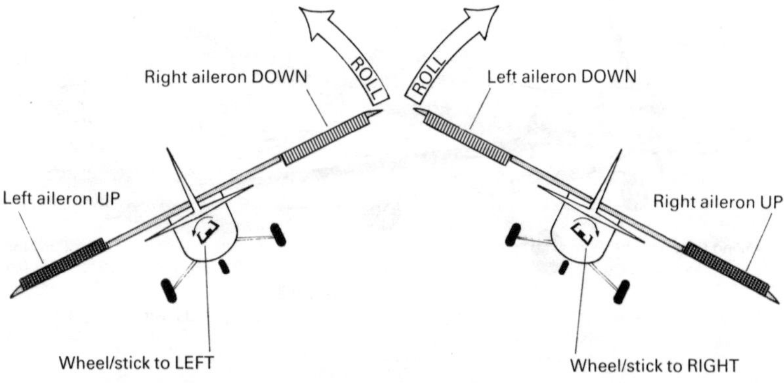

Fig. 3 The ailerons control the aircraft in the rolling plane. In this illustration the aircraft is seen from behind.

The rudder is moved by the **Rudder Pedals** and it controls the aircraft in the **Yawing Plane** (nose to left or right wing tip). These movements are illustrated in Fig. 4.

In practice the movements of the wheel/stick are easily remembered. In whichever direction the wheel/stick is moved that part of the aircraft will move away from the pilot, e.g. wheel/stick to left—left wing down, wheel/stick forward—nose down, etc. The movements and effects so far outlined are known as the **Primary Effects of Controls.**

The primary effects of rudder and aileron create additional or **Further Effects of Controls.**

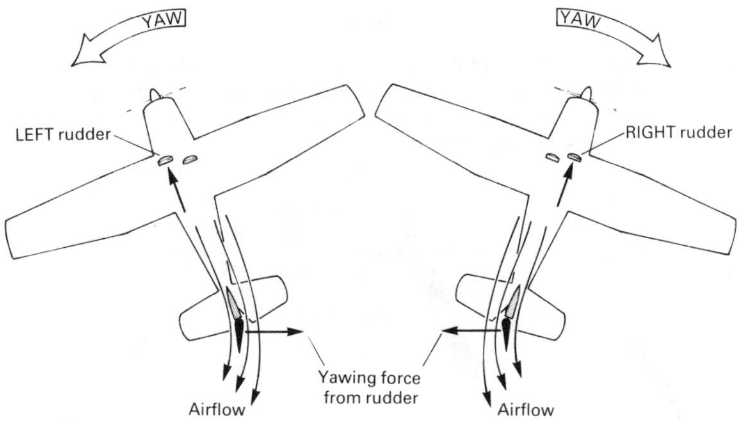

Fig. 4 The rudder controls the aircraft in the yawing plane.

Further Effects of Ailerons

If the wheel is turned to the left (or the stick is moved to the left) the aircraft will bank to the left. Reference to Fig. 5 will show that the lift and weight forces are now out of line, thus causing the aeroplane to sideslip towards the lower wing which in this case is the left one. During this slip the side of the fuselage together with the fin and rudder will be subjected to the air flowing up to meet the aircraft as indicated in Fig. 6. There is considerably more area behind the centre of gravity than in front, consequently the aeroplane will behave like a large weathercock, the nose turning towards the left wing *although the*

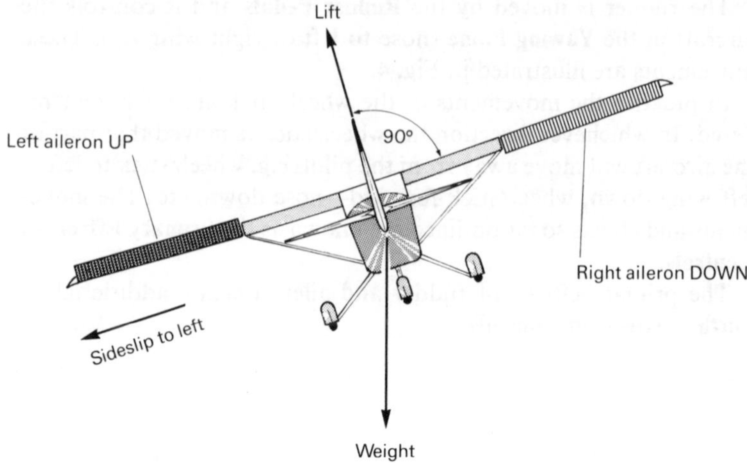

Fig. 5 Further Effects of Aileron, stage 1. In a bank lift remains at 90 degrees to the wing but weight acts vertically. The now out-of-line forces of lift and weight cause the aircraft to sideslip towards the lower wing.

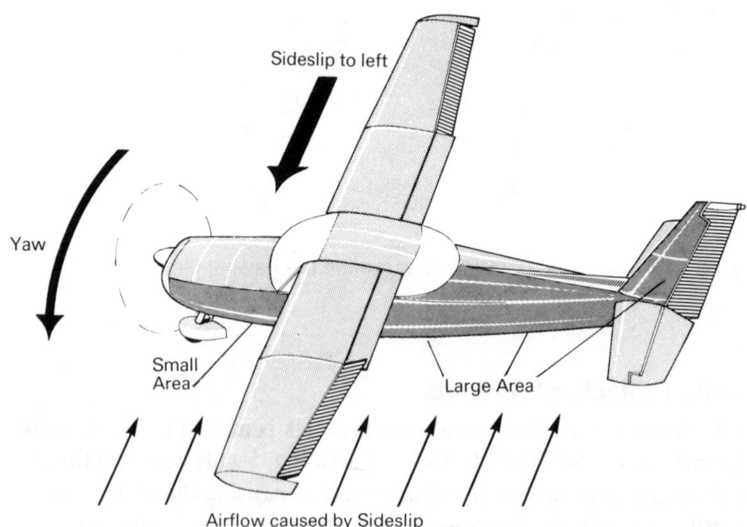

Fig. 6 Further Effects of Aileron, stage 2. With more side area behind its centre of gravity than in front the sideways airflow caused by the sideslip shown in Fig. 5 makes the aircraft 'weathercock' and a yaw towards the lower wing occurs *although no rudder has been applied.*

rudder is held in the central position. Similarly if the wheel/stick is moved to the right, the right wing will go down and the nose will swing towards the lower (right) wing. If the manoeuvre is left to develop, a spiral dive to the left or right will result as the nose follows the lower wing down below the horizon.

Further Effects of Rudder

If the left rudder pedal is pressed forward the nose of the aircraft will swing towards the left wing tip. The outer (right) wing will move faster than the inner one (similarly the outer man in a wheeling column of troops must step long while the inner man marks time). Because of the faster airflow, the right wing (in this case) will produce more lift than the left and a bank will occur *although the control column is held in the central position* (Fig. 7).

In the same way right rudder will produce a yaw to the right followed by a bank to the right. Because the nose follows the lower

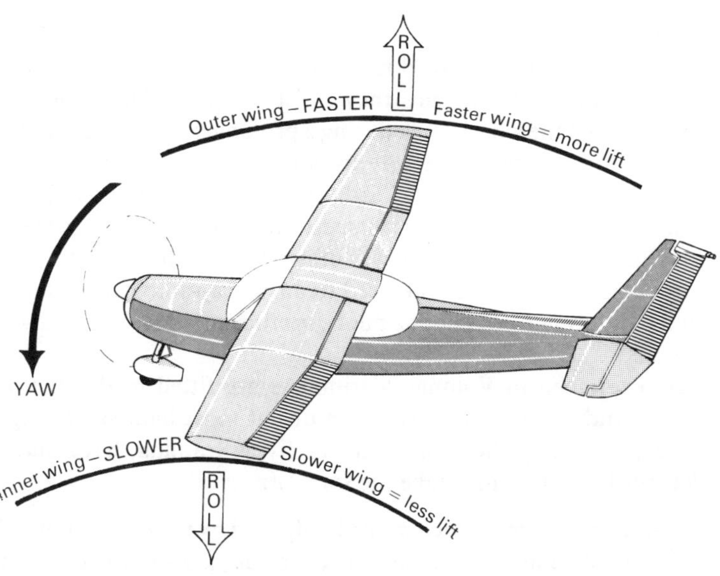

Fig. 7 Further Effects of Rudder. Yaw induced by (in this example) left rudder results in a faster right wing and a slower left wing and unequal lift causes the aircraft to roll *although no aileron has been applied.*

wing down below the horizon, a spiral dive in the direction of turn will develop.

From the foregoing it can clearly be seen that roll and yaw are closely related, and that the further effect of aileron is yaw and the further effect of rudder is roll.

Effects of Airspeed, Slipstream and Power

To provide the desired effect on the aircraft, the elevators, ailerons and rudder are dependent upon the airflow over them and it therefore follows that they will be more effective at high speed than low. Furthermore in most single-engined aeroplanes the rudder and elevators are situated within the **Slipsteam** from the propeller, consequently they are more effective with the power on than, for instance, during a glide, when the engine is throttled back. The ailerons being outside the slipstream are under the influence of the airflow only; therefore on many aircraft their 'feel' and effectiveness provide a good indication of airspeed.

Trim Controls

Trim controls are explained in Chapter 5. Volume 2 of this series, however, in practical terms they assist the pilot, for example, by holding back the wheel/stick during a prolonged climb without need to maintain backward pressure to obtain a required airspeed. This would be tiring for the pilot. Furthermore without a trim control the attitude of the aircraft and its airspeed would probably alter when at any time the pilot's attention was drawn to, say, the indications of a particular instrument or the need to answer a radio call. There is a limit to how many functions a pilot can efficiently perform at any one time.

As explained in Volume 2, trimmers are fitted to the ailerons, rudder and elevators of large aircraft but most light single-engine designs, certainly those intended for flying training, have only an elevator trim. This may take the form of:

1. A simple spring device which is adjusted by the pilot via the trim control to apply a tension force in the nose-up or nose-down sense.
2. An adjustable **Trim Tab** which is likewise controlled by the pilot via the elevator trim wheel or lever. The tab, which is located within or behind the elevator trailing edge, supplies an up or

down force to the elevator, holding it at the correct angle for the required condition of flight. When an aircraft is fitted with an **All Moving Tailplane** (known as a **Stabilator** in the USA) the trim tab surface is arranged to move in the same direction as the control surface. Such a device is known as an **Anti-balance Tab** and its principle is described on page 164 of Volume 2.

Whatever the type of trimming device its method of operation remains the same. It should be used in the following manner.

1. Attain the correct attitude and power setting for the required mode of flight (climb, descent, cruise, etc.)
2. Adjust the trim wheel until no control force is required to maintain the required flight condition.
3. Release the wheel/stick and check the trim for accuracy. Speed/attitude should remain steady if the aircraft has been properly trimmed.

Correct use of the trimmer requires a fine touch and to obtain the best results it is essential that the wheel/stick should be lightly held. A strong grip with the fist can only mask the feel of the control and make it impossible to recognize when the elevators are in proper trim for the required condition of flight.

A correctly trimmed aircraft makes fewer demands on the pilot, allowing the attention to be devoted to other important tasks such as navigation, instrument scan and LOOKOUT.

Effect of Flaps

The need for flaps and the different types of flap available to the aircraft designer are described in Chapter 4, Volume 2.

In so far as the pilot is concerned the main considerations related to use of flap are:

1. *Limiting Speed.* Primarily for structural reasons a maximum lowering speed is published for the aircraft type. It will appear in the aircraft manual and there may be a placard, giving the maximum flap lowering speed, on the instrument panel. It is common practice in modern aircraft to mark the dial of the airspeed indicator with a white arc. Provided the airspeed is within that arc flap may be lowered.
2. *Trim Changes.* Usually the first ten degrees or so of flap lowering have little effect on trim but in most aircraft there will be a need to adjust the trim as more flap is applied. Depending on the

design of the aircraft the effect of lowering flap is to cause a nose-down trim although there are exceptions in which the trim change is nose-up.

The most significant trim change occurs when full power is applied during a 'go around' procedure with full flap. In some aircraft it is necessary to apply a powerful forward pressure on the elevator control to prevent a pronounced nose-up attitude. Immediate use of the trim control will deal with the situation.

Likewise, when, for any reason, the flaps are raised, particularly from the fully-down position, there will be a change of trim which, in some aircraft, can be pronounced.

3. *Pitch Attitude*. For any particular airspeed the effect of lowering flap is to lower the nose. A by-product of this is that, at lower airspeeds, and later in the flying course exercises will be introduced where flying at reduced speed is a requirement, the view ahead is greatly improved because of the lower nose attitude.
4. *Increase Drag*. Although drag is the aircraft designer's enemy there are times when the ability to add drag at will is of assistance to the pilot. Whereas the major lift increase occurs with most flaps during the first 25 degrees of lowering (along with a small drag increase), further depression provides a more modest lift increase accompanied by a larger increase in drag. Because of the increase in drag when flap is lowered more power will be required to maintain any particular airspeed compared with the amount of power required when the flaps are not in use.

Different flap designs exhibit differing characteristics but when some types are raised there is a tendency for the aircraft to sink and lose height. In a light aicraft this is rarely very pronounced but it is a consideration to bear in mind when raising the flaps near the ground, for example following a 'go around' from a missed approach.

Operation and use of the Mixture Control

The purpose of the mixture control is explained in Chapter 6, Volume 2. Its use must be understood by the pilot:

(a) To obtain the best fuel economy and,
(b) To avoid damaging the engine.

Carburettor-type engines (as opposed to fuel injected engines) of the type fitted to most light training aircraft are confined to a simple mixture control in the form of a push-pull knob or a lever located near the throttle. Some aircraft, particularly those used for touring, are fitted with an **Exhaust Gas Temperature (EGT) Gauge** which is used for accurate setting of the mixture, the temperature of the exhaust gas being largely determined by the ratio of fuel/air mixture being supplied to the engine.

Operation of the mixture control, assuming that no EGT is fitted, is explained in the FLIGHT PRACTICE section of this chapter but the following general notes are for pilot guidance:

1. *During Take-off.* Most airfields in Europe are located no higher than 1000 feet above sea-level where the air is dense and there is no need to weaken the mixture for take-off. In those parts of the world where airfields are situated at 3000–6000 feet it will be necessary to adjust the mixture and obtain smooth running of the engine before take-off.
2. *During the Climb.* When operating into and out of the high airfields mentioned in the previous paragraph it will be necessary to lean the mixture until maximum RPM are achieved.

   During operations from airfields situated at up to 1000 feet the mixture should remain in RICH (i.e. fully forward) until the aircraft has passed 3000 feet when, even at full power, the mixture control may be moved towards weak to obtain the highest RPM. These are typical figures for a light trainer. They should be checked with the aircraft manual.
3. *In the Cruise.* Provided the power setting does not exceed 75 per cent the mixture may be leaned at any time to provide the best fuel economy. Here again this is a typical figure and it should be checked with the aircraft manual. Some manuals recommend leaning the mixture until the RPM decrease, then moving the mixture towards RICH until RPM are restored. Others advise leaning the mixture until there is a decrease of 25–50 RPM when a 10–13 per cent saving in fuel will occur.

Care must be exercised not to over-lean the mixture. Generally a lean mixture results in a hotter engine than one running on a rich mixture. Over-lean running will result in a rise in temperature (when a cylinder head temperature gauge is fitted there will be a red line indicating the maximum permitted running temperature), the engine will overheat and, as a result, the oil may thin causing further overheating leading to serious engine damage at worst and reduced engine life at best.

Careful use of the mixture control saves fuel. Careless use leads to expensive damage.

Operation and use of Carburettor Heat Control

The need for ice protection of the carburettor and induction system is described on page 227 of Volume 2. Although it is unlikely that a student pilot will be exposed to airframe icing conditions during training it should never be assumed that, because the day is bright and there are few clouds, carburettor ice will not occur.

The conditions of temperature and humidity that are likely to produce induction icing as explained in Volume 2 are notable for the wide range within which ice will form in the carburettor. In aircraft handling terms these are the factors to be remembered:

1. Never assume that, because it is a nice day, carburettor ice will not occur.
2. Learn to recognize the symptoms. A persistent decrease in engine RPM should not be dealt with simply by opening the throttle to restore power. Eventually the point will be reached where although full throttle has been applied RPM continue to decrease and rough running, a sure sign of over-rich mixture, is bound to follow.
3. Any decrease in RPM from the figure previously set should be regarded as a potential carburettor ice hazard. FULL carb heat should immediately be applied.
4. When carburettor ice has been allowed to develop the initial effect of applying carburettor heat may well result in an apparent worsening of the situation as ice melts and a mixture of ice and water enters the engine. This should not be allowed to provoke the removal of full carburettor heat which must be given time to deal with the situation. Usually carburettor heat acts quickly.

It should be remembered that the carburettor heat system relies upon engine temperature to provide heat. If the system is not used in time there may be insufficient engine heat to remove the ice.

The side effects of carburettor heat are:

(*a*) A reduction in power at any given throttle setting due to a reduction in air density entering the cylinders (pages 227–8, Volume 2).

(*b*) An increase in mixture strength (rich) due to (*a*).

(*c*) Prolonged use of the carburettor heat can, on a hot day, increase engine temperature to the point where detonation may occur. Consequently, it should only be used for brief periods in the course of removing induction ice.

Because of consideration (*a*) carburettor heat should not be used during the take-off. For the same reason, its use during final approach could prevent a climb away with full flap in the event of an overshoot.

The carburettor heat control should be used on an 'all or nothing at all' basis; part carburettor heat brings with it a risk of raising the inlet temperature into the icing range when the outside air temperature is too cold to allow formation of carburettor ice. Full heat, on the other hand, avoids this risk.

Cabin Heating and Ventilation

In light training aircraft, and single-engine touring aircraft the heater is a simple heat exchange built around the exhaust system. The position of the HOT/COLD control for comfortable cabin temperature can only be found by trial and error. In this respect it is much the same as a family car heater.

Separate fresh air vents are usually provided and these should always be used when the cabin heater is in operation, thus guarding against risk of exhaust fumes in the event of a fault in the heat exchanger.

The cabin heat may also be directed to the inside of the windscreen and, while this is primarily intended to be a demister, when all available heat is confined to the windscreen it is capable of defrosting small areas when such conditions are encountered in flight.

Instrument Indications as Appropriate

Modern flying training is based upon the early integration of visual and instrument flying techniques. Instrument flying, at one time something of a novelty in light aviation, now assumes a position of prime importance. In consequence, the student pilot will be introduced to the instruments at the earliest stages of training, even while being shown the basic handling manoeuvres.

Notwithstanding the importance of instrument flying student pilots must guard against devoting a disproportionate amount of their attention to the flight panel. Constant lookout is essential.

Exercise 4. Effects of Controls

AIM — *To learn the effects of the primary controls in flight.*

AIRMANSHIP — *Height: sufficient for exercise*
Airframe: flaps up
Engine handling: temperatures and pressures 'in the green'
Location: not in controlled airspace, etc
Lookout: use of clock code
Handover of controls
Suitability of weather

AIR EXERCISE

Actions	*Primary effects*	*Secondary effects*	*Banked attitudes*
Wheel/stick to LEFT	*Roll to LEFT (Att. Indicator)*	*Yaw to the LEFT, then spiral dive to LEFT*	*N/A*
Wheel/stick to RIGHT	*Roll to RIGHT (Att. Indicator)*	*Yaw to the RIGHT, then spiral dive to RIGHT*	*N/A*
Wheel/stick FORWARD	*Nose below horizon (Att. Indicator)*	*Airspeed increases Aircraft descends (AI., ASI., VSI., Altimeter)*	*Relative to pilot all effects and movements remain unchanged when in banked attitude*
Wheel/stick BACK	*Nose above horizon (Att. Indicator)*	*Airspeed decreases Aircraft climbs (AI., ASI., VSI., Altimeter)*	
LEFT rudder	*Yaw to LEFT (DI., Turn and Bal.)*	*Roll to LEFT, then spiral dive to LEFT*	
RIGHT rudder	*Yaw to RIGHT (DI., Turn and Bal.)*	*Roll to RIGHT, then spiral dive to RIGHT*	

High airspeed = more airflow = heavier and more effective controls
Low airspeed = less airflow = lighter and less effective controls
With slipstream = more effective rudder and elevators.No effect on ailerons
No slipstream = less effective rudder and elevators. No effect on ailerons

More power = climbing turn to left
Less power = descending turn to right

Pre-flight Briefing

Exercise 4. Effects of Controls (Trim, Flaps, Mixture, Carb. and Cabin Heat)

AIM — *To learn the effects of the secondary controls in flight.*

AIRMANSHIP — *Height: sufficient for manoeuvres*
Airframe: flaps as required
Engine handling: temperatures and pressures 'in the green'. Mixture/carb. heat use
Location: not in controlled airspace, etc
Lookout: use of clock code
Handover of controls
Suitability of weather

AIR EXERCISE

Actions	*Effects*	*Precautions/notes*
Elevator Trim		
Towards nose-heavy	*Back pressure on wheel/stick to hold attitude*	*Hold wheel/stick lightly* *Feel pressures*
Towards tail-heavy	*Forward pressure on wheel/stick to hold attitude*	*Move trimmer SLOWLY to remove control loads* *Release control – Check trim*
Flaps		
Lower, say 10° of flap	*Change in pitch attitude* *Change in trim*	*Below flap limiting speed before lowering flap* *Note lower nose attitude for any particular speed*
Lower full flap	*Bigger pitch change* *Bigger trim change*	*Note more power required to hold any particular speed*
Raise flaps	*Reverse pitch change* *Reverse trim change*	*Re-trim in stages* *Be prepared: aircraft may sink*
Mixture Control		
Move mixture to LEAN	*Decrease in RPM and engine will stop running if moved to ICO*	*Watch engine Ts and Ps*
Carburettor Heat		
Apply FULL heat	*Decrease in RPM*	*Avoid prolonged use* *Watch cylinder head temperature*
Cabin Heat/Demister		
Adjust temperature and distributor	*Cabin temperature increases* *Warm air to windscreen*	*Open fresh air vents while cabin heat is in use*

Flight Practice

COCKPIT CHECKS

(*a*) Set cruising power.
(*b*) Trim for level flight.

OUTSIDE CHECKS

(*a*) Altitude: sufficient for manoeuvre.
(*b*) Location: not over aerodromes or towns or in controlled airspace.
(*c*) Position: check in relation to a known landmark.

AIR EXERCISE

Primary Effects of Controls

(*a*) Move the wheel/stick to the left and the left wing goes down. Centralize control and level the wings.
Repeat to the right and notice instrument indications.

(*b*) Press the wheel/stick gently forward; the nose falls below the horizon. Notice airspeed increasing and a loss of height. Move the stick back and the nose rises towards the cockpit and above the horizon. Check instrument indications and note airspeed decreasing and a gain in height.

(*c*) The elevator movements are the same irrespective of the attitude of the aircraft. Now bank gently to the right and move the wheel/stick back and then forward. The aircraft's pitching behaviour will be the same in relation to the pilot.

(*d*) From straight and level flight move the left foot forward on the rudder pedals and notice the nose swings towards the left wing tip. At the same time a skid to the right will be felt and this will be indicated on the **Balance Indicator**.
Repeat to the right.

(*e*) Now bank gently to the left and apply right rudder. The nose will swing up towards the high (right) wing and left rudder will cause the nose to yaw towards the left wing and go below the horizon. In other words the rudder produces the same movement irrespective of the attitude of the aircraft.

(*f*) Notice that control effect is proportional to control movement.

Further Effects of Controls

(*a*) *Rudder*. Hold the wheel/stick in the central position and apply left rudder. The nose will swing towards the left wing tip and a roll will commence to the left developing into a spiral dive in the same direction.

Resume straight and level flight and repeat to the right.

(*b*) *Ailerons*. With the rudder in the central position move the wheel/stick to the right. A bank to the right will result followed by a yaw towards the right wing tip which will continue into a spiral dive in that direction.
Revert to straight and level flight and repeat to the left.

Effect of Airspeed and Slipstream

(*a*) At a high power setting, climb at 70 kt* and notice how sensitive are the elevators and rudder. Also note the feel of the ailerons.

(*b*) Now close the throttle and glide at 70 kt. The elevators and rudder will be less sensitive and a larger movement will be required to produce the same effect as before (with slipstream). Aileron feel will remain unchanged.

(*c*) Now fly at a low airspeed and note the feel of the ailerons. Increase to maximum speed and the ailerons will become heavier and more sensitive giving a good indication of airspeed.

Effect of Power

(*a*) In cruising flight trim the aircraft and remove the hands and feet from the controls.

(*b*) Open the throttle and very soon the nose will rise and the aircraft climb. At the same time a turn will develop in the opposite direction to propeller rotation. Rudder and forward elevator pressure are required to maintain balanced level flight.

(*c*) Return to cruising flight then close the throttle slightly. The nose will drop, a loss in height will occur and a turn will develop in the same direction as propeller rotation. Rudder and back pressure on the elevators are required to keep the aircraft in balanced level flight.

Effect of Trim

(*a*) From straight and level flight move the trim control to a 'nose heavy' position. Notice that backward pressure on the stick is required to maintain the same airspeed. Should the stick be allowed to ride in the 'hands off' position the nose will drop and speed will increase.

(*b*) Repeat this procedure moving the trimmer to a 'tail heavy' position when a forward pressure will be required to maintain the original airspeed. when left to fly 'hands off' the nose will rise and the airspeed will decrease.

*Use speed appropriate to type of aircraft.

Effect of Flaps

(*a*) Reduce speed until the IAS is within the flap operating range (white arc). Re-trim and notice the attitude of the nose relative to the horizon.

(*b*) Lower 10 degrees of flap and notice the change in trim. Maintain the same airspeed and compare the new nose position. It is lower in relation to the horizon. The aircraft is now descending and more power is required to maintain height. Notice the improved response of the rudder and elevators.

(*c*) Lower full flap and note the change in trim and reduction in airspeed. Trim the aircraft to hold approach speed.

(*d*) Now raise the flaps and feel the considerable change in trim as a result. Notice the aircraft sinks and there is a loss of height (*When applicable to aircraft type.*)

Operating the Mixture Control

(*a*) At a safe height set cruising power.

(*b*) Slowly bring back the mixture control until there is a slight drop in RPM (further leaning of the mixture will cause rough running and eventually a position is reached when the engine will stop producing power).

(*c*) During cross-country flights the best fuel economy for any particular power setting will be obtained by moving back the mixture control until the RPM decrease, then easing it forward to the point where the RPM are just restored to the original setting (Note: *teach as recommended in the manual.*)

Operating the Carburettor Heat

(*a*) Note the RPM, then operate the carburettor heat control. Notice the drop in RPM.

(*b*) When carburettor ice is present (denoted by a decrease in RPM and eventual rough running) use of the heat control will first cause a further drop and, when the icing is severe, more pronounced roughness before the RPM increase. At this stage the carburettor heat control should be returned to cold when the RPM should have returned to the previously set engine speed.

(*c*) Do not use part heat since under certain weather conditions this could raise the temperature into the icing range and cause carburettor icing.

Cabin Heat and Ventilation

(*a*) Note the position of the heat and cabin air controls. (Some aircraft allow heat to be directed to the windscreen.)

(*b*) Apply heat and notice that it may be controlled according to the needs of the occupants.

(*c*) When applying heat it is good practice to open a fresh air vent as a guard against possible fumes.

(*d*) Should ice form on the windscreen in flight, direct all heat to the windscreen using the control provided.

(*e*) If necessary open the clear vision panel (when fitted).

Exercise 5
Taxying

Background Information

Most aircraft of modern design have **Nosewheel Undercarriages** (sometimes known as **Tricycle** undercarriages). Compared with **Tailwheel** aircraft they are more stable in direction, easier to handle in crosswind conditions and visibility, particularly ahead of the aircraft, is greatly improved. Tailwheel management demands special techniques and these are described in Chapter 20.

When an airfield has its own air traffic control service, taxi, take-off and landing clearances will be passed over the radio. Nevertheless it is the pilot's final responsibility to ensure that it is clear to taxi irrespective of the permission he has received. For example, the tower may have cleared the aircraft for take-off but it is the pilot's duty to double check that no other aircraft is on the approach or occupying the runway.

During taxying these considerations should be borne in mind:

1. To avoid damage from stones or other debris the flaps must be raised.
2. The carburettor heat control must be in COLD. In the HOT position the air intake filter is by-passed leaving the engine unprotected from the risk of inhaling grit and abrasive dust.
3. Prolonged idling at very low engine speeds must be avoided to prevent sparking plug fouling and to ensure that the generator/alternator can meet electric load demands (radio, anti-collision beacon/strobes, etc.). 1000 RPM is an ideal minimum engine speed.
4. When flying training is being conducted on an airfield handling mixed traffic the greatest caution must be exercised when taxying behind larger aircraft, particularly jets. There have been cases where light aircraft manoeuvring too closely behind have been blown over and severely damaged.
5. Because a modern aircraft is easy to taxi this should not be allowed to encourage over-confidence. The smallest light trainer

is approximately five times wider than a family car and during tight turns the tail surfaces swing out some twenty feet behind the cabin. This must be uppermost in the mind while manoeuvring in confined spaces.

At busy airfields a ground marshal is sometimes employed to guide aircraft into and out of the parking areas. Standard hand signals are used and pilots should acquaint themselves with these.

Brake Check

At this stage of training the student will be mindful that flying is based upon a concept of 'check before you act'. This applies even while on the ground and because of their vital role during taxying one of the first actions to be taken after moving away from the parking area will be to test the brakes. In most aircraft these are operated by pedals attached to the rudder control, an arrangement which also allows braking effort to be applied to one wheel as an aid to turning in confined spaces. When Toe Brakes are fitted there will be a separate Parking Brake to lock on the foot pedals while, for example, the aircraft is stopped for power checks. In aircraft with toe brakes the rudder pedals of both the student and the instructor will be so equipped and each pair of brake pedals must be tested. The throttle must be fully closed while testing the brakes.

Some aircraft have hand-operated brakes which are applied to both wheels via a lever which may be locked on for parking. While hand-operated brakes usually apply equal braking effort to both mainwheels a number of aircraft incorporate a differential arrangement which is linked to the rudder pedals. When the brake lever is pulled back with rudder applied, for example, to the left, the left wheel will provide more retarding effort than the right.

Use of Power and Brake, separately and inter-related

An aircraft is controlled and manoeuvred on the ground by the use of power, rudder and brakes, either independently or interrelated. Since the throttle will constantly be adjusted to compensate for gradient, wind, for turns when brake is required and while integrating with other traffic, the throttle friction must be slackened before moving away from the parking ramp.

Moving off from a Standstill

After a final check that there is room to move forward without risk of striking persons or property the handbrake is released, and the throttle is opened until the aircraft starts to roll. On hard, level surfaces very little power is required to get the aircraft moving but on grass areas, particularly when they are soft, considerable power will be needed.

When the aircraft starts to move power should be reduced to the point where a safe taxying speed is being maintained for the conditions at the time. This can only be demonstrated by the flying instructor but after the inertia of the aircraft has been overcome less power will be required to keep moving.

At the first opportunity the throttle is closed and the brakes must be checked as described in the previous section.

Control of Speed – effect of Surface and Gradient

Speed is controlled by adjusting the throttle but when fully closing the throttle fails to slow the aircraft sufficiently, for example while rolling down an incline, brake must be applied. Disc brakes are less prone to fade than the old drum brakes that used to be fitted to aircraft but unnecessary disc, pad and tyre wear can be avoided by considerate treatment of the brakes. **Avoid applying the brakes without first closing the throttle**

While taxying it should be remembered that, like a car, an aircraft has inertia; effort is required to start moving – effort is needed to stop or reduce speed. These efforts are provided by engine power and brake respectively but no vehicle can instantly change speed; time is required to accelerate, reduce speed or stop. Obviously stopping distance is affected by taxying speed; the faster the speed the greater the stopping distance required. Taxying speed must therefore be matched to the conditions at the time.

Rolling friction on grass, particularly following heavy rain, can entail the use of very high power settings but when the ground is soft to the point where the wheels are making deep ruts care must be taken not to over-heat the engine by prolonged use of near-full throttle. The situation should be dealt with by manhandling the aircraft onto harder ground. These considerations relating to ground surface should be remembered:

1. It is sometimes necessary to make the transition from grass to hard surface, or hard surface to grass. Often there is a difference

in height between the two surfaces rather like mounting a low pavement. Light-aircraft wheels are small and propeller tip clearance is limited so the transition from one surface to the other must be made slowly and one wheel at a time (Fig. 8). A fast transition made head on, both mainwheels together, imposes unnecessary strain on the nosewheel assembly and the aircraft may pitch forward striking the propeller on the ground.

2. Whenever possible areas of loose stones must be avoided since these may be lifted by the propeller to cause extensive blade damage.
3. Wet grass can seriously diminish braking efficiency, adversely affecting an aircraft's stopping distance. This is particularly important during the landing roll – another application of taxying.

When a rising gradient is encountered, and few airfields are completely level, more power will be required to maintain speed. Conversely, while running down a slight incline the throttle will have to be completely closed and speed is then controlled on the brakes.

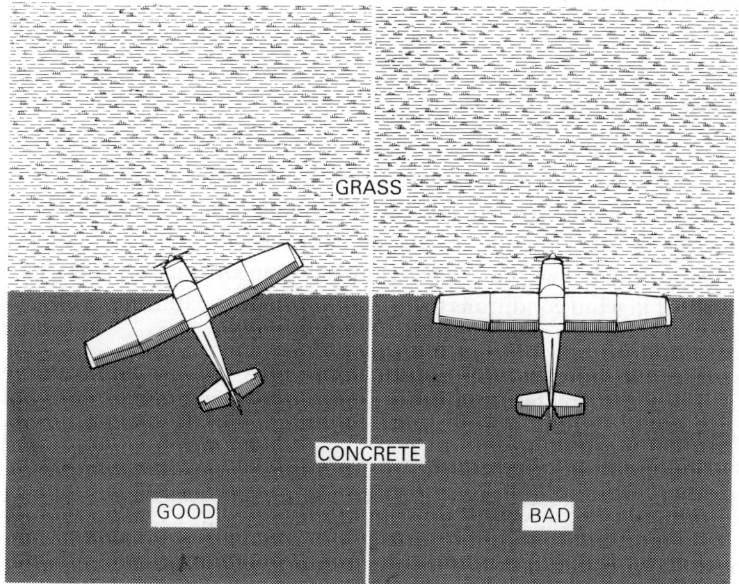

Fig. 8 Transition from grass to hard surface or hard surface to grass should be made one wheel at a time to safeguard the undercarriage and avoid risk of propeller damage.

Control of Direction

Several aircraft designs feature a castering nosewheel and these require differential use of brake while controlling direction and making turns on the ground. When the winds are light moderate turns can be made with rudder assisted with slipstream.

The most common arrangement is to link the nosewheel strut to the rudder pedals, either directly or through tension springs. In either case the nosewheel will turn when rudder pressure is applied – left rudder LEFT TURN : right rudder RIGHT TURN.

When manoeuvring in confined areas brake may be needed to tighten the turn radius but it should never be allowed to lock one wheel and cause unnecessary strain on the undercarriage together with a risk of tyre damage.

While turning on a small radius, with the aid of brake on the inside of the turn, additional power may be required to keep the aircraft moving. This is an occasion when power against brake is permitted.

Generally the modern light aircraft is easy to steer on the ground provided the good view ahead is not allowed to encourage over-confidence. Sudden, fast turns put unnecessary strain on the aircraft, are unpleasant for passengers and are a potential hazard to other pilots.

Effect of Wind

Nosewheel aircraft are less affected by wind than tailwheel designs but it should always be remembered that training aircraft are very light. During the flying course student pilots will not be exposed to marginal wind conditions but students usually qualify, and the private pilot left without supervision has been known to operate light aircraft in wind conditions that are beyond the capabilities of human or machine.

Fuselage and fin area behind the aircraft's centre of gravity is several times greater than that in front, consequently, in strong wind conditions there is a tendency to 'weathercock' into the prevailing wind direction. Over-confidence must not be allowed to erode the advantages of nosewheel aircraft over tailwheel designs.

To prevent a wing lifting, or perhaps the tailplane being caught by a following gust and then lifting the rear fuselage sufficiently to drive the propeller tips into the ground, the flying controls should be held as illustrated in Fig. 9.

A strong following wind will encourage an increase in taxying speed. This will require closing the throttle and the use of brake.

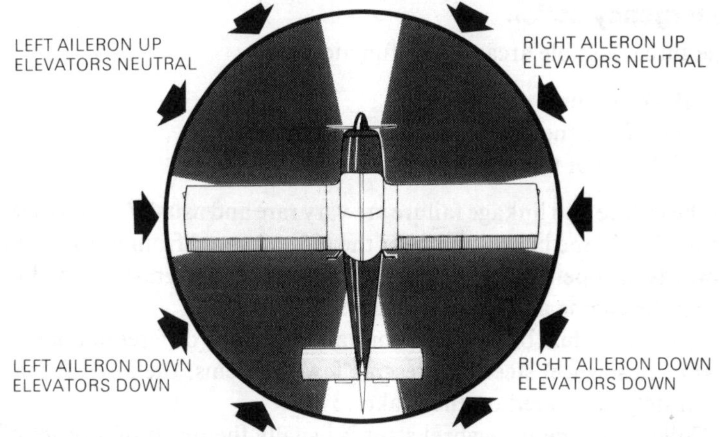

Fig. 9 Use of the flying controls to safeguard the aircraft while taxying in a strong wind.

Rudder Function Check

While some aircraft have a fully caster nosewheel the majority incorporate the nosewheel steering arrangement previously mentioned. When this takes the form of spring linkage between rudder pedals and nosewheel strut it is possible to check the rudder for full movement, left and right, while the aircraft is stationary (this is part of the pre-starting checks).

Many aircraft have a direct linkage between the rudder pedals and the nosewheel assembly. In these it is only possible to check for full and free rudder movement while taxying. This check must be done at low speed.

Instrument Function Checks

At a later stage in training the student pilot may progress to instrument flying and learn, among other exercises, how to enter cloud and climb to a safe level on instruments. Clearly the time to discover an instrument fault is on the ground, not in the air, and the instrument checks to be made while taxying are described under GROUND PRACTICE.

Emergency action

The three possible areas of malfunction are:

1. Brake failure.
2. Failure of the nosewheel steering linkage.
3. Collapse of the nosewheel strut.

Brake failure and linkage failure are very rare and usually easy to deal with. Should the brakes fail steer the aircraft away from objects and towards an open space. If possible turn onto the grass, then shut down the engine and await assistance.

With individual brake pedals or hand-operated differential brakes loss of nosewheel steering presents few problems; the aircraft may accurately be steered on the brakes.

Collapse of the nosewheel strut is usually the result of a series of heavy landings. Although injury to the occupants of an aircraft involved in this kind of incident is practically unknown, damage to the propeller, crankshaft, engine cowlings and firewall that follows such an incident can be extensive.

It is not unknown for the collapse to occur while an innocent party is handling the aircraft, suffering the experience because the pilot(s) responsible failed to report previous heavy landings.

Should the nosewheel strut collapse immediately turn off the fuel, ignition and battery master switch, exit the aircraft and await assistance.

Pre-flight Briefing

Exercise 5. Taxying

AIM To manoeuvre the aircraft on the ground, along the taxiways and in confined places.

AIRMANSHIP Pre-taxi checks (flaps up, carb heat COLD, slacken throttle friction)
Airfield discipline
Surface area assessment
Suitability of weather
Lookout

AIR EXERCISE

Speed control	*Control of direction*	*Instrument function checks*
LOOKOUT *Power to move off* *Brake check* *Power/Brake control of speed* *Power on hard surfaces* *Power on grass* *Power up an incline* *Brake down an incline* *Grass/concrete transition* *Avoiding bad surface*	*LOOKOUT before turns* *Nosewheel steering* *Adjusting power* *Turns with brake* *Effect of Wind* *Elevator/Aileron use* *Wingtip/tail unit clearance while turning* *Rudder Check at LOW SPEED* *full LEFT* *full RIGHT*	*Turn Left* *DI and compass readings DECREASING* *Turn needle to LEFT: Ball to RIGHT* *Turn Right* *DI and compass readings INCREASING* *Turn needle to RIGHT: Ball to LEFT* *Attitude indicator level while turning* *VSI shows no climb or descent*

Exercise 5E. Emergencies

Brake Failure *Avoid obstacles* *Steer for open space, grass for preference* *Shut down engine* *Await assistance*	*Nosewheel Steering Failure* *Reduce taxying speed* *Control direction with brakes* *Avoid confined areas, park in suitable position* *Shut down engine*	*Nosewheel Strut Collapse* *Shut down engine:* *ICO, Ignition* *Fuel OFF* *Exit aircraft* *Await assistance*

Ground Practice

COCKPIT CHECKS

Flaps UP
Carburettor heat COLD
Throttle friction SLACK

OUTSIDE CHECKS

No obstructions ahead of the aircraft, wingtips and tail surfaces clear.
LOOKOUT

GROUND EXERCISE

Brake Check

(*a*) When certain it is clear to move ahead release the parking brake and add power sufficient to start the aircraft rolling.
(*b*) Close the throttle and apply the brakes. Note: when toe brakes are fitted both sets, student and instructor, must be checked.

Use of Power and Brakes, separately and inter-related

(*a*) Release the parking brake, open the throttle, start the aircraft moving then reduce power to prevent excessive taxying speed. Close the throttle and, on level ground the aircraft will slow down slightly. Increase power and the aircraft will accelerate.
(*b*) Close the throttle, gently apply the brakes and note the retarding effect. Now bring the aircraft to a halt and apply the parking brake.

Control of Speed – effect of Surface and Gradient

(*a*) The aircraft is now moving up an incline. Note that more power is required to maintain speed.
(*b*) While taxying down an incline the throttle must be fully closed and the speed controlled on the brakes.
(*c*) Taxi from a hard surface onto the grass and make the crossing slowly at about 45 degress – one wheel at a time. Note that more power is required to taxi on soft ground than on tarmac or concrete.
(*d*) Avoid areas of loose stones, etc.

Control of Direction (*Aircraft with Nosewheel Steering*)

(*a*) At a slow taxi speed LOOKOUT in the direction of turn, then apply rudder in that direction. Note that the turn radius may be tightened by applying more rudder.
(*b*) If necessary add power to maintain forward movement.
(*c*) Now try a turn in the opposite direction.

(*d*) To tighten the radius of turn apply full rudder, then add brake in the same direction. On no account lock the wheel on the inside of the turn. Extra power will probably be required.
(*e*) Repeat a tight turn in the opposite direction.

Effect of Wind (*To be demonstrated under moderate wind conditions*).

(*a*) The aircraft is now taxying into wind. Hold the elevators and ailerons in the neutral position.
(*b*) The wind is now coming from ahead but from the left. Prevent the wing from lifting by raising the left aileron (wheel/stick to the left). The elevators should be neutral.
(*c*) When the wind is from ahead but from the right, right aileron should be applied to prevent the right wing from lifting.
(*d*) The wind is from directly behind the aircraft. with the ailerons neutral apply down elevator (wheel/stick forward) to prevent the tail lifting with risk of the propeller striking the ground.
(*e*) We have now turned to the left and the wind is from the rear left quarter. Maintain down elevator and apply down aileron on the left wing (wheel/stick to the right) to prevent the wing lifting.
(*f*) Now the wind is from the rear right quarter. Move the wheel/stick to the left, lower the right aileron to prevent the wing rising. Keep the elevators down to safeguard the propeller.
(*g*) Notice that when taxying downwind there is a tendency for the speed to increase. Prevent this with use of brake.

Rudder Function Check (*Aircraft with fixed linkage nosewheel steering*)

(*a*) Make sure there is plenty of room ahead, reduce speed, taxi along the right of the taxiway, then look left to ensure it is safe to turn.
(*b*) Apply full rudder to the left and check for full and free movement.
(*c*) Now repeat the check to the right, first checking that it is clear to turn.

Instrument Function Checks

(*a*) At a low taxying speed move along the right hand side of the taxiway and check it is clear to turn left.
(*b*) Turn left and check – DI and compass readings decreasing, turn needle to the left and ball to the right.
(*c*) Now turn right; check DI and compass readings increasing, turn needle to the right and ball to the left.

(*d*) The attitude indicator should show no bank or pitch and the VSI should indicate no climb or descent during these checks.

Emergencies

Brake Failure

(*a*) If the brakes fail steer away from obstacles, make for an open grass area to increase rolling friction, shut down the engine, turn off the fuel, ignition and master switch. Then await assistance.

Nosewheel Steering Failure

(*a*) The nosewheel steering has failed. Reduce taxying speed and steer with use of differential brake.

(*b*) Do not attempt to park in a confined space. Taxi to an open position near the maintenance hangars, apply the parking brake, shut down the engine and report the malfunction.

Nosewheel Strut Collapse

(*a*) Imagine the nosewheel strut has collapsed and the aircraft is standing tail high with the engine resting on the ground.

(*b*) Carry out the emergency actions:
Ignition OFF
Fuel OFF
Advise ATC
Battery master OFF
Mixture to Idle cut-off
Parking brake ON

(*c*) Exit the aircraft, taking care not to fall from the tilted cabin, and await assistance.

Exercise 6

Straight and Level Flight

Background Information

When an aircraft is flying at a constant speed, height and direction the forces acting on it are in equilibrium, i.e. lift equals weight and thrust equals drag (Fig. 10). A change in any one of these forces will bring about a change in the others. For example, if power is increased, the thrust will be greater than the drag and the aircraft will accelerate. The increase in speed will produce an increase in lift and since the lift component will now be greater than the weight, the aircraft will begin to climb. Conversely, when power is reduced, thrust becomes less than drag and the aircraft will slow down until thrust and drag are in balance. The lower speed will reduce lift and the aircraft will descend.

Changes in throttle setting have an effect on the directional stability of the aircraft in addition to altering the amount of thrust.

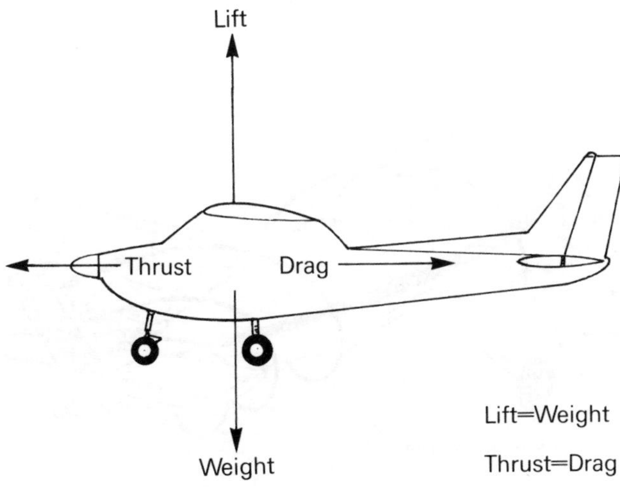

Fig. 10 Simplified diagram showing that in straight and level flight Lift equals Weight and Thrust equals Drag.

The tendency to yaw when the power setting is changed results from **Slipsteam Effect**. This is illustrated in Fig. 11. Influenced by the rotation of the propeller, the slipstream travels back around the fuselage in a helical path. Able to pass freely under the rear fuselage, the helical slipstream makes contact with one side of the fin and rudder, thus deflecting the tail of the aircraft to one side and causing a swing. The direction of swing will depend upon the rotation of the engine, most modern power plants turning in a clockwise direction when viewed from behind.

To overcome slipstream effect, the designers may use one or more of the following methods:

(a) Arrange a spring (sometimes adjustable in flight) to hold on rudder in the desired direction.

(b) Attach a small metal tab to the trailing edge of the rudder. This is set on the ground by bending in the direction necessary to hold on rudder.

(c) Offset the fin at a slight angle in order to give a permanent force in the opposite direction to the swing.

(d) Install the engine out of line with the fuselage so that the airscrew pulls to one side and counteracts the swing.

(e) Fit an adjustable trim tab to the rudder.

Unfortunately, with the exception of (d) and (e) and the adjustable version of (a), all of these arrangements only compensate perfectly at

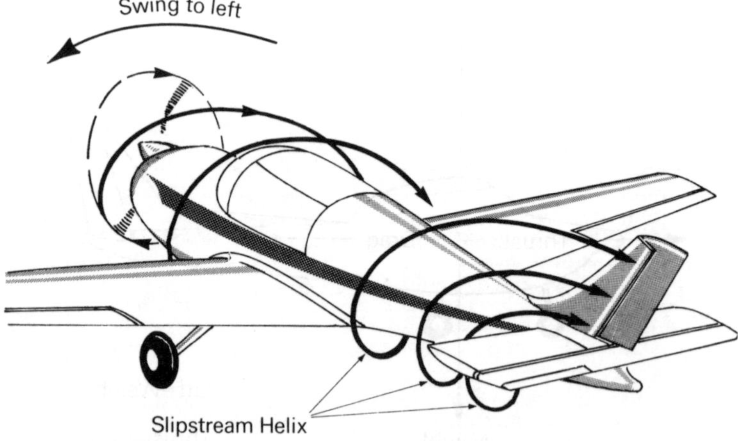

Fig. 11 Slipstream Effect, a cause of swing due to the propeller slipstream striking one side of the fin and rudder.

the cruising speed for which they are set, and changes of throttle upset the balance of the system. A rudder trim, which can be adjusted from the cockpit is of course the best arrangement.

Good straight and level is the basis of cross-country and instrument flying and the ability to fly in a straight line at a pre-determined height and airspeed will take the student pilot some time to master.

Attaining Level, Balanced Flight

When straight and level flight is to be attained after climbing to the required cruising level the nose is first lowered to the correct position relative to the horizon.

Climbing power is left on to help accelerate the aircraft and, as cruising speed is approached, the throttle should be brought back to the correct RPM. The correct order of actions for setting up straight and level flight is:

ATTITUDE - POWER - TRIM

Adjusting the Throttle

Unless the accelerator is depressed to provide more power when a car encounters a hill it will slow down and the engine RPM will decrease. Likewise, unless the driver reduces pressure on the accelerator a car will run faster downhill and its engine RPM will increase accordingly.

An aircraft behaves in much the same manner as a car in this respect. *At a fixed throttle setting*, when the nose is raised and the airspeed decreases load on the propeller/engine will increase causing a reduction in RPM. Conversely, if the aircraft is placed in a shallow dive the airspeed will increase, load on the propeller/engine will decrease and the RPM will increase.

More advanced aircraft are fitted with a constant speed propeller which automatically compensates for changes in airspeed and engine power but training aircraft have fixed pitch propellers. These are affected by airspeed in the manner described.

From the foregoing it follows that when setting the throttle to provide a particular engine speed the ASI must be indicating within a few knots of cruising speed, and not, for example, before levelling out and while still in the climb. Such a technique would result in the engine running at excessive RPM as the aircraft accelerated into the cruise.

Hunting the Airspeed is a common fault. As the term implies the student pilot will go from 'too slow' to 'too fast' in a series of prolonged pitching movements. This is caused by not allowing the airspeed to settle before making any corrections. Like any vehicle an aircraft has inertia and time is required to accelerate or decelerate. Over-correction and incorrect use of the trim control are perhaps the commonest faults of all.

For any condition of flight it is correct practice to think in terms of attitude in relation to the horizon and the position for cruising speed should be learned. Once the aircraft has settled, small corrections to the speed are made by raising or lowering the nose. In other words the airspeed is controlled by the elevators (angle of attack) and the purpose of the throttle is to determine whether the aeroplane will gain height, maintain height or descend at any particular airspeed. Small height corrections may be made with the elevators, trading speed for height.

Lateral Level, Direction and Balance

In Exercise 4 the inter-relationship between roll and yaw was explained. It therefore follows that unless the wings are level during straight and level flight the aircraft will turn in the direction of bank: left wing down, LEFT turn; right wing down, RIGHT turn. When, due to flying with one wing lowered, the aircraft has turned, say, five or ten degrees off heading a return to the correct direction is made by banking in the corrective sense. Every effort must be made to keep the wings level.

Inexperienced pilots sometimes fly with a wing down and, seeking to prevent a turn, opposite rudder is then applied. The out of balance condition that follows will be indicated by the balance indicator; the ball will move towards the lower wing. Out of balance flying of this kind is uncomfortable for the passengers and it also carries a speed penalty of several knots.

Effect of Power Changes and related Attitudes

Obviously, if speed is increased above normal cruising, more power will be needed to maintain height. On the other hand a reduction of speed by moving the wheel/stick back will cause the aeroplane to climb unless the power is decreased, and the method of adjustment is outlined under flight practice.

The aeroplane differs from other vehicles in a manner which is perhaps unique. It may be flown at a speed which requires the least amount of engine power for level flight. If an attempt is made further to reduce this speed by holding up the nose the aeroplane will begin to sink and *more* power will be required to maintain height. Indeed, because of the rapid increase in drag which is experienced at high angles of attack, and high drag requires high thrust to balance the forces, the lowest possible flying speed requires *full* throttle. At the other end of the scale, maximum speed also demands full power in order to balance the high drag resulting from the faster airflow (see graph 3, page 346).

Use of Trim

During a cross country flight a pilot is faced with a number of tasks which, in total, may be regarded as impossible to handle by some students. The pilot is required to study maps and charts, keep to a pre-arranged flight plan, use the radio equipment, monitor the instruments and maintain a good lookout. It therefore follows that flying straight and level must become second nature, as indeed it can with practice.

Much of the workload described can be minimized by good planning before flight while other tasks will become easier with experience. However, a properly trimmed aircraft will allow the pilot more time for other things; there are more important matters to be dealt with than constantly correcting a gain or loss or height due to a tail or nose heavy trim situation.

A little time devoted to accurate trim and careful power setting will more than repay the pilot in reduced workload.

Effect of Change of Configuration

Normally, straight and level flight is conducted with the flaps raised and the engine set to the recommended power for the cruise. In general terms, 75 per cent power will produce the fastest maximum continuous cruise for a light trainer and this will occur at 8000 feet at which altitude the throttle will probably be fully open. For maximum range a low power setting in the region of 45–55 per cent would be used but a good compromise between speed and range is usually achieved at 65 per cent power.

Power settings and all performance figures are listed in section 5 of the Flight/Operators Manual for the aircraft.

When, for any reason, there is a need to fly slowly, forward visibility will be improved by lowering 10–15° of flap. For any given speed the nose will adopt a lower position relative to the horizon and, since more power is required when the flaps are lowered, more slipstream will be generated. This will result in better rudder and elevator response.

Effect of Load

Each person represents approximately 10 per cent of the total weight of a light trainer. When flown at the maximum take-off weight it will fly slower than the same aircraft with perhaps half full tanks and only one person in the cabin. Speed difference is usually 3–5 kt.

Instrument Indications

Throughout the flight the aircraft's performance will be monitored on the instruments. Their interpretation and significance will be explained during the air exercise.

Exercise 6. Straight and Level Flight

AIM — *To fly the aircraft at a constant speed and height in the required direction and to fly at different airspeeds.*

AIRMANSHIP — *Height: sufficient for the exercise*
Airframe: flaps up
Engine handling: temperatures and pressures 'in the green'
Location: not in controlled airspace, etc
Lookout: use of clock code
Suitability of weather

AIR EXERCISE

Attaining level (from climb)	*Lateral level, direction and balance*	*Power/speed changes*	*Configuration changes*
LOOKOUT *Lower nose to correct position on horizon* *Airspeed increasing* *Hold attitude* *At expected speed:* *Cruising RPM* *Re-trim* *Check:* *Altitude* *Airspeed* *Trim* *Adjust* *Carb. Heat* *ATTITUDE-POWER-TRIM*	*LOOKOUT* *Wings level* *Balance; ball in centre* *Check heading on DI* *Correct heading with gentle bank* *OUT OF BALANCE DEMO.* *Check ASI reading while in balanced flight* *Apply: Left bank right rudder* *NOTE: Ball to left* *Lower airspeed*	*LOOKOUT* *At Constant Attitude* *Add Power – CLIMB* *Less Power – DESCEND* *Maintaining Altitude* *Add Power – ASI INCREASE* *Less Power – ASI DECREASE* *At Minimum Power Setting* *Note speed* *Raise nose and note:* *airspeed lower* *loss of height* *Add power – maintain height* *Max. Power = Min. Speed*	*LOOKOUT* *NOTE speed* *attitude* *RPM* *Lower 15° flap* *Hold speed and altitude* *NOTE attitude* *RPM* *Rudder and elevator feel* *Lower full flap* *Hold speed and altitude* *Return to normal Cruise and raise flaps* *LARGE TRIM CHANGE*

Flight Practice

OUTSIDE CHECKS

(*a*) Altitude: sufficient for manoeuvre.
(*b*) Location: not over aerodrome or towns or in controlled airspace.
(*c*) Position: check in relation to a known landmark.

AIR EXERCISE

Attaining Level Balanced Flight relative to the Horizon

(*a*) At the required altitude place the nose of the aeroplane on the horizon in the approximate straight and level attitude.
(*b*) Hold the aeroplane in this position and as the required speed is approached set the throttle to cruising RPM.
(*c*) Allow the speed to settle and move the elevator trimmer until no pressure is required on the wheel/stick. Make any adjustments to the airspeed which may be necessary by adjusting the attitude with the wheel/stick and re-trim.
(*d*) With the RPM correct check the height. If the aeroplane tends to climb lower the nose a little. On the other hand with a gradual loss of height the nose must be raised slightly. Re-trim after each correction.
(*e*) Larger height corrections require power adjustments.

Lateral Level, Direction and Balance

(*a*) Keep the wings level to prevent the aeroplane from turning off heading.
(*b*) With the wings level prevent the aircraft from yawing by correct use of rudder.
(*c*) Should the aeroplane swing off heading, move the wheel/stick in the direction necessary to bring it back into line. This movement should be co-ordinated with a little rudder in the same direction to maintain balanced flight.
(*d*) Avoid flying with 'crossed controls'; for example left wing down and right rudder to maintain heading. Notice the decrease in speed when flying out of balance.

Effect of Power Changes, Related Attitudes and Use of Trim

(*a*) From cruising speed open the throttle another few hundred RPM and notice that if the attitude is held constant the aircraft will climb.
(*b*) Prevent the climb by forward pressure on the wheel/stick and

re-trim. The airspeed will now be higher than cruising speed. Note the lower nose attitude.

(*c*) Now reduce power below cruising RPM. Notice that if the attitude is held constant the aircraft loses height. Prevent loss of height by backward pressure on the wheel/stick and re-trim. The aircraft will now fly at a lower speed than cruising. Note the higher nose attitude.

(*d*) Progressively decrease the power until the aeroplane just maintains height at a low airspeed (the speed will depend upon the aircraft type). Reduce the speed by moving back the wheel/stick. When the airspeed has settled the aeroplane will lose height slowly although the power has not been altered.

(*e*) Progressively reduce the speed step by step and note that power must be increased to maintain height until full throttle is required at the lowest possible flying speed. The aircraft is now flying in a very steep nose-up attitude.

(*f*) Now practise straight and level flight at selected airspeeds, trimming the aircraft to fly 'hands-off'.

Effect of Change of Configuration

(*a*) Imagine the weather has deteriorated and you require to fly at Low Safe Cruising Speed. Reduce speed to within the flap limiting arc.

(*b*) Lower part flap (according to type but usually 10–15°), add power to maintain height and re-trim. Notice the lower nose attitude for the reduced speed, giving improved visibility ahead and the increase in rudder/elevator effectiveness due to slipstream effect.

(*c*) Lower full flap, maintain speed and height by adding power. Note the very low nose attitude and high power setting.

(*d*) Return to normal flaps-up straight and level. Notice large trim change as flaps are raised. In some aircraft there will be a tendency to sink and lose height as the flaps are raised.

Instrument Indications

When fitted the following instruments will give these indications during straight and level flight. At this stage of training they should be used to supplement outside visual references –

(a) *Airspeed Indicator (ASI).* Because time is required by the aircraft to change speed there is a slight delay before the instrument settles to a new airspeed. This lag is appreciable and can cause inexperienced pilots to 'hunt the airspeed'.

(b) *Altimeter.* Under certain conditions vertical currents may cause height to vary although the airspeed and RPM are correct and the pilot must compensate for these variations during prolonged straight and level flying. In the case of a sudden gain in height the nose should be lowered and height lost by increasing the speed at that power setting. Similarly a small loss in height can be regained by holding up the nose. In extreme cases power setting adjustments will also be needed.

(c) *Engine Speed Indicator* (sometimes referred to as the Tachometer). Notice that a change in airspeed will alter the RPM, although the throttle has not been moved. Like a car running downhill a gentle dive will decrease the load on the propeller and cause the RPM to increase. A climb will produce the reverse effect. For this reason always re-check the RPM when the airspeed has settled.

(d) *Turn and Balance Indicator.* This is two instruments in one, so arranged because of the interrelation between directional and lateral movements of the aircraft. Notice that, if a wing is deliberately lowered while keeping straight on rudder, the ball (slip and skid) will move towards the lower wing, indicating a slip in that direction. If the wings are held level and the rudder made to yaw the aircraft, the needle (turn) will show a turn in the appropriate direction while the ball will displace away from the turn, indicating a skid outwards. Some Turn and Slip indicators make use of two needles, the lower pointer indicating Rate of Turn while the upper needle shows Slip or Skid. In practice the two-finger presentation is similar in use to the more modern needle and ball instrument. Yet another presentation provides turn information in the form of a small, banking aircraft. Such an instrument is known as a **Turn Co-ordinator**. No pitch indications are given on these instruments although at first glance they resemble an Artificial Horizon.

(e) *Vertical Speed Indicator (VSI).* The slightest vertical movement of the aircraft up or down is indicated and becomes apparent as the wheel/stick is moved back or forth. The VSI indicates in feet per minute rate of climb or descent.

(f) *Direction Indicator (DI).* This gyro-operated instrument must be synchronized with the compass before it can be used for navigational purposes. Turn to the left and right when the DI will instantly measure the angular change of heading.

(g) *Atttitude Indicator (AI).* Notice at low airspeeds the model plane is above the horizon bar whereas the wings of the model cover the horizon line at cruising speed. A gentle dive places the symbol below the horizon line. Bank the aeroplane slightly and, in addition to the attitude of the symbol in relation to the horizon bar, notice how the pointer indicates the angle of bank. Like the Direction Indicator the AI is a gyro-operated instrument.

(h) *The Magnetic Compass*. See Exercise 9, page 103.

Exercise 7
Climbing

Background Information

Imagine a motor-car moving steadily along a level road. When it comes to a hill more power will be required if it is to ascend. Similarly the aeroplane requires more power while climbing.

In addition to height and airspeed it is now necessary to introduce a further measurement – **Rate of Climb**. As the term implies this refers to the rate at which the aeroplane gains height and it is expressed in feet per minute. The Vertical Speed Indicator (VSI) referred to in the previous chapter is arranged to read zero in level flight. The instrument will also register **Rate of Descent** in feet per minute and the dial is marked in the manner shown in Fig. 12.

Maximum rate of Climb

It is sometimes necessary to attain a specific altitude as quickly as possible and to attain the maximum rate of climb two conditions must be satisfied:

1. Aerodynamic efficiency.
2. Engine efficiency.

Fig. 12 The Vertical Speed Indicator showing a constant height.

If the maximum amount of engine power is to be devoted to producing the highest rate of climb it follows that the wing must provide the most lift possible for the least drag. For example, a small angle of attack would result in high speed but a poor climb rate. Conversely, a climb attempted at too high an angle of attack would create excessive drag which would be paid for in terms of wasted power. Forward speed would be low, there would be a risk of overheating the engine due to the lack of cooling air and, because of the demands on power occasioned by excessive drag, rate of climb would be poor.

Ideally, the maximum rate of climb would occur when the wing is flown at its best lift/drag angle of attack (page 139, Volume 2), but the airspeed corresponding to this angle is rarely ideal for aircraft with fixed pitch propellers since, because of the relatively low speed, loads imposed on the propeller prevent the engine developing its maximum power.

The importance of engine power during a maximum rate climb is easy to visualize in motoring terms – a car climbing a hill, the need to change to a lower gear and maintain engine RPM/power and so forth. In an aircraft the combined factors of aerodynamics and engine power are closely interrelated in the following manner. Imagine that aircraft X requires 60 hp to fly level at its best lift/drag ratio angle of attack and that the speed corresponding to that angle is 65 kt. If the engine has a maximum power of 100 hp there would be 40 surplus horsepower available to climb the aircraft at that speed. It therefore follows that the more surplus horsepower available, the better the rate of climb. In other words, the best rate of climb will occur when the biggest difference exists between **Horse Power Available** and **Horse Power Required** and a typical graph showing these two curves is included in Appendix 1. However to ensure that the engine is capable of developing its maximum power the airspeed quoted in the aircraft manual for maximum rate of climb is usually slightly faster than the best L/D speed.

In essence, there is an optimum speed which, at maximum power, will produce the fastest rate of climb. Any attempt to climb at a faster or slower speed will reduce climb rate (Fig. 13).

Cruise Climb

Although for training purposes maximum rate of climb is often used to gain height quickly for the upper air exercises a more common technique, one adopted in cross country/transport flying, is the **Cruise Climb**. As the term implies the climb is flown at an airspeed

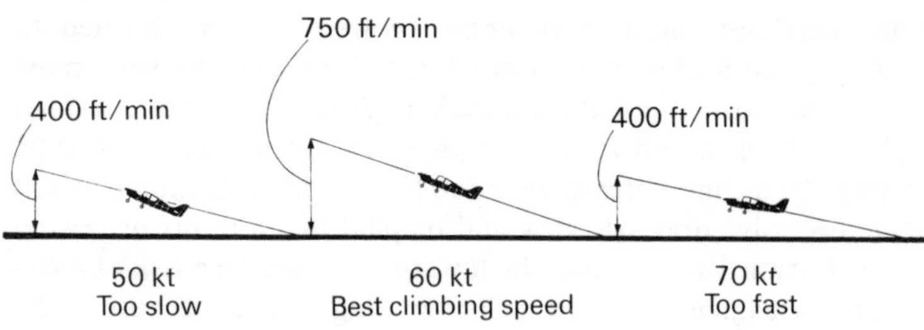

Fig. 13 The relationship between Airspeed and Rate of Climb.

somewhere between that for normal cruise and the previously described best rate of climb speed.

The purpose of the cruise climb is to gain height while covering distance more effectivley than would be the case when all the surplus horsepower is being used to produce the fastest rate of climb.

Maximum Angle of Climb

At some time in the future, after gaining a PPL, the pilot may be faced with a need to operate into and out of small private airstrips. Techniques to be adopted during the take-off and landing will be described in the relevant sections of this book but at this stage it is sufficient to mention that when obstacles (trees, power lines, etc.) have to be cleared **Maximum Angle of Climb** is more important than achieving the maximum rate of climb. The latter is expressed in terms of height gained in feet/minute. Maximum angle of climb, on the other hand, entails gaining height on the steepest ascending flight path possible.

To achieve the maximum angle of climb maximum power is required and the airspeed will be lower than that required for best rate of climb. Some aircraft, but certainly not all designs, benefit from the use of flap. Whether or not flap should be used will be stated in the aircraft manual. It should be noted that during this type of climb rate of height gain will be lower than that achieved when flying at the speed recommended for best rate of climb. However, airspeed is lower and the change in relationship between feet/minute climbed and feet/minute forward is such that a steeper climb path is attained with the advantage of improved obstacle clearance.

Initiating the Climb

Assuming the aircraft is in cruising flight and it is intended to gain, say, 2000 feet as quickly as possible the first action is, as always, to look left, ahead, right and above before attempting to change cruising level. Only after the pilot is satisfied that it is safe to climb should the throttle be opened to provide maximum engine power. Sequence of actions for entering the climb may be remembered by the words:

POWER ATTITUDE TRIM

The nose is raised to adopt the climbing attitude, something that will be demonstrated by the flying instructor who will align a convenient part of the aircraft (top of the instrument panel, edge of windscreen etc., according to aircraft type) with the horizon. Speed must be allowed to settle before fine adjustments on the wheel/stick are made to attain the airspeed recommended for maximum rate of climb. The aircraft should then be accurately trimmed.

While adopting the climbing attitude it is essential to remain laterally level, otherwise a turn will develop towards the lower wing.

Balanced Flight

Although some of the more powerful single-engine tourers and all twin-engine aircraft are fitted with a rudder trim which may be adjusted from the cabin, light trainers rely on a fixed metal trim tab, affixed to the trailing edge of the rudder, which is set to provide balanced flight at cruising speed. Changes in power and/or airspeed from that speed will entail use of the rudder to restore balanced flight.

During a maximum rate climb the airspeed is relatively low and power is at or near maximum, slipsteam and torque effect are at a high level and unless the pilot applies corrective rudder the aircraft will fly out of balance and rate of climb will suffer.

When the engine turns clockwise when seen from behind there will be a tendency during the climb for the aircraft to yaw to the left, causing a skid to the right which will be indicated by the ball in the balance indicator. Therefore right rudder should be applied to maintain balanced flight. In some aircraft considerable rudder pressure is required and this calls for determination on the part of the pilot if corrective action is to continue while attending to other matters during the climb.

Power, Speed and Rate of Climb

During the early stages of the exercise the instructor will demonstrate the effects on rate of climb of various power settings, from low power/level flight at best L/D speed up to maximum power. The effects of adopting speeds above and below the recommended figure for maximum rate of climb will also be shown.

Lookout during the Climb

Modern light aircraft tend to climb in a somewhat steep nose-up attitude and this can limit the view ahead. Bearing in mind than an aircraft changing altitude is in a similar position to a car changing lanes on the road the importance of lookout will be apparent.

To ensure that there is no danger of climbing into another aircraft, or even a large cloud, two methods of clearing the area ahead are open to the pilot:

1. The nose may *gently* be lowered at regular intervals and then returned to the climbing attitude.
2. The aircraft may be turned slightly left and right of the climb path so that the area ahead is revealed.

Returning to Straight and Level Flight at Selected Heights

Larger and more powerful aircraft than trainers initiate the return to straight and level flight some 50–150 feet below the required altitude, thus allowing time for the transition from one mode of flight to the next. This technique does not apply to light aircraft with relatively low rates of climb. Sequence of actions is:

ATTITUDE POWER TRIM

At the required altitude the pilot must look in all directions to ensure that it is safe to level out. Then the nose is lowered to the cruising attitude while climbing power is retained to help accelerate the aircraft.

As the speed increases the load on the propeller will decrease, engine RPM will, in consequence, increase and as the ASI approaches cruising speed the throttle will be moved back, quite considerably in most aircraft, to attain the correct RPM for cruising flight. During the transition from climb to level flight it is essential that the wings remain level and that the new nose attitude is maintained by forward pressure on the wheel/stick while the airspeed increases. The aircraft should then be accurately trimmed. At this stage there will be no need to hold on rudder (balance).

Engine Handling Considerations

During prolonged climbs the engine is delivering near maximum power at a time when forward speed is relatively low. While this should present no cooling problems in temperate climates the cylinder head temperature gauge (when fitted) should be watched when flying in countries where high temperatures are normal. If the red line is approached the nose must be lowered and power should be reduced while the engine cools. The climb can then be resumed.

Because air density decreases as height is gained so engine power decreases and this in turn results in a gradual reduction in rate of climb from take-off to the absolute ceiling of the aircraft. More advanced aircraft are sometimes fitted with a **Turbocharger** for the purpose of:

(*a*) Providing extra power during take-off and the climb.
(*b*) Maintaining maximum power up to, typically, 12,000 feet.

Low-powered trainers are fitted with simple **Normally Aspirated** engines which suffer from the power loss with height already mentioned. The effect on rate of climb, assuming standard temperature conditions, for a light trainer in widespread use is as follows. Note that the recommended speed for maximum rate of climb is reduced slightly as height is gained.

Altitude (ft)	*Indicated airspeed (kt)*	*Rate of climb (ft/min)*
Sea-level	67	715
1000	66	675
2000	66	630
3000	65	590
4000	65	550
5000	64	505
8000	62	380
10,000	61	300
12,000	60	215

Whereas full throttle is used continuously for the climb when flying light trainers, in more advanced aircraft a setting for maximum continuous power is often quoted in the aircraft manual.

Use of the Mixture Control

Above 3000 feet the mixture control should be moved towards LEAN (i.e. eased back) until the maximum RPM are indicated thus confirming that the fuel/air ratio that altitude. Further leaning will be required as height is gained but over-leaning will cause over-heating of the engine. This will be confirmed on the cylinder head temperature gauge (when fitted).

Exercise 7. Climbing (Maximum Rate of Climb)

AIM — *To gain height at the fastest rate while maintaining a constant heading and to revert to straight and level flight at the required altitude.*

AIRMANSHIP — *Height: suitable for the exercise*
Airframe: flaps up
Engine handling: temperatures and pressures 'in the green'. Lean mixture above 3000 ft
Location: not in controlled airspace, etc
Lookout
Suitability of weather

AIR EXERCISE

Entering the climb	*During the climb*	*Return to straight and level*
LOOKOUT *Maximum power* *Check yaw (rudder)* *Raise nose to climbing attitude* *Hold attitude* *Wings level* *Adjust attitude for correct speed* *Re-trim* *Order of actions* *POWER ATTITUDE TRIM*	*LOOKOUT (lower nose or turn at intervals)* *Instrument Checks:* *Airspeed (Ref: ASI)* *Balance (Ref: Ball)* *Heading (Ref: DI)* *Rate of Climb (Ref: VSI)* *Engine Handling:* *Above 3000 ft, lean mixture* *Check engine Ts and Ps*	*At required altitude:* *LOOKOUT* *Lower nose to S & L attitude* *Maintain balanced flight* *Hold attitude* *Wings level* *As cruising speed approaches:* *Power for crusing* *Adjust attitude/speed* *Re-trim* *Order of actions* *ATTITUDE POWER TRIM*

Maximum Angle of Climb (Flap as recommended in the Aircraft Manual)
NOTE: Lower Airspeed: Steeper nose Attitude: Lower Rate of Climb: Steeper Climb Path
WARNING: Flaps must not be raised before all obstacles are cleared

Flight Practice

COCKPIT CHECKS

(*a*) Trim for straight and level flight.
(*b*) Power set for cruising conditions.

OUTSIDE CHECKS

Look around and ascertain that no other aircraft are in the intended path of climb.

AIR EXERCISE

Initiating the Climb (maximum rate)

(*a*) Check wings are level and open the throttle fully. Be prepared for the aircraft to swing due to slipstream and torque effect and check the yaw with rudder.
(*b*) Move the wheel/stick back and assume the climbing attitude using a point on the aircraft in relation to the horizon.
(*c*) Hold this attitude until the airspeed settles, then if necessary move the wheel/stick slightly backwards or forwards until the correct climbing speed is reached. Hold this attitude and re-trim so that there is no load on the wheel/stick.
(*d*) Notice the altimeter readings which should indicate a steady gain in height while the VSI will show the rate of climb. Notice the indication of the attitude indicator.
(*e*) Keep a good lookout. During a long climb turn gently from side to side or at intervals lower the nose in order to see ahead.

Balanced Flight

(*a*) During the climb maintain heading by keeping the wings level. If the aircraft has turned while setting up the climb regain the required heading by banking gently in the corrective direction.
(*b*) Check that the aircraft is in balance and if necessary correct by applying rudder in the direction indicated by the ball in the balance indicator.

Demonstration of Power, Speed and Rate of Climb

(*a*) From straight and level flight reduce power and adopt speed to that recommended for maximum rate of climb. Check that the aircraft is maintaining height and re-trim.
(*b*) Add, say, 200 RPM, maintain the same speed and note the rate of climb.

(*c*) Now apply full power, adjust attitude to hold the same speed and note the higher rate of climb.
(*d*) Hold up the nose and reduce speed. Note that rate of climb decreases.
(*e*) Lower the nose and increase to above maximum rate of climb speed and note that the VSI is showing a lower than maximum rate of climb.

Returning to Straight and Level Flight at Selected Heights

(*a*) At the required altitude move the wheel/stick forward and bring the aircraft into the level attitude.
(*b*) As cruising speed is approached reduce the throttle setting to cruising RPM, checking any tendency for the aircraft to swing. Finally, when airspeed settles to that required, check engine RPM and adjust if necessary. Retrim for level flight.

Engine Handling Considerations

(*a*) Always make final RPM adjustments at the correct airspeed.
(*b*) At intervals check the engine temperatures and pressures, particularly during prolonged climbs in hot conditions.
(*c*) Above 3000 feet slowly move back the mixture control until maximum RPM are achieved. The mixture is now correct for that altitude. As height is gained continue adjusting the mixture to maintain the correct fuel/air ratio.
(*d*) Avoid over-leaning which can over-heat the engine.

Effect of Flaps

(*a*) Trim the aircraft at normal climbing speed and note the rate of climb on the VSI.
(*b*) Lower optimum flap (15°–25°, according to aircraft type), hold the previous speed and re-trim. Note the reduced rate of climb.
(*c*) Now reduce speed to that recommended for the climb with flap. There will be a slight improvement in the climb angle and a reduction in maximum rate of climb.
(*d*) Raise the flaps and note the trim change.

En Route Climb (Cruise Climb)

(*a*) At maximum continuous climbing power trim the aircraft at the recommended speed for the en route climb.
(*b*) Note the slight reduction in normal climb rate and the larger increase in forward speed.

(*c*) For economy use the mixture control in accordance with the aircraft manual. This is the climb technique usually adopted during navigational flights.

Maximum Angle of Climb

(*a*) At maximum power select a higher than usual nose-up attitude and trim the aircraft at the recommended best climb gradient speed.
(*b*) If recommended for the type, lower optimum flap.
(*c*) Note the high nose attitude, low forward speed and rate of climb which will be slightly less than for a maximum rate climb. The aircraft is now climbing at an angle giving maximum obstacle clearance.
(*d*) In a prolonged climb at low airspeed check the engine instruments for signs of overheating and either reduce power or increase the airspeed if this is evident.

Exercise 8
Descending

Background Information

The descent is not only used to change from a higher to a lower altitude; it is also the final manoeuvre immediately prior to landing. There are two methods of descending: *(a)* **Gliding** and *(b)* **Powered Descent**. Each method has its own particular advantages and applications.

In 'Straight and Level Flight' the function of the throttle control was explained. When it is closed completely the aeroplane is left without the thrust required to overcome drag and a force must be provided from another source in order to keep the aeroplane moving forward. The 'weight' component takes over this function and Fig. 14 shows how it is able to fill the gap left by thrust. Taking the extreme case, if the aeroplane were dived in a vertical position, 'weight' would be opposed and balanced by 'drag'. There would be little lift because of the small angle of attack during the manoeuvre, but the arrangement is unusual since under normal conditions thrust and not weight balances drag. In this position the aeroplane would lose height at its maximum rate of descent and very quickly exceed the maximum speed limit for the airframe (**Never Exceed Speed**).

Applying these principles to a more normal if less extreme case, Fig. 15 shows the aeroplane in a 45° dive. It can be seen that weight is balancing both lift and drag. Although the rate of descent will be less than that experienced in a vertical dive, the aeroplane will still move forward under the influence of its own weight, or putting the situation into motoring terms, it will 'coast downhill'.

The rate of descent would be high in a 45° dive but at its optimum angle of attack the aeroplane will have the flattest glide path in relation to the ground. Such a condition will only occur when the weight is called upon to oppose the most amount of lift for the least amount of drag. In other words, the flattest glide will occur at the best lift/drag ratio angle of attack. This corresponds to a particular speed for the type of aircraft and any attempt to glide at a higher or lower

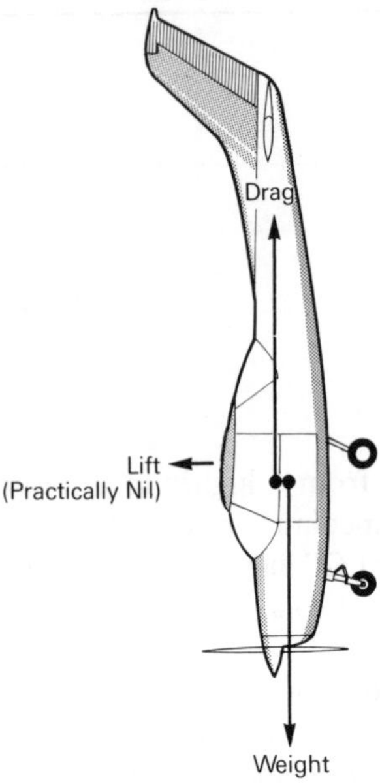

Fig. 14 Without Thrust, when the engine is throttled back, the aircraft continues to move forward under the influence of Weight. In the extreme case of a vertical dive Weight will be balanced by Drag when the speed has settled.

airspeed will result in a steeper **Glidepath** (measured in relation to the ground).

If too high a gliding speed will steepen the glide path, for the foregoing reasons, an attempt to stretch the glide by holding up the nose will cause the aeroplane to sink because of the increased drag resulting from the higher angle of attack. This can be misleading to the pilot because, although the aeroplane seems to be in the level attitude, the glidepath is steep. Fig. 16 illustrates the effect of airspeed on the glidepath.

Initiating the Descent

When changing from level to descending flight the pilot is, as in the climb, in a similar position to the motorist changing lanes. Lookout is therefore the first action before closing the throttle. Naturally, the lack of slipstream and engine torque, hitherto balanced during cruising flight by the fixed rudder trim tab, will cause a swing (to the

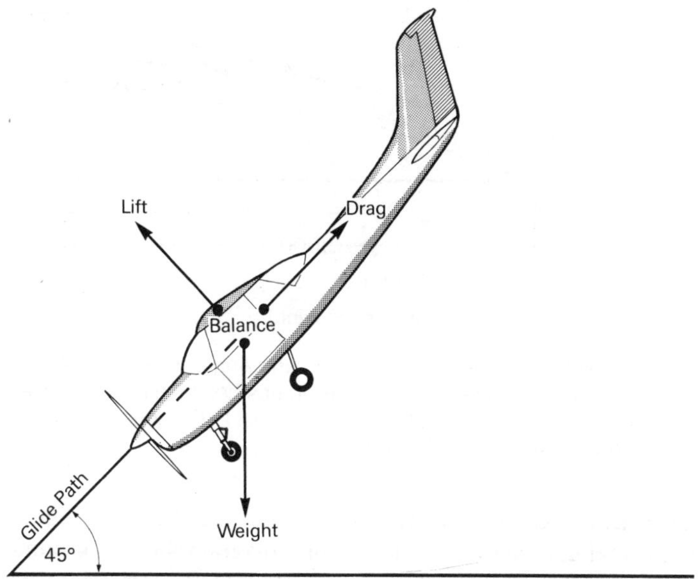

Fig. 15 In this steep glide Weight is balanced by Lift and Drag. As in Fig. 14, Weight is the force pulling the aircraft forward.

right in the case of engines rotating clockwise when seen from behind). This will have to be contained with left rudder.

To guard against carburettor icing the carburettor heat control must be fully applied *before* closing the throttle, thus checking its correct function by the small decrease in RPM that follows and guarding against inadvertent use of the mixture control. Such action would operate the idle cut-off and stop the engine.

Gliding, Flaps up

Like maximum rate of climb the speed for best gliding performance (i.e. maximum distance covered per unit height loss) is achieved at or near best L/D ratio angle of attack and the speed relating to this condition will be quoted in the aircraft manual.

Usually a descent is made with power and this will be described later in the exercise. However gliding is important and it should be practised by the pilot with a view to estimating gliding distances. Engine failure is, these days, very rare but later in the course the action to be taken in these circumstances will be described. Without power the aircraft must glide to the selected forced landing area and

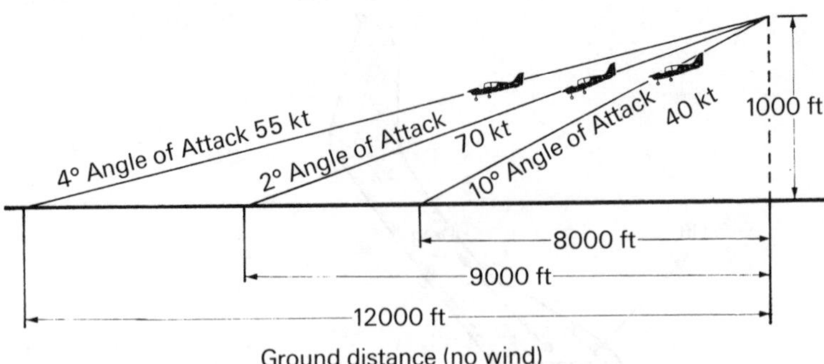

Fig. 16 The relationship between Airspeed and Angle of Glide. At 40 kt the aircraft sinks rapidly. 70 kt dissipates height in a gentle dive. For this particular aircraft 55 kt is the best L/D speed which produced the flattest descent path.

Section 3, EMERGENCY PROCEDURES of the Owner's/Flight/Operating manual usually includes a simple diagram showing ground distance covered for every 2000 feet of height loss. (In some manuals the information is in Section 5, PERFORMANCE.) The figures quoted relate to conditions of no wind, a headwind reducing ground speed and steepening the glide path and a tailwind having the reverse effect.

An average training aircraft will cover approximately 1·5 nm over the ground for every 1000 feet of height loss, assuming no wind.

During prolonged glides the engine must be warmed at intervals by opening the throttle. This will be more fully explained under 'Engine Considerations'.

Resuming Straight and Level Flight

As the aircraft descends towards the new cruising altitude the pilot must carefully check that it is clear to level out without risk of flying near to another aircraft or entering cloud.

As the nose is raised to the level attitude power is applied, the carburettor heat is moved to COLD and the aircraft will then accelerate to cruising speed. Small adjustments to height, speed and power will then be made and the aircraft will be accurately trimmed.

Flaps and Trim

The effect of lowering flap is to steepen the glide path, the steepest descent path occurring with maximum flap. Most aircraft exhibit a nose-down trim as the flaps are lowered and this is easily compensated for with the trim control.

Of more importance is the change in trim which occurs when maximum power is applied following a descent with full flap. A powerful nose-up pitch can result, requiring considerable forward pressure on the wheel/stick until the aircraft is re-trimmed. Flaps are described on page 170, Volume 2 of this series.

Effect of Power on Rate of Descent

There are times when it is necessary to control the rate of descent within fine limits, e.g. instrument flying or while approaching to land. The glide is unsuitable under these conditions and a **Powered Descent** must be used for the purpose. A speed below that related to best lift/drag ratio is used so that with the engine throttle closed there is a high rate of descent. Throttle is added sufficiently to give the desired rate of descent which will be indicated on the VSI. This instrument was mentioned in the chapter on 'Climbing'. When the descent is too fast, more power is added but if a higher rate of descent is required power is reduced. During these throttle adjustments the airspeed must be kept constant on the elevators, i.e. rate of descent is controlled by the throttle and airspeed is controlled with the wheel/stick.

Maintaining a constant Rate of Descent and Approach Simulation

During the **Landing Approach** a flat glide path can be a disadvantage for several reasons. Obstacle clearance is poor when crossing the airfield boundary prior to landing and forward vision is bad because of the flat, nose-up attitude of the aeroplane. Flaps are employed to:

(*a*) Reduce the stalling speed thus providing additional safety while flying at low airspeeds.
(*b*) Increase drag, allowing a more nose-down attitude to be adopted for any given airspeed, thus improving forward visibility.
(*c*) Because of (*b*), provide a larger range of descent rates as the throttle is adjusted.

The increased power required when flaps are lowered will provide more effective rudder and elevator control. In most cases the nose tends to drop when the flaps are lowered because the centre of pressure moves back as they are applied. There are a few aircraft which have a nose-up tendency but these are the exception rather than the rule.

As explained in Volume 2 the first 15°–25° of depression gives the largest lift increase and a modest increment in drag while further lowering increases the drag to a marked degree with little further improvement in lift. If it is necessary to raise the flaps while flying near the ground this should be done gradually since some aircraft tend to sink as the flaps are lifted.

The powered descent is the standard technique used in aviation for changing cruising levels and approaching an airfield prior to landing. In preparation for the landing approach, which is the subject of an exercise to be described later, the pilot should practise maintaining constant descents at varying rates, using part and full flap. In all cases the airspeed is adjusted on the elevators while the throttle (power) is used to control the rate of descent.

Because of the advent of jet aircraft, in which the pilots directly control thrust when the throttles (known as thrust levers in jets) are moved, there has been a tendency among some instructors to reverse the long-established 'speed with elevator – descent rate with power' technique. However, the pilot of a piston/propeller-driven aircraft is adjusting power, not thrust, on the throttle and the technique described in this book will prove more convenient than trying to adapt jet handling to light aircraft. In practice the accurate control of descent rate towards a runway entails an intelligent combination of attitude and power adjustments. These will be described more fully in Exercise 13.

Engine Considerations

When a descent is commenced from altitudes where the mixture has been leaned it should first be moved into RICH (i.e. fully forward) before the descent is initiated.

Mention has already been made of the need to apply carburettor heat before closing the throttle at the start of a glide but in prolonged gliding descents the engine must on no account be allowed to become cold. At intervals of 1000 feet (500 feet in cold weather) the throttle should be fully opened for a few seconds to prevent over-cooling.

A cold engine can be slow to respond when power is needed at the bottom of the descent and allowing the engine to over-cool will materially shorten its life.

The Sideslip, Entry and Recovery

Before the introduction of flaps the **Sideslip** was the prime method of increasing the rate of descent without diving and thus increasing the airspeed, possibly at a time when it is undesirable.

In modern aviation the main application of the sideslip is to counter drift while landing (Exercise 13). In a sideslip a wing is lowered and yaw is prevented by applying rudder in the opposite direction, e.g. wheel/stick to the left, left wing down. Prevent left yaw by holding on right rudder. During a sideslip the pilot will be flying with 'crossed controls'. The aeroplane will now have a path of descent which is somewhere between the nose and the lowered wing and not straight ahead (Fig. 17).

When in order to speed up the descent the angle of bank is increased so must opposite rudder be increased if the aeroplane is to keep straight. Maximum angle of bank will be reached when full rudder is applied to maintain direction and any attempt to increase the bank still further will cause a yaw.

It is convenient to practise sideslips at several thousand feet and to save time lost in regaining height after each descent some power can be left on to reduce the high rate of sink.

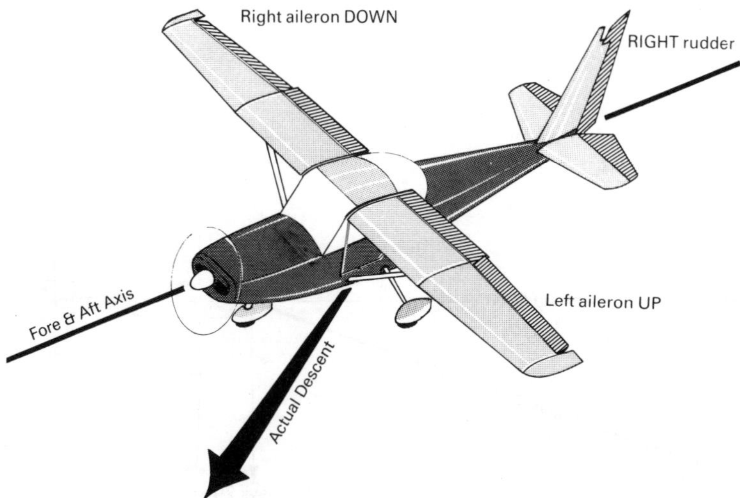

Fig. 17 The Sideslip. Illustration shows a sideslip to the left using left aileron for the bank and right rudder to prevent yaw to the left. Note direction of the descent.

A velocity additional to the forward speed occurs during a sideslip. This is illustrated in Fig. 18 where it can be seen that there is a 'sideways' speed towards the lower wing tip. The actual path of descent is shown as a solid bold line and it will be appreciated that, since it represents the resultant of two speeds (one forward and one sideways), it must therefore be faster than the gliding speed before sideslipping is commenced. For this reason there is a tendency for the airspeed to increase during a sideslip and the nose must be held up slightly if the airspeed is to remain constant.

The sideways speed is greater than many experienced pilots realize and it is interesting to line up with a runway at say 500 feet and slip off the excess height. The distance travelled to one side of the runway during the brief descent is considerable. It is therefore impossible to effect a landing off a sideslip when the aeroplane is lined up with the landing direction, (*a*) because of the very considerable drift and (*b*) because at the end of the sideslip the aeroplane would more likely be over the edge of the airfield than on the landing area.

Most aircraft manuals recommend restricting the amount of flap to be used in a sideslip, half being typical. The sideslip as a means of dealing with crosswind conditions during an approach and landing is explained in Exercise 13.

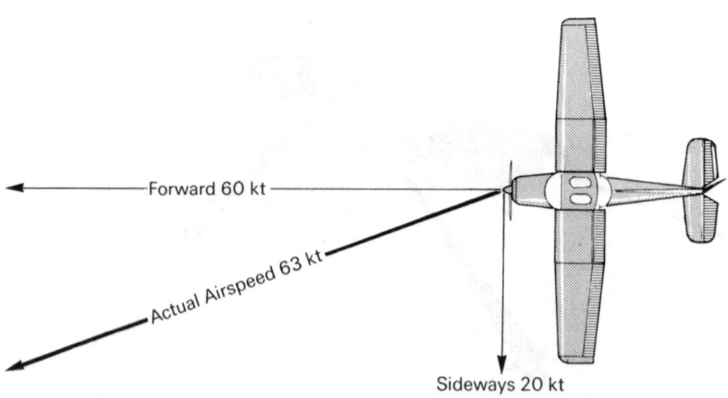

Fig. 18 The airspeed tends to increase during a sideslip because it is the product of two velocities, one forward and the other sideways towards the lower wing.

Recovering from a Sideslip

To stop a sideslip and return the aircraft to a straight glide the wings are levelled and at the same time opposite rudder is removed. The nose, which was held up slightly to prevent an increase in airspeed, is allowed to adopt the correct gliding attitude.

Instrument Indications

During a descent of any kind the altimeter will indicate a steady decrease in altitude and the VSI will show the rate of descent.

The attitude indicator will depict the aircraft symbol slightly below the horizon by an amount that depends on the power (RPM) being used and whether or not flap has been selected. In a correctly balanced descent the ball will remain in the centre of the balance indicator. The usual check of engine temperatures and pressures should be made at intervals.

Pre-flight Briefing

Exercise 8. Descending (Gliding, Flaps up).

AIM — *To lose height at a constant rate while maintaining both heading and airspeed, and to revert to straight and level flight while at the required altitude.*

AIRMANSHIP — *Height: sufficient for the exercise*
Engine handling: temperatures and pressures 'in the green'. Carb. heat
Location: outside controlled airspace, etc
Lookout
Suitability of weather

AIR EXERCISE

Entering the descent	*During the descent*	*Return to straight and level flight*
LOOKOUT *Carb heat: Mixture RICH* *Close throttle* *Check yaw* *Wings level* *Balance* *Adjust attitude/speed* *Re-trim* *Order of Actions* *POWER ATTITUDE TRIM*	*LOOKOUT* *Instrument Checks:* *Airspeed* *Balance* *Heading* *Rate of descent* *Engine Handling:* *Open throttle and warm every 500–1000 ft*	*At required altitude LOOKOUT* *Open throttle to cruise RPM* *Carb heat COLD* *Raise nose to S & L attitude* *Wings level* *Balance* *Adjust power/attitude* *Re-trim* *Order of Actions* *POWER ATTITUDE TRIM*

Exercise 8. Descending (Engine Assisted with Flap)

AIM — *To lose height at a controlled rate while maintaining both heading and airspeed, and to revert to straight and level flight at the required altitude.*

AIRMANSHIP — *Height: sufficient for the exercise*
Airframe: flap as required
Engine handling: temperatures and pressures 'in the green'
Location: outside controlled airspace, etc.
Lookout
Suitability of weather

AIR EXERCISE

Entering the descent	*During the descent*	*Return to straight and level flight*
LOOKOUT *Mixture RICH* *Reduce power and speed* *Check yaw* *Wings level* *Balance* *Lower part flap (LIMIT SPEED)* *Adjust attitude/speed* *Re-trim* *Order of Actions* *POWER ATTITUDE TRIM*	*LOOKOUT* *Instrument Checks:* *Airspeed* *Balance* *Heading* *Rate of descent* *Adjusting descent rate* *Add Power: = slower descent rate* *Reduce Power: = faster descent rate* *Add Flap:* *= nose-down trim* *=faster descent rate* *Reduce Flap:* *=nose-up trim* *=slower descent rate*	*At required altitude LOOKOUT* *Open throttle to cruise RPM* *Raise flaps* *Adopt S & L attitude* *Wings level* *Balance* *Adjust power/attitude* *Re-trim* *Order of Actions* *POWER ATTITUDE TRIM*

Pre-flight Briefing

Exercise 8. Descending (Sideslipping)

AIM — *To increase the rate of descent in a sideslip and practise the exercise for future use in countering drift before landing.*

AIRMANSHIP — *Height: sufficient for the exercise*
Airframe: flaps UP
Engine handling: temperatures and pressures 'in the green'. Part power to reduce height loss during practice
Location: outside controlled airspace, etc.
Lookout
Suitability of weather

AIR EXERCISE

Entering the sideslip	*During the sideslip*	*Return to straight glide*
LOOKOUT *Carb heat: Mixture RICH* *Set up glide and trim* *LEFT aileron* *RIGHT rudder to check yaw* *Hold airspeed (nose up)* *NOTE: Intrument indications* *High rate of descent* *Changed flight path* *Relate glide path to line feature*	*LOOKOUT* *Increase bank slightly* *Add rudder: check yaw* *NOTE: Higher descent rate* *More drift – (check line feature)* *Increase bank until FULL rudder is required to hold yaw* *Maximum sideslip* *Warm engine every 500–1000 feet*	*LOOKOUT* *Level wings* *Centralize rudder* *Establish glide attitude* *Check:* *Wings level* *Balance* *Airspeed* *Heading* *Trim* *SAFETY HEIGHT*

Flight Practice

COCKPIT CHECKS

(*a*) Trim for straight and level flight.
(*b*) Engine set for cruising RPM.
(*c*) Carburettor heat control as required.

OUTSIDE CHECKS

(*a*) Altitude: sufficient to begin a descent.
(*b*) Location: not near other traffic, airfields or controlled airspace.
(*c*) Position: check in relation to a known landmark.

AIR EXERCISE

Initiating the Descent

(*a*) Check below the aircraft. Apply carburettor heat. From cruising power close the throttle fully, at the same time keeping straight. Using the elevators, prevent the nose from dropping below the horizon.
(*b*) As the airspeed decreases to near best recommended gliding speed, allow the aeroplane to settle in the gliding attitude and re-trim. Note the attitude of the nose in relation to the horizon.

Gliding, Flaps Up

(*a*) Make any airspeed adjustments which may be necessary by slight backward or forward movement of the wheel/stick and re-trim. The aeroplane is now descending in a straight line at a steady rate and airspeed. Note the instrument indications.
(*b*) Maintain a good lookout and during prolonged glides open the throttle at 500/1000 feet intervals (according to temperature) in order to warm the engine.

Resuming Straight and Level Flight

(*a*) At the required altitude open the throttle to cruising power, keep straight and adopt the straight and level attitude in relation to the horizon. Set the elevator trim control in the approximate position. Return the carburettor heat to COLD.
(*b*) Allow the airspeed to settle and check the RPM. Re-trim.

Gliding: Flaps Down

(*a*) Check below the aircraft. Apply carburettor heat. Close the throttle from cruising power and keep straight as before. Maintain straight and level attitude by backward pressure on the wheel/stick and hold this position until the speed has reduced to below maximum for flap operation.

(*b*) Lower flap to the required number of degrees and allow the aeroplane to take up the 'flaps down' gliding attitude which will be steeper than before.

(*c*) Re-trim at the correct airspeed and the glide path is now steeper. The aircraft symbol on the altitude indicator will indicate the steeper attitude. Also the airspeed is lower than before although there is a steeper nose-down attitude. Forward visibility is now much improved.

Resuming Straight and Level Flight

(*a*) At the required altitude open the throttle to cruising power while keeping straight and bring the aeroplane into the level attitude. Return the carburettor heat to COLD.

(*b*) Before the aeroplane accelerates beyond flap limiting speed, raise the flaps gradually and re-trim. If preferred trimming may be carried out in stages as the flaps are raised. When the airspeed has settled to that required check RPM and re-trim.

Effect of Power on Rate of Descent

(*a*) Check below the aircraft. From straight and level flight select a lower power setting, say, 1500 RPM.

(*b*) Reduce the speed some 10 kt below best gliding speed. Re-trim when the airspeed has settled and note the attitude of the nose in relation to the horizon.

(*c*) The rate of descent is less than before and the glide path flatter. Because of the slipstream both rudder and elevators are more effective than in the glide.

(*d*) Control the rate of descent by increasing or decreasing the RPM while the airspeed is kept constant with the elevators.

(*e*) Resume straight and level flight as before, check that it is clear ahead of the aircraft.

(*f*) Now practise the powered descent, using various degrees of flap.

(*g*) With full flap notice the steep descent path when power is reduced and that considerable power is needed to maintain height.

Maintaining a Constant Rate of Descent and Approach Simulation

(*a*) Check below the aircraft, then reduce power and bring the speed back to 10 kt below best gliding speed.

(*b*) Lower half flap (according to aircraft type) then adjust the throttle to provide a 500 ft/min rate of descent. Re-trim.
(*c*) Controlling the airspeed on the elevators and the rate of descent with the throttle maintain the descent rate and airspeed. If descent rate is too high increase power slightly and raise the nose a little to maintain speed. If the descent rate is too low reduce power slightly and lower the nose a little to maintain speed as before.
(*d*) Now practise the exercise at the same speed but at various rates of descent. This is the basis of a landing approach.
(*e*) Check the altimeter and ensure that the descent does not continue below a safe height.

Engine Considerations

(*a*) Before closing the throttle completely always apply full carburettor heat. Note the decrease in RPM – this confirms that the control is working correctly.
(*b*) Before descending always ensure that the mixture is RICH.
(*c*) During prolonged power-off glides warm the engine every 500–1000 feet (according to temperature) by opening the throttle fully for a few seconds. Leave the carburettor heat in HOT to help maintain engine temperature.
(*d*) At intervals check the engine temperatures and pressures.

The Sideslip – Entry and Recovery

(*a*) At several thousand feet (with a little power on to reduce the rate of descent during practice), move the wheel/stick to the left and apply sufficient right rudder to prevent a yaw.
(*b*) Correct the tendency for the airspeed to increase by backward pressure on the wheel/stick. Notice the path of descent is between the nose and the lower wing.
(*c*) Steepen the bank and increase the rudder to keep straight and the rate of descent will increase.
Bank still further until full opposite rudder is required to keep straight. When any further bank will cause a yaw this is the maximum rate of sideslip.
(*d*) To resume a straight descent, level the wings, centralize the rudder and move the wheel/stick gently forward to maintain the airspeed.
(*e*) Now repeat the exercise to the right.

Instrument Indications

(*a*) During a gliding descent note the position of the aircraft symbol in the attitude indicator. The altimeter is showing a constant height loss and the vertical speed indicator is showing the rate of descent. In turbulence this instrument will tend to fluctuate.

(*b*) Lower full flap and note that to maintain IAS the aircraft symbol has to be moved further below the horizon bar, the rate of descent has increased and the altimeter is decreasing in reading more rapidly.

(*c*) Now add power in stages until the aircraft is maintaining height with full flap. Retrim at intervals. Note the varying instrument readings.

Exercise 9
Turning

Background Information

The Mechanics of Turning

Unless it is forced into a turn a moving body of any kind will proceed in a straight line. For example a stone swung around on the end of a length of string will describe a circle only because the string is held at one end and it is therefore pulled towards the centre. Should the string break the stone will immediately fly away in a straight line.

An aeroplane behaves in exactly the same way as a stone in resisting a circular path. If an attempt is made to hold the wings level while applying rudder, the aircraft will skid in the opposite direction, rather like a car on ice. To avoid skidding it is necessary to provide a force towards the centre of the turn to act like the string in the case of the experiment with the stone. This is accomplished by inclining the lift (which acts at right angles to the wings top surface) towards the centre of the turn. To do this it is necessary to **Bank** the aircraft.

Lift is now made to provide two forces, a vertical one performing the usual function of supporting the weight of the aeroplane, and the other in a horizontal plane which acts as the 'string' pulling the aircraft towards the centre of the turn. This is called **Centripetal Force** and the tighter the radius of turn for any particular airspeed the more powerful must be this force.

Entry into Level Medium Turns and the Return to Straight and Level Flight

A Medium Turn is used for changing direction during general flying such as manoeuvring around the airfield circuit. Before altering heading it is vitally important to ensure that there is no risk of turning into the path of another aircraft and the pilot must look thoroughly around the aircraft, starting from the opposite side to the direction of intended turn, through the windscreen and then to the side and behind. In the case of a left turn the visual check starts on the right of the aircraft.

Aircraft of high-wing design suffer from the disadvantage of blocking visibility in the direction of turn after the wing has been lowered and it is prudent first to lift the wing slightly and clear the area into which the aircraft will be turning.

The turn is initiated by rolling on bank to an angle of some 25°–30°. Some aircraft require a little rudder in the same direction to maintain balanced flight and there is a tendency to lose height in the turn unless the nose is raised slightly. In aircraft with side-by-side seating the pilot sits off centre and nose position relative to the horizon varies slightly when seen during left and right turns.

To stop turning the pilot must look in the intended direction of flight and ensure that there are no other aircraft in the flight path. The wings are then rolled level, rudder (if it has been applied) is centralized and the nose is allowed to adopt the straight and level attitude.

Control of Angle of Bank and Radius of Turn

Radius of turn for any aircraft is dependent upon:

(*a*) Airspeed.
(*b*) Angle of bank.

Both factors are under the control of the pilot and they allow the aircraft freedom to turn on a wide or tight radius. Turns at shallow angles of bank are used for instrument flying purposes; steep turns (i.e. 45° of bank or more) enable the pilot to orbit a point on the ground which is more or less directly below the aircraft.

Fig. 19 shows that at any particular airspeed to tighten the radius of a turn it is necessary to increase the angle of bank so making more of the lift available as a turning force, i.e. centripetal force. If height is to be maintained during the turn more lift will be required and this is obtained by backward pressure on the wheel/stick, thereby increasing the angle of attack. This will, of course, result in a decreased airspeed and the steeper the angle of bank (tighter the turn) the lower the airspeed. It will be seen from the diagram that in a 60° banked turn twice the normal amount of lift is required to maintain height. During medium turns, however, (25°–30° of bank) the decrease in airspeed only amounts to some 5 kt.

When the aircraft is turning the outer wing will move faster than the inner wing, causing unequal lift. Consequently in practice it will

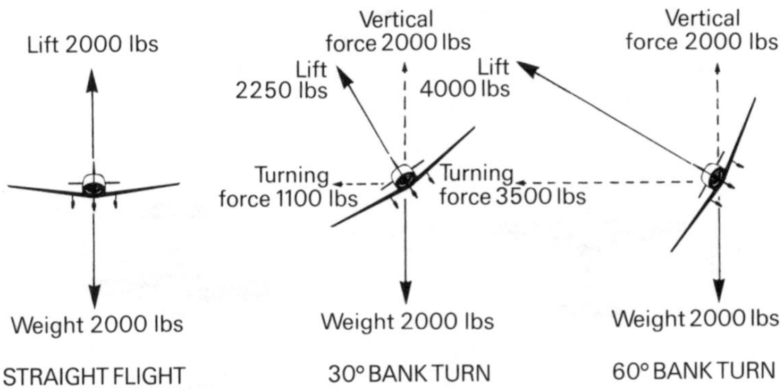

Fig. 19 The forces in a turn. Lift must serve two functions and provide both a turning and a lifting force.

be found necessary to move the stick away from the direction of turn and back to the neutral position in order to prevent a steeper angle of bank developing than that intended.

Accurate angles of bank are shown on the attitude indicator.

The number of degrees an aircraft changes heading in a given time is known as the **Rate of Turn**. A change in heading of 3° per second (equivalent to a complete reversal of direction in one minute) is known as a Rate 1 Turn.

The rates explained in Fig. 20 are shown on the **Turn Co-ordinator** or, when fitted, **Turn and Balance Indicator** in the aeroplane.

Rate of turn is dependent upon the airspeed and the angle of bank. The higher the airspeed the greater the angle of bank required for any given rate of turn. At low airspeeds the angle of bank will be less for any given rate of turn than at faster speeds. The approximate angle of bank for a Rate 1 turn may easily be calculated – 10% of the Indicated Airspeed in knots + 7 or 10% of the Indicated Airspeed in MPH + 5, e.g. Rate 1 turn at an IAS of 90 kt = 9 + 7 = 16° Angle of Bank. Rate 1 turn at an IAS of 500 MPH = 50 + 5 = 55° Angle of Bank.

Slipping and Skidding

Aircraft with well designed ailerons which provide additional drag on the inside of the turn rarely need the use of rudder during turns. When balanced turns cannot be made on the ailerons alone, and a number of aircraft exhibit this characteristic, there will be a tendency to slip in towards the centre of the turn. Rudder in the same direction as the

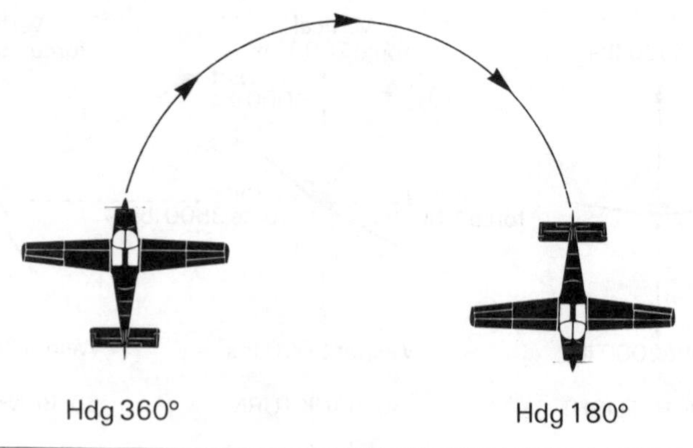

Rate 1=180° Turn in 1 min

Rate 2=180° Turn in ½ min

Rate 3=180° Turn in ¼ min

Rate 4=180° Turn in 7½ sec

Fig. 20 The Rate of Turn.

turn will then have to be applied to restore balance. In practice the pilot will enter the turn by co-ordinating aileron and rudder in the same direction.

The amount of rudder required is usually quite small and too much rudder will induce a skid outwards, i.e. in the opposite direction to the turn. These out of balance faults are dealt with in the next section.

Instrument Indications

During visual flight (i.e. flight other than by sole reference to the instruments) the instruments are used to confirm the accuracy of a turn and to correct faults that may occur as a result of faulty handling.

Slipping and Skidding as mentioned in the previous section will be indicated by the balance ball in the turn and balance indicator or, when one is fitted, the turn co-ordinator. Corrective action is simply a matter of applying rudder in the direction indicated by the ball until it centres. Bank angle must be maintained with aileron during these corrections.

Rate of Turn, the number of degrees change of direction per unit of time, will be shown on the turn needle of the turn and balance indicator or, in the case of a turn co-ordinator, the degree of bank on the aircraft symbol. These instruments will provide rate of turn information even when slip or skid is present. However, in a properly balanced turn the ball will be in the centre and a full glass of water in the cabin would not spill, even in a steeply banked turn. The turn and balance indicator and the turn co-ordinator are shown in Fig. 21.

Angle of Bank is shown pictorially on the attitude indicator. There is also a pointer and scale giving the number of degrees of bank.

Nose attitude, too low or too high, will also be shown on the attitude indicator but this instrument is incapable of displaying aircraft pitch angles (nose-up, nose-down) sufficiently accurately to ensure that height is not gained or lost while in a turn. To obtain this information one must refer to the vertical speed indicator for short-term climb–descend information while the altimeter will show a gain

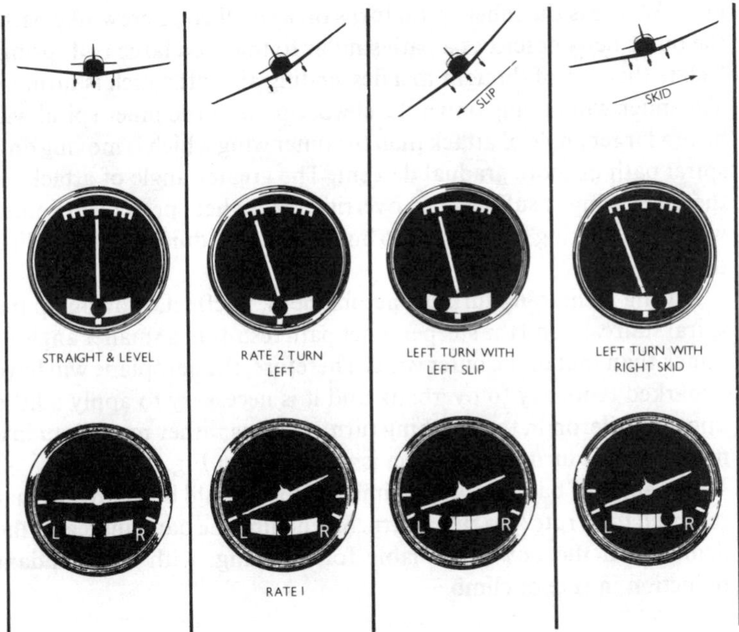

Fig. 21 Indicating Rate of Turn and Balance. Top row of instruments shows a typical Turn and Balance Indicator capable of reading up to a Rate 4 turn. Below them are shown the same readings on the Turn Co-ordinator. These instruments are only calibrated up to Rate 1.

or loss of height less instantly but more accurately since the VSI is subject to fluctuations in turbulent conditions.

Climbing and Descending Turns

At this stage of flying training, the student pilot will have practised climbing, descending and turning. It is often necessary to climb over a specified area in a series of turns or perhaps descend towards an airfield which is to one side of the aeroplane. In either case a combination of the three exercises mentioned will be required.

In a medium level turn it is necessary to move the wheel/stick away from the direction of the turn in order to prevent the angle of bank becoming too steep (page 96). During a descending turn there is no such tendency and bank must be held on throughout. The reason for this behaviour will be understood if a spiral staircase is imagined. A little thought will reveal that both the inner and outer spirals of such a staircase will descend through the same distance from one level to the next. Whereas the inner spiral turns on a small corkscrew-like path, the outer helix describes a path similar to that of a large coil spring. Relate this line of thought to a descending aircraft which is turning. The inner wing flying down the steeper path of the inner spiral will have a larger angle of attack than the outer wing which is moving on a spiral path of more gradual descent. The greater angle of attack on the inner wing is sufficient to override the higher speed of the outer wing thus making it necessary to hold on aileron during a descending turn.

During a climbing turn the aeroplane is, in effect, moving up the spiral staircase and the steeper inner path results in a smaller angle of attack than that of the outer wing. Therefore, the aeroplane will have a marked tendency to overbank and it is necessary to apply a little opposite aileron in the climbing turn, perhaps rather more than in a medium level turn. Spiral effect is illustrated in Fig. 22.

The angle of bank during a climbing turn should be limited to that required for a rate ½ to rate 1 turn. Steepening the bank increases the demand on the power available for climbing, with an attendant reduction in rate of climb.

Stall Prevention during Descending Turns

Figure 19 showed that in a 60° bank turn double the amount of lift was required to maintain height compared with straight and level flight. As the angle of bank increases so lift requirements multiply.

These lift increases during turns are matched by a reaction in the

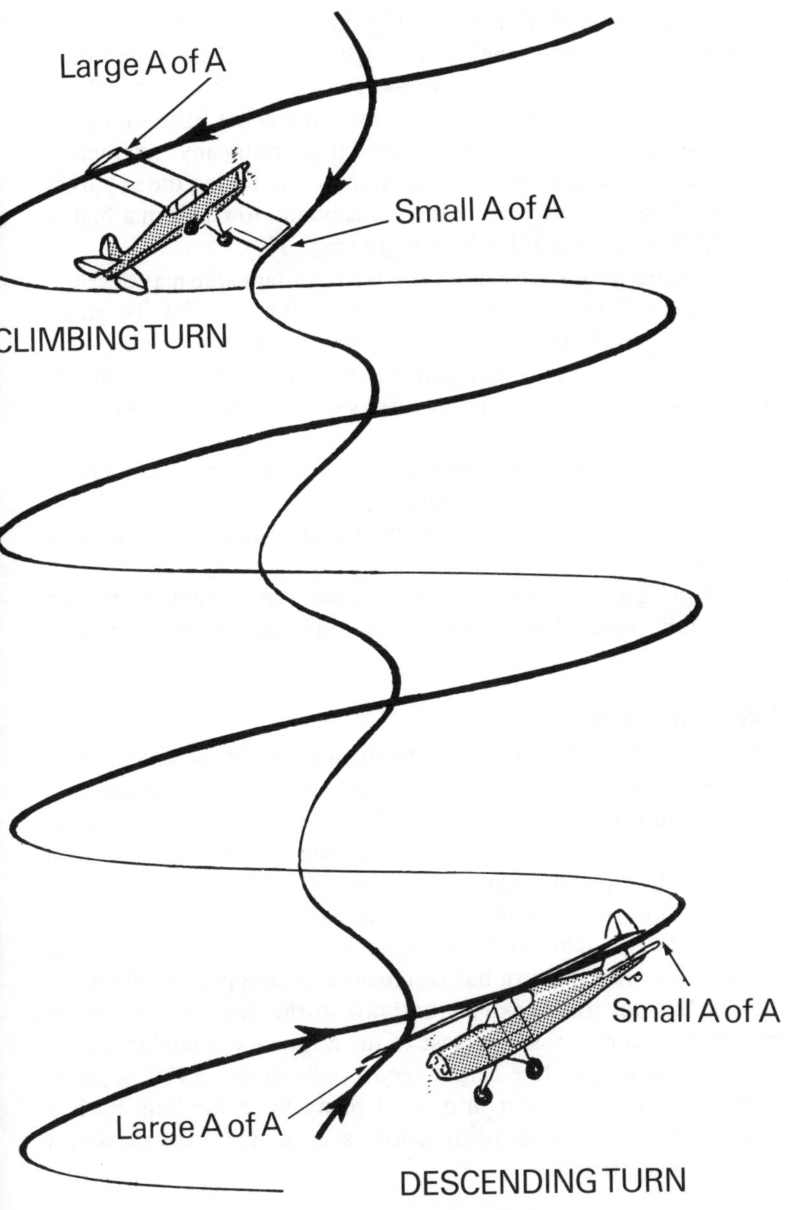

Fig. 22 Climbing and Descending Turns showing the spiral paths followed by the inner and outer wings and the effect on their angles of attack.

opposite direction which can be felt by the occupants of an aircraft as an apparent increase in their own weight, forcing them more firmly into their seats. This is the well-known 'g' force and it will be discussed in greater detail when the student reaches Exercise 15.

At this stage it should be understood that during any turn there is an increase in 'g' and this has the effect of increasing the apparent weight of the aircraft, which, in turn is bound to result in a higher stalling speed than that for level flight (page 117).

If a stall occurs during a turn there is a risk that a spin may develop. This topic will be dealt with in Exercises 10A and 10B. To guard against the possibility it is the practice before entering a gliding turn to lower the nose slightly and gain an additional 5 kt for bank angles of up to 30° or so. Steeper gliding turns should be entered 10 kt above normal gliding speed.

During descending turns with power there is no need to increase airspeed. Instead the power should be increased slightly since this will reduce stalling speed and improve elevator and rudder control response.

Modern light aircraft are less prone to inadvertent spinning but the precautions outlined in this section should be taken nevertheless.

Slipping Turns

Under certain conditions, it is convenient to slip off height during a descending turn rather than in a straight glide. Such a manoeuvre is called a **Slipping Turn** and, because like any turn, the outer wing travels faster, a steeper angle of bank can be attained than in a straight side-slip and height loss can be very considerable.

In principle the slipping turn is a gliding turn with too much bank for the rate. It is achieved by reducing the backward pressure on the wheel/stick after the turn has commenced and applying a little top rudder but not to the extent that yaw in the direction of turn is prevented. From a position close to the downwind boundary of the field it is possible to lose height very rapidly during a 180° slipping turn prior to rolling out into wind ready for a landing. Such a procedure is ideal when obstructions such as tall trees prevent a straight-in approach.

Turning onto Selected Headings

In the early stages of training the student will be expected to turn at a constant height with the aircraft in balance. The next stage is to turn towards a prominent feature (large cloud, lake, etc.). It will soon become clear that roll out from the turn should be commenced 5–10° before reaching the feature. Time is required to level the wings and, during the process, turning continues, albeit at a reducing rate.

Unless the new heading is anticipated there can be only one result; the aircraft will turn through the chosen point and it will be necessary to alter heading in the opposite direction. When such corrections have to be made, or when the change in direction is limited to a relatively few degrees, a gentle angle of bank should be adopted so that the turn is at a slow rate.

Use of the Direction Indicator

The Direction Indicator is described in Chapter 9, Volume 2. Being a gyro-operated instrument with no inherent means of seeking magnetic north it must be synchronized with the magnetic compass before its readings may be applied to any form of navigation. The same text also explains why the instrument must be re-synchronized with the magnetic compass at intervals of 10–15 minutes.

The instrument is simple to use and, unlike the magnetic compass, its readings are not affected by acceleration–deceleration or the lateral level errors that occur while flying on northerly and southerly headings (compass errors are described on pages 47–50, Chapter 2, Volume 2).

When using the DI for turns onto selected headings precautions similar to those previously described for turns onto prominent features apply, namely, the roll out should commence some 5°–10° before reaching the new heading.

During visual flight the pilot must maintain a good lookout while turning. The temptation to 'bury the head in the cockpit', concentrating on the instruments at the expense of collision avoidance must be resisted. The correct technique to adopt when using the DI is to monitor its readings at intervals during the turn, spending the final moments during roll-out confirming heading.

Compass Turns

Compass errors, as explained on pages 47–50, Chapter 2, Volume 2, render the instrument difficult to use in an aircraft unless certain rules are observed.

1. When reading the compass during straight and level flight the wings must be level and the airspeed must be steady.
2. While making turns angle of bank must be limited to that required for a Rate 1 change in heading. Turns at faster rates render the instrument unreadable.

Compass errors, turning and acceleration, are the result of magnetic **Dip**, the tendency for the magnet system to be pulled down towards the magnetic poles. As explained in Volume 2, the effects of dip are at their maximum at the magnetic poles, becoming less pronounced as the equator is neared. The effects of dip act in reverse directions, north and south of the equator. The following rule of thumb corrections for compass errors relate to flying in the northern hemisphere. South of the equator all effects and corrections apply in the reverse direction.

Turning on to North

Roll out of the turn some 25°–30° before the compass reads north. Because the compass magnets will have dipped towards the lower wing as the northerly heading is approached, the action of levelling the wings will swing the magnet system back through the last twenty-five or so degrees. It may be necessary to correct the heading with gentle turns.

Turning on to South

Roll out of the turn some 25°–30° after the compass reads south. Here again the action of levelling the wings will bring the magnets up from the lower wing on to south.

When on north or south the wings must be level while checking the compass.

Turning on to East or West

As there is no lateral level error on east or west it is not necessary to overshoot or undershoot the turn by 25°–30° but allowance should be made for the time required to stop the turn. The roll out should be commenced 5°–10° before east or west.

When checking compass heading on east or west the airspeed must be steady otherwise acceleration or deceleration error will affect the reading.

The errors described are at their maximum on N, S, E and W and they change progressively as the aircraft turns. For example, midway between north and east (NE) there will be some lateral level error and some acceleration and deceleration error.

Pre-flight Briefing

Exercise 9. Turning (Level Medium Turns)

AIM — *To turn the aircraft at a constant rate and height and roll out on a required heading, either with reference to an outside visual reference or the direction indicator.*

AIRMANSHIP — *Height: suitable for the exercise*
Engine handling: temperatures and pressures 'in the green'
Location: outside controlled airspace, etc.
Lookout
Suitability of weather

AIR EXERCISE (turn to the LEFT)

Entering the turn	*During the turn*	*Roll-out from the turn*
LOOKOUT (high-wing precautions) *Bank to left (aileron)* *Slight left rudder (balance)* *At 25–30 degrees:* *Back pressure on wheel/stick to maintain height* *Maintain bank angle (ailerons neutral)* *S-M-O-O-T-H use of controls* *Co-ordination* *Avoid over-reaction to faulty bank angle/nose attitude*	*LOOKOUT* *Instrument Checks:* *Bank angle (Ref: AI)* *Constant height (Ref: VSI & Alt)* *Progress of turn (Ref: Visual/DI)* *Turn Rate 2–2 (Ref: Turn needle)* *Balance (Ref: Ball)* *Correct:* *Bank/Rate with aileron* *Balance with rudder* *Height with elevator*	*LOOKOUT (particularly in intended new direction)* *ANTICIPATE by 5–10 degrees:* *Visual roll-out point* *DI heading* *Level wings (aileron)* *Centralize rudder* *Relax elevator pressure* *Instrument Checks:* *Height* *Heading* *Balance* *Adjust as necessary*

Pre-flight Briefing

Exercise 9. Turning (Climbing Turns)

AIM	*To gain height at a constant rate while turning.*
AIRMANSHIP	*Height: suitable for the exercise* *Airframe: flaps as required* *Engine handling: temperatures and pressures 'in the green'* *Location: outside controlled airspace, etc.* *Lookout* *Suitability of weather*

AIR EXERCISE (climbing turn to the RIGHT)

Entering the turn (from the climb)	*During the turn*	*Roll-out from the turn*
LOOKOUT (high-wing precautions) *Gentle bank to right (aileron)* *Slight right rudder (balance)* *At 15–20 degrees:* *Slight back pressure on wheel/stick to maintain airspeed* *Maintain bank angle (ailerons)*	*LOOKOUT* *Tendency to over-bank* *Instrument Checks:* *Bank angle (Ref: AI)* *Constant climb rate (Ref. VSI)* *Progress of climb (Ref. Alt.)* *Progress of turn (Ref: Visual/DI)* *Max. Rate 1 turn (Ref: Turn needle)* *Balance (Ref: Ball)* *Correct:* *Bank/Rate with aileron* *Balance with rudder* *Airspeed with elevator*	*LOOKOUT (particularly in intended new direction)* *ANTICIPATE by 5–10 degrees:* *Visual roll-out point* *DI heading* *Level wings (aileron)* *Centralize rudder* *Instrument Checks:* *Height* *Rate of climb* *Heading* *Balance* *Adjust as necessary*

Pre-flight Briefing

Exercise 9. Turning (Descending Turns)

AIM — *To lose height at a constant rate while turning in a glide, with power and with or without flap.*

AIRMANSHIP — *Height: sufficient for the exercise*
Airframe: flaps as required
Engine handling: temperatures and pressures 'in the green' Carb. heat during glides
Location: outside controlled airspace, etc.
Lookout
Suitability of weather

AIR EXERCISE (gliding turn to the LEFT)

Entering the turn (from the glide)	*During the turn*	*Roll-out from the turn*
LOOKOUT (high-wing precautions) *Increase gliding speed 5 kt by lowering nose slightly* *Bank to left (aileron)* *Slight left rudder (balance)* *At 25–30 degress:* *Maintain bank angle (hold on aileron)* *Maintain airspeed*	*LOOKOUT* *Need to hold on bank* *Instrument Checks:* *Bank angle (Ref: AI)* *Descent rate (Ref: VSI)* *Descent progress (Ref: Alt.)* *Progress of turn (Ref: Visual/DI)* *Balance (Ref: Ball)* *Correct:* *Bank angle with aileron* *Balance with rudder* *Airspeed with elevators* *WARM ENGINE every 500–1000 ft*	*LOOKOUT (particularly in intended new direction)* *ANTICIPATE by 5–10 degrees:* *Visual roll-out point* *DI heading* *Level wings (aileron)* *Centralize rudder* *Return to gliding speed (elevator)* *Instrument Checks:* *Height* *Heading* *Airspeed* *Balance* *Adjust as necessary*

Gliding Turn with Flap — *NOTE: Lower nose attitude: Lower speed: Higher rate of descent*

Descending Turn with Power and Flap. — *NOTE: Rate of descent adjusted with power: More responsive elevator and rudder = Better control*

Pre-flight Briefing

Exercise 9. (Slipping Turns)

AIM — *To increase the rate of descent in a gliding turn by introducing sideslipping*

AIRMANSHIP — *Height: sufficient for the exercise*
Airframe: flaps up or at maximum recommended in the aircraft manual
Engine handling: temperatures and pressures 'in the green'. Carb. heat during glides
Location: outside controlled airspace, etc.
Lookout
Suitability of weather

AIR EXERCISE (slipping turn to the LEFT)

Entering the turn	*During the turn*	*Roll-out from the turn*
LOOKOUT (high-wing precautions) *Increase gliding speed 5 kt by lowering nose slightly* *Bank to left (aileron)* *At 30–35 degrees:* *Slight opposite rudder (right)* *Slight forward elevator* *Avoid excessive airspeed*	*LOOKOUT* *Need to hold on bank* *Low turn rate for bank angle* *High sink rate* *Instrument Checks:* *Bank angle (Ref: AI)* *Descent rate (Ref: VSI)* *Descent Progress (Ref: Visual near ground)* *Progress of Turn (Ref: Visual near ground* *Sideslip (Ref: Ball to Left)* *Correct:* *Bank angle with aileron* *Airspeed with elevator* *Sideslip with rudder*	*LOOKOUT (particularly in intended new direction)* *ANTICIPATE by 5–10 degrees* *ANTICIPATE height above ground (high sink rate)* *Level wings (aileron)* *Centralize rudder* *Return to gliding speed (elevator)* *CHECK PROXIMITY TO GROUND* *Instrument Checks:* *Height* *Airspeed* *Balance*

Flight Practice

COCKPIT CHECKS

(*a*) Trim for straight and level flight.
(*b*) Engine set at cruising RPM.

OUTSIDE CHECKS

(*a*) Altitude: sufficient for the manoeuvre.
(*b*) Location: not over towns or airfields, or in controlled airspace.
(*c*) Position: check in relation to a known landmark.

AIR EXERCISE

Entry into Level Medium Turn and return to Straight and Level Flight

(*a*) LOOK OUT! Start on the opposite side to the turn, continue around the nose of the aircraft and end by looking behind in the intended direction of turn. Make sure no other aircraft are in the vicinity.
(*b*) Bank by moving the ailerons in the direction of turn at the same time applying a little rudder in the same direction in order to maintain balance. Correct any tendency for the bank to steepen by moving the wheel/stick away from the direction of the low wing and maintain the nose of the aircraft on the horizon by slight backward pressure on the stick. The aircraft is now turning at a steady rate without gaining or losing height.
(*c*) To come out of the turn, look ahead, level the wings by applying aileron and rudder in the opposite direction to the turn. As the wings become level, maintain the nose in the correct position by returning the wheel/stick to its trimmed position.

LOOK AROUND FOR OTHER AIRCRAFT BEFORE COMMENCING THE TURN, DURING THE TURN AND PRIOR TO COMING OUT OF THE TURN. THE IMPORTANCE OF THIS CANNOT BE OVER-STRESSED.

Control of Angle of Bank and Radius of Turn

(*a*) During a medium turn the bank angle should be 25–30°. Practice assessing this angle by relating the horizon to a part of the aircraft.
(*b*) Check the bank angle with the attitude indicator. If the bank is too shallow apply more aileron to increase the rate of turn. At the correct bank angle centralize the ailerons to prevent over-banking. If the bank is too steep, rate of turn will be high and the nose will tend to drop below the horizon. Reduce bank with the ailerons and maintain the correct angle.

Slipping and Skidding

(*a*) If the ball in the turn and balance indicator has moved towards the centre of the turn the aircraft is slipping in. Apply rudder in the direction indicated by the ball to regain balanced flight.

(*b*) If the ball has moved away from the direction of turn the aircraft is skidding out. Apply rudder in the direction indicated by the ball as before.

Control of Height

(*a*) If the aircraft is gaining height too much back pressure has been applied on the wheel/stick. Lower the nose slightly.

(*b*) Should the aircraft lose height the nose must be raised by slight back pressure on the wheel/stick.

Instrument Indications (to be demonstrated during a turn)

(*a*) During a properly balanced medium level turn the instruments will read as they are now indicating:

Attitude Indicator	showing 25–30° of bank, nose slightly above the horizon
Turn needle	reading Rate 2–2½
Ball	in the centre confirming balance
Direction Indicator	giving a steady change in heading
Altimeter	showing a constant height
Vertical Speed Indicator	at zero
Airspeed Indicator	reading some 5 kt less than for straight and level flight

(*b*) Instrument scan during visual flight must not inhibit LOOKOUT while turning.

Climbing Turns

(*a*) While climbing look around in the usual way before commencing the turn.

(*b*) Initiate a turn in the required direction but limit the angle of bank to that for a Rate ½ to Rate 1 turn. Be prepared to 'hold off bank' rather more than for a level turn and maintain the correct climbing speed with the elevators.

(*c*) Keep a constant lookout and note the instruments indicate a turn with an increase in height.

(*d*) To return to the straight climb, look ahead and roll out of the turn in the usual manner while holding the climbing attitude.

Descending Turns: Gliding

(*a*) From the straight glide look around and increase the airspeed some 5 kt by depressing the nose slightly.

(*b*) Commence a turn in the required dirction but be prepared to 'hold on bank'. Maintain the higher airspeed.

(*c*) Notice the instruments indicate a turn together with a loss of height. Keep a good lookout while turning.

(*d*) To resume the straight glide, look ahead, roll out of the turn in the usual manner and slow down to the recommended gliding speed.

(*e*) Now repeat the exercise with the flaps down. The gliding speed is lower and to gain the extra 5 kt for the turn the attitude of the aeroplane is steeper.

Descending Turns: Engine Assisted

(*a*) From a powered descent open the throttle slightly and after the usual visual checks go into a turn without increasing airspeed.

(*b*) To tighten the turn open the throttle still further as the bank is increased.

(*c*) Resume straight descent by rolling out of the turn and reduce power to give the required rate of descent maintaining the airspeed steady throughout.

(*d*) Repeat the exercise using various flap and power settings.

The Slipping Turn (in suitable aircraft)

(*a*) Position the aeroplane in a glide on the base leg fairly close to the boundary of the airfield.

(*b*) Commence a gliding turn on to the approach in the usual way but steepen the angle of bank. Reduce the rate of turn by releasing some of the backward pressure on the wheel/stick and at the same time take off rudder but keep the aeroplane turning. Maintain usual speed for the gliding turn.

(*c*) Because of the high rate of descent allow plenty of height for the roll out from the turn.

Turns onto Selected Headings

(*a*) Imagine you wish to make a turn through 90° to the right. Find a prominent cloud or ground feature on the right wingtip then roll into a turn to the right.

(*b*) Maintain a good lookout throughout the turn and, at intervals, glance at the instruments to confirm angle of bank, balance and constant height.

(*c*) Look out for the selected reference point, then when the nose is within 5–10° of it check that the new intended flight path is clear and smoothly roll out of the turn.

(*d*) Now practise turns in both directions, through varying changes of heading.

Use of the Direction Indicator

(*a*) Check that the DI is synchronized with the magnetic compass and select a new heading. Remember that to reduce the heading a left turn is required; to increase heading turn right.

(*b*) After a careful lookout check, roll into the turn in the usual way.

(*c*) At intervals glance at the instruments and check angle of bank, balance, height and turn progress on the direction indicator. Resist the temptation of confining the attention to the instruments at the expense of LOOKOUT.

(*d*) 5–10° before the new heading appears on the DI check it is safe to fly in the new intended direction, then roll out of the turn. Adjust the new heading if necessary with gentle use of the ailerons in the appropriate direction.

Compass Turns

(*a*) Fly west and find a reference point on the horizon on which to keep straight.

(*b*) Increase the airspeed with a sudden burst of power or by depressing the nose. Keep straight on the reference point and notice the compass card has rotated during the acceleration indicating an apparent turn towards north.

(*c*) Now reduce speed by closing the throttle sharply or raising the nose. Keep straight on the reference point and notice the compass card has rotated and indicated an apparent turn towards south during the deceleration.

(*d*) Now turn on to east and notice the effects of acceleration and deceleration are the same as on west. In either case when the airspeed has settled the compass will return to its correct reading.

(*e*) Commence a Rate 1 turn to the left on to north and some 25°–30°

before the required heading, roll out of the turn. The compass will now be within a few degrees of north and small corrections should be made as required. The wings must be level to obtain a correct reading.

(*f*) Keeping the rudder in the straight-flight position, lower one wing and notice how the compass card turns towards it. Level the wings and the card will swing back again.

(*g*) Turn on to south at Rate 1 in either direction. Notice that at first the compass will give no indication of turn since the action of banking pulls the card in the same direction as the turn – towards the lower wing. Carry on turning until the compass card appears some 25°–30° past south then roll out of the turn. The card will swing to within a few degrees of south and small corrections can be made as required. The wings must be level to obtain a correct reading.

(*h*) Lower one wing as before and notice the compass card will move towards it as on north. Level the wings and the card will swing back again.

(*i*) Commence a Rate 1 turn on to east or west. Notice the card gives an immediate indication of turn since the action of banking pulls the card in the opposite direction to the turn – towards the lower wing. When within 5°–10° of east or west roll out of the turn. The compass will indicate within a few degrees of the desired heading. Small corections can be made as required but the airspeed must be steady to obtain a correct reading.

(*j*) Now try turning on the compass at a high rate. The instrument is of little use and on occasions indicates a turn in the wrong direction.

Note. When flying a straight heading, the pilot must not watch the compass continually. The aircraft should be kept straight on a distant object, and its heading checked by frequent reference to the compass.

Exercise 10A
Stalling

Background Information

Stalling occurs when the angle of attack is increased until the air can no longer flow smoothly over the wings. At this point, which is known as the **Stalling Angle**, the airflow breaks down into eddies with consequent loss of lift, and the situation together with the attendant large increase in drag is illustrated in Fig. 23.

As the angle of attack is increased the centre of pressure moves forward but at the stall there is a general collapse of the lift envelope and the centre of pressure moves sharply back behind the centre of gravity thus upsetting the balance of the aircraft and causing the nose to drop although the stick is held right back. Nose-down pitch at the

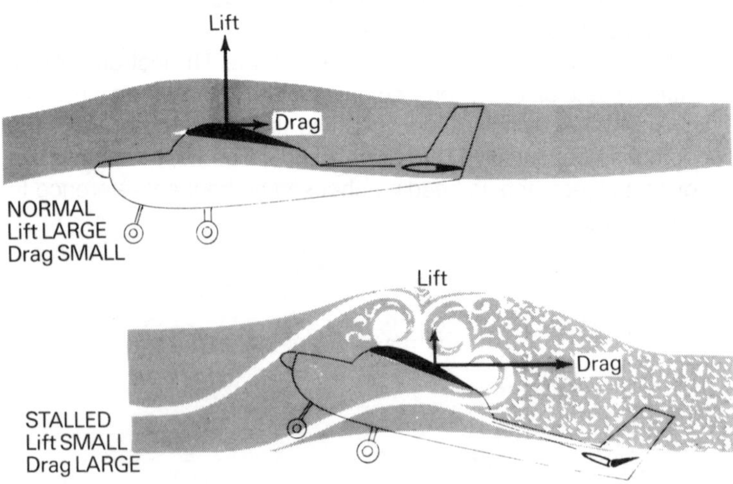

Fig. 23 Airflow, Lift and Drag in normal flight (top aircraft) and the same aircraft at the stall. Drag has increased due in the main to turbulence, Lift has decreased, the Centre of Pressure has moved back and the nose is about to drop.

stall is known as the **'g' break** (Fig. 24). Loss of elevator effectiveness, resulting from low airspeed, is accentuated by turbulent airflow from the wings moving back to the tail surfaces and over the elevators. Many aircraft give a prior warning of the stall by **Buffeting** (best described as shuddering). This will be felt through the elevators via the control wheel or stick. When no natural buffet exists the manufacturers may fit buffet inducing strips to the wing leading edges, located near the fuselage to cause turbulence over the tailplane at high angles of attack (Fig. 25).

It is the practice on modern light aircraft to fit a stall warning device, usually in the form of a small sensor positioned on the leading edge of the wing. At high angles of attack the sensor (which consists of a small vane-operated electric contact) switches on a stall warning light in the cockpit. Some installations include an audible warning, while another type in common use takes the form of a simple whistle which becomes vocal at high angles of attack. These devices are arranged to give warning some 5–10 kt before the stall. Larger and more complex aircraft have a **Stick Shaker** which behaves as the name implies. There may also be a **Stick Pusher** which automatically eases forward on the elevators when a stall is approached.

Stalling can occur in any attitude since the angle of attack is dependent upon the *relative* airflow. For example, if the aircraft is flown slowly with insufficient engine power it will commence to sink, the relative airflow will approach from below the wings and the stalling angle will be reached as depicted in Fig. 26. Stalling will occur

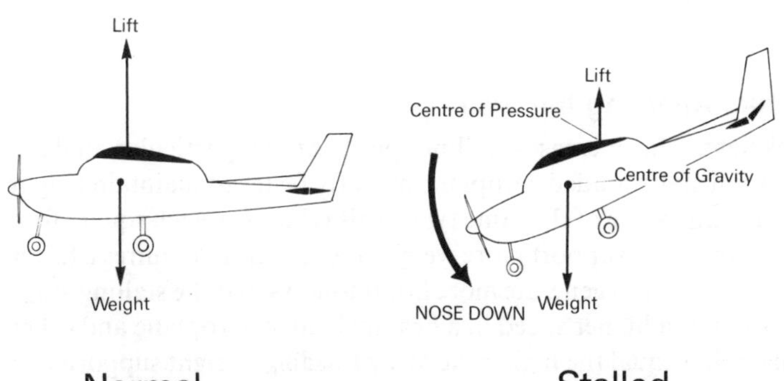

Fig. 24 Location of Lift and Weight before and after the stall. Note the reduction in Lift when stalled and the rearward movement of the Centre of Pressure which contributes to the nose drop.

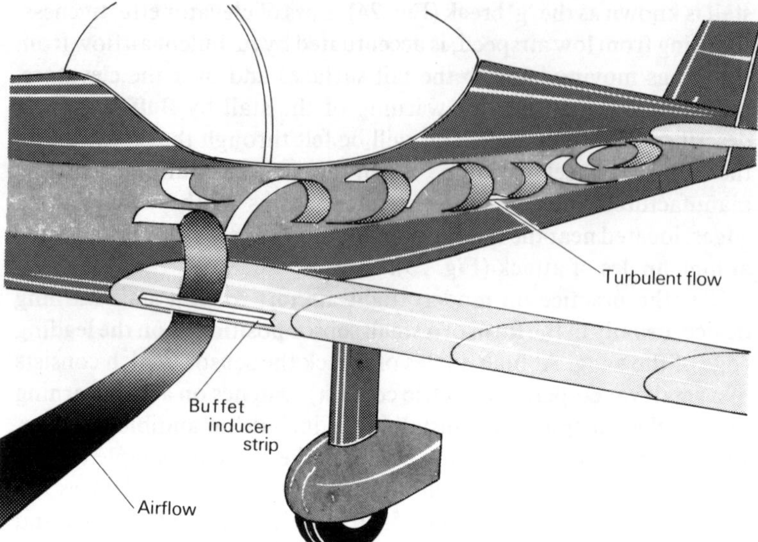

Fig. 25 The Buffet Inducer Strip. At large angles of attack these cause turbulent airflow which, in some aircraft, can be felt through the wheel/stick via the elevators. Buffet Inducers are sometimes employed to reduce wing drop at the stall by stalling the inner areas of the wing before the outer areas.

at any time when the wings are presented to the relative airflow at too great an angle.

Factors Affecting the Stall

(*a*) **Weight**. This increases stalling speed. At any particular angle of attack a heavily loaded aeroplane must fly faster to maintain height than an empty one of the same type. This is because each square foot of the wing must support more weight and therefore requires a faster airflow in order to generate more lift. It follows that the stalling angle will occur at a higher speed in a heavily loaded aeroplane and other things being equal the higher the **Wing Loading** (weight supported in pounds per square foot of wing area) the higher the stalling speed.

(*b*) **Power**. This decreases stalling speed. The reason is two-fold. First, with the engine running, the slipstream helps maintain a smooth airflow over that part of the wing which is within its

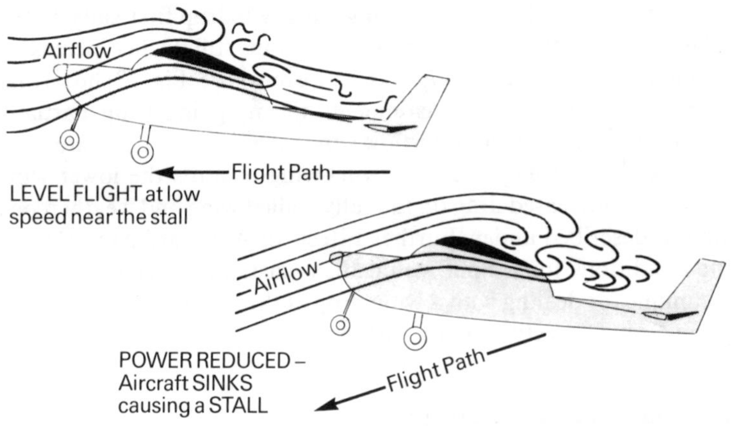

Fig. 26 Same Attitude, different Angle of Attack. Aircraft A is in level flight at low speed but not fully stalled. Aircraft B is in the same flight attitude but a descent has increased its Angle of Attack and caused a stall.

influence. Secondly, it will be appreciated that, with power on, shortly before the stall the aeroplane will normally be flying in a steep, nose-up attitude. In effect thrust will now be inclined upwards, contributing to lift and relieving the wings of part of their load.

(*c*) **Loading**. This increases stalling speed. Not to be confused with the weight of an aeroplane which may be increased by the addition of passengers or freight, loading occurs during certain manoeuvres such as steep turns, loops and the recovery from a dive. Under these conditions the effects of 'g' (which can be felt by the occupants of the aeroplane) temporarily increase the wing loading, and therefore the stalling speed is higher during these manoeuvres. Such a condition is called a **High-speed Stall.**

(*d*) **Flaps**. These are of many designs but broadly speaking first 15°–30° of lowering decreases the stalling speed. Further depression of the flaps will make a further small reduction to the stalling speed but the main aerodynamic change will be an increase in drag.

The Fully Developed Stall

Due to manufacturing inaccuracies and particularly on aeroplanes with highly tapered wings it is usual for one wing to stall slightly before the other. The early stalling wing will drop first causing the relative airflow to approach from below, thus still further increasing its angle of attack and stalling that wing even more. Fig. 27 shows that concurrent with these developments the up-going wing partially unstalls because its relative airflow comes from above. If the situation is allowed to develop the nose will swing towards the lower wing because of increased drag on the fully stalled wing and weathercock action (described under 'Further Effects of Ailerons,' page 31). All the conditions for a **Spin** would then exist. The correct handling technique for dealing with a wing drop at the stall is described under **Wing Drop Corrective Action**, later in this chapter.

Pre-stall Checks, HASELL

Any manoeuvre likely to entail loss of height must not be commenced until safety precautions have been taken. The most widely accepted airmanship check, which is enacted before stalling, spinning or aerobatics, is remembered by the word HASELL, the individual letters representing the following safeguards:

HEIGHT	Sufficient for the manoeuvre
AIRFRAME	Brakes off Flaps as required for the exercise
SECURITY	No loose articles in the cabin Harness and hatches secure
ENGINE	Temperatures and pressures 'in the green' Carb. Heat to HOT before closing the throttle Electric fuel pump ON (when fitted)
LOCATION	Outside controlled airspace Within sight of a known landmark
LOOKOUT	Not over towns, airfields or other aircraft.

The 'brakes off' requirement under AIRFRAME applies to aircraft with differential systems which limit the amount of rudder movement when they are applied. However 'brakes off' should be recited in *all* aircraft; airmanship checks are only of value if they are simple to remember and they can be applied to most aircraft types.

The LOOKOUT part of the check will be enacted during a clearing turn, either through 360° or, turns to the left and right through some 120° in each direction. Fuel pump ON is to assure fuel supply during the manoeuvre.

Symptoms and Recognition of the Stall

Many students approach the stall with some trepidation, possibly because others learning to fly have given dramatic accounts of what is in store.

It is certainly true that light aircraft of earlier design often exhibited decisive stalling characteristics, and military aircraft of high performance sometimes have a violent stall, however, light aircraft of modern design are, in the main, without vice in this respect. Indeed, in some trainers it is difficult for the instructor to produce a convincing 'g' break at the stall.

In the early stages of training the student will be shown the symptoms of the stall. Power off, the nose is raised to the approximate climbing attitude and attention will be drawn to the following developments:

1. Airspeed steadily decreasing.
2. Controls becoming less and less effective. This can be felt by the pilot.
3. Airframe noise becoming quieter.
4. Tendency to sink which must be corrected by more back pressure on the wheel/stick.
5. At about 5 to 10 kt before the stall the stall warning device (light and/or audible warning) will activate.
6. Shortly before the stall, low-wing aircraft will begin to buffet as the airflow breaks down and turbulence from the wings affects the tailplane and elevators. In high-wing designs buffet is rarely discernible because the turbulent airflow passes above the tail surfaces.
7. At stalling speed the nose will drop although the wheel/stick is held right back, there may be a slight tendency for a wing to drop and there will be a low airspeed allied to a high rate of sink.

The speed at which the aircraft stalls should be noted. In some light aircraft indicated speeds of only 35 kt or less may be seen but at high angles of attack there is considerable ASI reading error and the true airspeed may be 45 kt, flaps up and perhaps five knots less with full flap.

Stall and Recovery without Power

If moving the wheel/stick back until the stalling angle is reached results in a stall then logic suggests that moving the elevator control forward will un-stall the aircraft. This is, in fact, the case.

The simple act of moving forward the wheel/stick, reducing the angle of attack and gaining gliding speed, will effectively recover the aircraft from a stalled condition but at the price of excessive height loss. The instructor will draw attention to the altimeter immediately prior to the stall and compare its reading after the recovery just explained. Height loss will be in the region of 300–400 feet and clearly, had the stall occurred near the ground, such a loss of height could have caused a serious accident.

To minimize height loss during the recovery from the stall, a technique, known as the **Standard Recovery**, has been developed. This is now described.

Stall and Recovery with Power

When a deliberate stall has been entered as before, the various stages previously listed will occur. Immediately before the 'g' break the altimeter reading should be noted, then as the nose drops at the stall full power should be applied. Carburettor heat will have been applied before closing the throttle and this should be returned to the COLD position as the throttle is applied to ensure that full power is available.

As power is being applied the wheel/stick, previously held fully back to induce the stall, should be relaxed forward (as opposed to moved deliberately forward), allowing the nose to take up an attitude on or just below the horizon. At that stage the wings will have un-stalled and airspeed will be increasing. With practice, height loss can be limited to 50–100 feet or even less.

Wing Drop Corrective Action

To prevent or minimize wing drop at the stall the various design features explained in Chapter 4, Volume 2 (slats, washout, buffet inducers, leading edge droop) are sometimes incorporated by the manufacturers, either singly or perhaps several features together.

The pilot's contribution to preventing a wing drop at the stall is confined to preventing yaw while holding up the nose during practice and, in particular, at the stall itself.

Although the ailerons of modern aircraft remain active at speeds below the stall their use for the purpose of raising a dropped wing can induce a potentially dangerous situation. Taking a left wing drop as illustrated in Fig. 27 application of right aileron would lower the one on the left, effectively increasing the angle of attack on the already stalled wing and simultaneously increasing its drag. At the same time the up-going aileron on the raised, right wing would, in effect, decrease its angle of attack, partly un-stall it and cause a reduction in drag.

With more drag on the dropped wing and less drag on the raised wing producing a turning force in addition to that already being induced by the initial roll (roll/yaw effect was explained in further effects of aileron, page 31) a powerful turning force is generated at a time when the aircraft is stalled and all the ingredients for a spin will then be present.

If yaw at low speed is a prime requirement for the development of a spin (described in greater detail in Exercise 11) then clearly spin prevention must entail *preventing the yaw from developing*. Consequently, when a wing drops during a stall, accidental or intentional, the with-power recovery previously described will be carried out and yaw must be stopped by using sufficient rudder in the opposite direction to the dropped wing. In other words if the right wing drops left rudder will be applied along with the elevator and throttle inputs.

Use of rudder must be limited to no more than is necessary to check the yaw. The wings are then levelled with aileron *after* flying speed

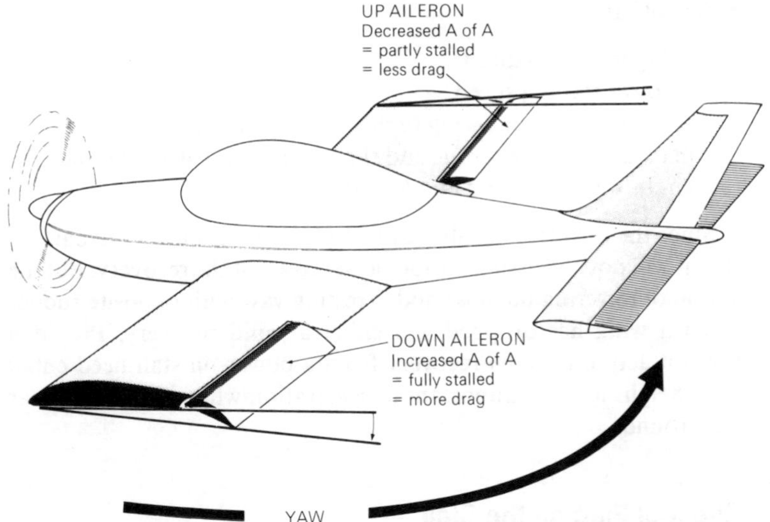

Fig. 27 Risks of using aileron to raise a wing when one has dropped at the stall. Increase in Drag caused by the lowered aileron on the dropped wing allied to a decrease in Drag on the raised aileron induces more Roll and Yaw. This could lead to a spin.

has been regained. On no account should opposite rudder be used in an effort deliberately to raise the dropped wing since there is a risk of inducing a spin in the opposite direction. In practice, most modern light aircraft will automatically roll level when the yaw is arrested with opposite rudder.

Stall with Power

Even with maximum power it is possible to stall an aircraft. Naturally, because power is inclined upwards thus contributing towards lift, the stall will occur in a higher nose-up attitude and at a lower speed than a power-off stall.

The purpose of practising the stall entry with power and its recovery is to reproduce what is sometimes called a **Departure Stall**, i.e. a stall that occurs during the climb-out soon after take-off.

Having regard to the very high nose attitude that must be reached before an aircraft will stall while under full power it seems hard to credit that such inadvertent stalls are, from time to time, the cause of serious accidents. It is to educate the inexperienced pilot and protect him from such accidents that the stall with power is demonstrated.

The main differences between power-on stalls and those made power-off are:

1. A higher nose attitude.
2. A lower stalling speed.
3. A tendency for one wing to drop sharply under the influence of maximum engine torque and slipstream effect at a time when the aircraft is stalled and at a low airspeed.

Since the throttle may already be fully opened during a departure stall it will not be possible to add power to assist the recovery. But the action of lowering the nose and arresting yaw with opposite rudder when a wing has dropped will effect a rapid recovery. Provided prompt action is taken recovery from a power-on stall need entail little height loss, an important consideration when it happens near the ground.

Effect of Flap on the Stall

The function of flaps and the different types of these high-lift devices is described in Chapter 4, Volume 2. In so far as the stall is concerned lowering of flap will result in a reduction of stalling speed according to the number of degrees applied.

Most light aircraft have relatively low stalling speeds, even in the **Clean** (i.e. flaps up) configuration and since flaps can only cause a percentage reduction in stalling speed the actual figure expressed in knots is bound to be small – a percentage of a small number will itself be small.

Generally light trainers exhibit a full flap stalling speed reduction of 5 to 8 kt. Larger business turboprops have a 20 knot reduction in stalling speed when the flaps are fully lowered and, to illustrate the point, a large jet transport might save 60 kt or more when the flaps and leading edge high-lift devices are brought into use.

When the flaps are fully lowered rate of speed decrease as the nose is held up to induce a stall is noticeably higher than without flap.

Stall in Approach Configuration and Recovery

Whereas the stall with power is taught to represent an unintentional stall soon after take-off this exercise is intended to help the student recognize the symptoms of a stall while on the approach prior to landing and teach the correct recovery action.

Being on the approach the aircraft will have part or perhaps full flap applied and it will be descending under power. If the pilot tries to correct a 'too low' situation by holding up the nose (instead of applying more power which is the correct action) a stall could occur. To recover with a minimum loss of height the nose must immediately be lowered to the level attitude, or slightly below, and power should be increased. Provided the nose has not been allowed to adopt a diving attitude full power should be used.

Recovery at the Incipient Stage

On the basis that prevention is better than cure there are obvious advantages to recovering from a stall situation before it has properly developed.

Practice of this exercise should be regarded as a valuable insurance against becoming involved in an inadvertent stall. Usually such incidents occur when the pilot's attention is concentrated on other matters. It is a well known fact that most stall/spin accidents in the past have occurred at a time when the pilot was under pressure. Possibly there was an engine failure and the consequent need to find a suitable field for a forced landing. Whatever the reason, such accidents need never happen and this exercise is intended to develop in the pilot an automatic recognition of an impending stall and the correct recovery action.

When the stall warning sounds the nose should immediately be lowered on to or just below the horizon and full power is applied, at the same time checking any tendency to yaw with rudder.

Stall at Various Speeds and Recovery

The purpose of this exercise is to demonstrate the effects of 'g' on stalling speed. In a turn, when lift is called upon to support the aircraft and provide a turning force (centripetal force) the 'g' that results, and which increases as the angle of bank becomes steeper, has an effect on the aircraft similar to adding weight in the cabin. This was mentioned under *Factors Affecting the Stall* on page 116. Figures relating to a well known four-seat tourer are shown in the following table.

Bank angle (degrees)	*Stalling speed (kt)*	*Percentage increase*
0	49	0
35	53	8
45	59	20
60	71	43
75	97	98

The demonstration entails entering a steep turn without adding power, gradually easing back on the wheel/stick, noticing the increased 'g' which can be felt as the occupants are pressed firmly into their seats and the stall, which will occur at a higher than normal speed. Recovery entails use of power and elevator as before. Some aircraft tend to roll sharply or **Flick** in one direction or the other when the stall occurs under 'g' loading and this must be prevented from developing by use of rudder in the opposite direction to roll.

As mentioned at the beginning of this chapter, stalling should not be approached with apprehension in a modern light aircraft. It should nevertheless be taken seriously and thoroughly understood.

Exercise 10A. Stalling (Standard Recovery, stall without Flap)

AIM — *To recognize the symptoms of the stall and to recover with a minimum loss of height*

AIRMANSHIP — *Height: sufficient for the exercise*
Airframe: flaps up, brakes off
Security: no loose articles in the cabin, harness and hatches secure
Engine: temperatures and pressures 'in the green', fuel pump on, carb. heat management
Location: not in controlled airspace, etc., in sight of known landmark
Lookout: clearing turn to avoid towns, airfields and other aircraft
Suitability of weather

AIR EXERCISE

Stall entry	*During the stall*	*Recovery from stall*
Fuel pump on (when fitted) *LOOKOUT* *Carb. heat HOT* *Close throttle* *Hold up nose (climbing attitude)* *SYMPTOMS* *Speed decreasing* *Controls less effective* *QUIET!!* *Progressive back pressure on wheel/stick required* *At stall warning:* *Check altimeter* *Buffet* *STALL*	*Wheel/stick hard back* *Nose down* *Low airspeed* *High sink rate* *Nose pitching up/down (some types of aircraft)* *Little noise*	*Wheel/stick forward* *If wing drop use opposite rudder to check yaw* *At gliding speed add power* *Level out* *Note height loss* *STANDARD RECOVERY* *From stall entry as before* *At 'g' break:* *Full power – carb. heat COLD* *Nose on or just below horizon* *Check wing drop with opposite rudder* *Note reduced height loss*

Stall with power — *NOTE: High nose attitude: lower stall speed: tendency to roll (some aircraft)*
Stall with flap — *NOTE: Faster speed reduction: lower stalling speed*

Pre-flight Briefing

Exercise 10A. Stalling (Incipient stage and stall at various speeds)

AIM — *To recover from the approach to the stall before it can develop and to recover from a High Speed Stall.*

AIRMANSHIP — *Height: sufficient for the exercise*
Airframe: flaps up, brakes off
Security: no loose articles in the cabin, harness and hatches secure
Engine: temperatures and pressures 'in the green', fuel pump on
Location: not in controlled airspace, etc., in sight of known landmark
Lookout: clearing turn to avoid towns, airfields and other aircraft
Suitability of weather

AIR EXERCISE (Incipient stall)

Stall entry	*During stall approach*	*Recovery from stall*
Fuel pump on (when fitted) *LOOKOUT* *Reduce power* *Hold up nose (climbing attitude)* *Recognize symptoms*	*Continue back pressure on Wheel/stick* *Note airspeed* *Note height* *Keep straight with aileron* *Check yaw with rudder*	*When stall warning activates:* *Lower nose to level attitude* *Apply full power* *Resume S & L flight* *Check altimeter reading*

AIR EXERCISE (High speed stall)

Stall entry	*During the stall*	*Recovery from stall*
Enter steep turn without extra power *Tighten turn* *Speed decreasing, more back elevator* *Notice 'g'* *Aircraft stalls*	*Note airspeed higher than wings-level stall* *Note tendency for aircraft to roll sharply into or out of turn (according to type)*	*Relax back pressure on Wheel/stick* *Check roll/yaw with opposite rudder* *Full throttle unless nose is well below horizon* *Level wings, return to S & L flight*

Flight Practice

Vital actions 'HASELL'

Height – sufficient for the exercise.
Airframe – brakes off, flaps as required.
Security – harness and hatches secure. No loose articles in the cockpit.
Engine – fuel pump 'on', temperatures and pressures normal.
Location – out of controlled airspace in sight of known landmark.
LOOK OUT – carry out an inspection turn. Not above towns, airfields or other aircraft.

Air exercise

Introduction

To induce confidence in his student the instructor will stall and recover thus demonstrating the gentle nature of the manoeuvre as it applies to light training aircraft.

Symptoms and Recognition of the Stall

(*a*) Apply carburettor heat then throttle back, keep straight and raise the nose just above the horizon.
(*b*) As the airspeed decreases note the progressive ineffectiveness of the controls. The aircraft is becoming very quiet.
(*c*) Continue the backward pressure on the wheel/stick until the pre-stall buffet is felt and the stall warning device operates. The aircraft is now close to the stall (Recover).

Stall and Recovery Without Power

(*a*) Apply carburettor heat, throttle back, keep straight, and raise the nose just above the horizon.
(*b*) Notice height and airspeed and continue backward pressure on the wheel/stick until the aircraft stalls. (Some types buffet shortly before the stall giving a clear warning.)
(*c*) Recovery is effected by easing the wheel/stick forward thus gaining flying speed, when the nose should be gently brought up to the horizon and power applied. Notice loss of height which may be considerable. Near the ground this could be dangerous.

Stall and Recovery with Power (Standard Method)

(*a*) Proceed into the stall as before, noting height shortly before the stall.

(*b*) As soon as the nose commences to drop, open the throttle fully and gently ease the stick forward. It should not be necessary for the nose to go far below the horizon. Note that very little height is lost.

Wing Drop Corrective Action

If a wing should drop, opposite rudder must be used to prevent a yaw developing. Level the wings after flying speed has been regained.

Stall with Power On

Practise stalling with a little engine power and progressively increase this. The more power used the steeper the attitude of the aircraft and the lower the speed before the stall. Recover as already explained adding extra power where possible.

Effect of Flap on Stall

With the flaps lowered and the engine throttled back carry out a stall and notice how rapidly the speed decreases as the wheel/stick is brought back. The stall will occur at a lower airspeed than from a glide without flap and the nose is likely to drop more severely. Recover by using the standard technique.

Stall in Approach Configuration and Recovery

This exercise simulates the possibility of a stall during an engine assisted landing approach. As the wheel/stick is brought back the airspeed will not decrease as rapidly as in the previous exercise and stalling speed will be lower because of the engine power. Recover by using the standard technique.

Recovery at the Incipient Stage

(*a*) Approach the stall in the usual manner.
(*b*) Check: yaw, airspeed and height.
(*c*) When the stall warner sounds or (if applicable) buffet is felt, apply full power, lower the nose to the horizon and prevent yaw with rudder. Note the height loss which should be minimal.
(*d*) Now practise this exercise from the turn, power on and power off, also with and without flap.

Stall at Various Speeds and Recovery

(*a*) Go into a steep turn without increasing the power.
(*b*) Tighten the radius of turn by backward movement of the wheel/stick.

(*c*) Shortly before the stall buffeting may be felt. Notice the stall is at a higher speed than usual when power is on.

(*d*) There may be a tendency to roll sharply into or out of the turn at the stall.

(*e*) To recover, ease the wheel/stick forward and increase the power. Prevent roll/yaw from developing with rudder applied in the opposite direction. The ailerons will again become effective and capable of correcting any alteration in angle of bank caused by one wing stalling before the other.

Exercise 10B
Spin Awareness and Avoidance

Background Information

An examination of aircraft accidents related to light aircraft clearly indicates that a very high proportion are the result of stall or spinning incidents. Statistics from various countries show a remarkable uniformity in this respect.

Until 1 August 1984 entry and recovery from a spin formed part of the training for a UK Private Pilot's Licence. After that date the UK Civil Aviation Authority no longer required spin training and it has been replaced by the exercise to be described in this chapter. Spinning is therefore included in this manual as a non-mandatory exercise.

Although spin entry and recovery is no longer a requirement for the purpose of PPL training it has been retained for professional pilots and flying instructors.

Slow Flight and the Effects of Yaw

Few pilots have spent more than a few moments flying an aircraft at low speed, power on and in various configurations, and while the setting up of slow flight demands no more than patience (time is required while the aircraft settles into very slow flight and trim changes have to be made) there are certain areas that must be treated with respect. These are:

1. **Yaw**. Other than the waste of fuel and possibly a degree of discomfort for the occupants of an aircraft that is flown out of balance, yaw is not of any significance in safety terms when it is introduced at normal operating speeds.

   At airspeeds near the stall yaw can present a hazard because, as previously explained in the section dealing with a dropped wing stall (Exercise 10A) the roll that follows a yaw will increase the angle of attack and drag on the down-going wing and have a reverse effect on the rising wing. These are ideal conditions for the onset of a spin (Exercise 11).

2. **High Power/Low Speed**. At low airspeeds near the stall control effectiveness is reduced because of the slower airflow. At high power settings engine torque and slipstream effect are at their maximum and a combination of these factors means that the pilot must make a deliberate effort to hold on considerable rudder deflection if balanced flight is to be maintained.

   If a stall does occur while the engine is at a high power setting, and if the pilot fails to maintain balanced flight by checking the tendency to yaw, there will be a pronounced roll in the opposite direction to propeller rotation. Some aircraft will, in these conditions, enter a spin in a manner that can be both sudden and pronounced.

3. A slow flying aircraft is more affected by wind than one at normal cruising speed. If, for operational reasons, it is necessary to fly slowly at low levels (*Operation at Minimum Level* is described in Exercise 16) the effects of wind will become more than usually apparent and it is easy for an inexperienced pilot to gain the impression that the aircraft is slipping or skidding. An incorrect rudder correction at low speed could induce a dangerous yaw.

The various manoeuvres forming part of this exercise are intended to demonstrate the need for special care while flying at low airspeeds, the potential danger of yaw under these conditions and the need for stall/spin awareness so that the pilot will adopt an instinctive, corrective response. Stall/spin accidents need not happen if pilots understand the subject and develop their low speed handling skills.

Pre-stall/spin checks

The HASELL check was described on page 118 and although many of the exercises that follow should entail no height loss it must be used nevertheless.

Slow flying, particularly in the early stages, requires of the pilot some concentration and by adopting the HASELL check the exercise can proceed in the knowledge that there is sufficient altitude below the aircraft, airspace is not being violated and there is no risk of collision with other aircraft.

Recognition and Recovery at Incipient Stage and from various Flight Attitudes and Configurations

For flying training purposes it is generally agreed that the incipient stage is that part of the spin entry up to the time when the wings become vertical.

The instructor will demonstrate how a spin can develop from various flight attitudes when these are adopted at a low airspeed. Particular areas for attention are:

1. Maximum power climbs at low speeds with uncompensated yaw (i.e. out of balance flight).
2. Descending turns at low speeds which are out of balance.
3. Risks of incorrect use of rudder because of misleading impressions of slip/skid while flying at low levels in strong winds.
4. Failure to deal with powerful nose-up changes of trim following a 'go around' at maximum power in the landing configuration. This could lead to a low airspeed and excessive yaw due to slipstream and torque effect.

The instructor will demonstrate a spiral dive, drawing attention to the differences between this manoeuvre and a spin. (Explained in *Exercise 11, Spinning*).

Spinning with Flaps Lowered

Although there are occasions, such as the overshoot from a landing mentioned in item 4 above, when flap will be lowered at a time when there is a risk of spinning, intentional spins must never be attempted in this configuration.

Flaps disturb the airflow over the elevators and rudder, degrading the effectiveness of these controls at a time when they are needed for spin recovery. If an inadvertent spin does occur the flaps must be raised while recovery action is being made.

Spinning with Power

Some aircraft must not be allowed to enter a spin with power while others will only make the entry when a little power has been left on to increase the effectiveness of the rudder and the elevators, both controls being used to initiate the spin.

When an inadvertent spin has started the throttle must be closed during the recovery procedure.

Spinning with Aileron

Like the use of power, aileron may not be used for spin entry in some aircraft while others need pro-spin aileron (i.e. aileron down on the inside of the spin to increase the angle of attack and drag on that wing) for satisfactory entry. Whether or not aileron or power are permitted while entering a deliberate spin will be stated in the aircraft manual.

During any spin recovery and whatever the stage of development (incipient or fully developed as described in Exercise 11) aileron must be neutral while the other actions are being carried out.

Recovering from an Incipient Spin

A spin can be induced from a glide at low speed when too much rudder has been applied. It could also occur during a powered descent, say, on the base leg, when the aircraft would be at circuit height.

Imagine the situation. The aircraft is descending with part flap and perhaps 1500 RPM. The pilot has carelessly let the airspeed drop to a low figure 5–10 kt above that for the stall. He becomes aware that he has flown past the runway centreline and in an effort to rectify the situation an excessive amount of rudder is applied and, because of the low height, an attempt is made to keep the wings level with opposite aileron. These are, of course, a misuse of the controls and thoroughly poor handling techniques.

The fact that a little power has been left on will make the rudder and elevators more effective and, suddenly, a stage is reached where the aircraft stalls just as a yaw had developed. Very prompt action will be required to stop the aircraft from rolling past the wings-vertical position and the recovery action will entail easing forward the wheel/stick, applying rudder in opposition to yaw and centralizing the ailerons. Provided the nose has not dropped to a diving attitude power may be left on to assist the recovery but when a steep nose-down attitude has developed the throttle must be closed. There is little point in using power to increase the rate of descent.

Flight at V_{so} + 10 kt and V_{s1} + 10 kt reducing to V_{so} + 5 kt and V_{s1} + 5 kt.

Because a large family of key speeds has grown up alongside the introduction of jet aircraft it has been necessary to codify them in the form of an internationally agreed list of definitions. The **V Code**, as it is known, is mainly related to aircraft larger than light trainers and it takes the form of velocities applicable to such matters as **Never Exceed Speed** (V_{ne}), **Maximum Flap Extension Speed** (V_{fe}), and a number of others of importance to multi-engine aircraft. V_s is the stalling speed and this is further qualified by the two velocities mentioned in the heading to this section:

V_{so} Stalling Speed or minimum steady flight speed at which the aircraft is controllable in the landing configuration.

V_{s1} Stalling Speed or minimum steady flight speed at which the aircraft is controllable other than in the landing configuration.

The reference to stalling speed or minimum steady flight speed may at first seem confusing but the demarcation between these two conditions of flight would be difficult to express in terms of speed. V_{s1} relates to stalling speed/minimum steady flight speed in any configuration (flaps up, half flap, etc.) but not in the landing configuration when full flap is assumed to be deployed.

The purpose of this part of the exercise is to develop skill in flying safely at low airspeeds and in varying configurations. In the early stages of training the student will be expected to fly accurately at V_{so} and V_{s1} plus 10 kt. As confidence and skill are gained these speeds will be reduced to V_{so} and V_{s1} plus 5 kt.

Initiation and Recovery (Low-speed Flight)

Slow flight will be practised in straight and level, while climbing and descending and during turns. Because of the nature of these exercises certain precautions must be taken:

1. The process of settling into the low speeds must not be hurried. The aircraft has inertia and time must be allowed for it to settle. Also, while speed is reducing and power is being adjusted to maintain, gain or lose height it will be necessary constantly to re-trim.
2. At low airspeeds, balance being particularly important, it will be necessary to apply *and maintain* rudder pressure keeping the ball centred at all times. When high power settings are in use this rudder pressure can be considerable.
3. While climbing at low airspeeds and high power settings engine cooling will be less effective than normal. Engine temperatures and pressures must be monitored at intervals more frequent than in flight at higher cruising speeds. Rate of climb will be lower than at the correct climbing speed.

4. Because of the higher than usual nose attitude associated with slow flight lookout is impaired and pilots must be aware of the need for special care.
5. While turning at low speeds there will be a further and quite rapid speed reduction as bank is applied. Power must be added to maintain both speed and altitude.

Transition from Slow to Normal Speed Flight

During the change from V_{so} + 5 kt to normal cruising speed flaps will have to be raised, causing a considerable change in trim. Flaps should be raised in stages, trimming at intervals, and care should be exercised to maintain heading, height and balance throughout. The engine will also have to be adjusted as cruising speed becomes established. It will then be necessary to make a final trim check.

Instrument Indications

By now it will be clear that Exercise 10B is in two parts:

(*a*) Spin awareness and avoidance.
(*b*) Low speed flight.

Aircraft instruments assume varying importance according to phase of flight and the pilot should develop a talent for automatically seeking relevant information from the correct instrument(s).

During Incipient Spin Entry

ASI	Low, reducing airspeed
Altimeter/VSI	According to whether the approach to the spin is from level, climbing or descending flight
Turn needle	A turn or yaw to left or right
Ball	A skid away from the turn/yaw
DI	Heading change to left or right
AI	(*a*) High nose attitude when aircraft is in low speed, level flight and while climbing (*b*) Lower nose attitude during descents (*c*) Bank during turns

During Incipient Stage of Spin

ASI	Low airspeed
Altimeter/VSI	Loss of height
Turn needle	Maximum rate of turn in yaw direction
Ball	Large deflection away from yaw
DI	May topple
AI	May topple

After Recovery

ASI	Airspeed moving towards normal
Altimeter	Considerable height loss
VSI	Near zero position
Turn needle	Central
Ball	Central
DI	May need re-setting with magnetic compass
AI	May require up to 10 minutes to erect (some AIs do not topple)

During Slow Flight, 10 kt and 5 kt above V_{so} and V_{s1}

The only significant instrument indications during this phase of flight are.

Ball	A tendency, particularly at high power settings, for the ball to move away from centre. This MUST be corrected by applying rudder in the direction indicated by the ball (RIGHT rudder when the engine rotates clockwise seen from behind).
AI	Because of the low airspeed smaller than usual angles of bank are required to attain any particular rate of turn.

The Spiral Dive

Although the spiral dive is a totally different manoeuvre to the spin, with no low speed element involved, an understanding of it is required because, superficially at least, the spiral dive may appear to be a form of spin.

It is important to understand the difference between the two manoeuvres because recovery action demands totally dissimilar techniques.

The spiral dive is a steep turn with the nose in a low or diving attitude. Airspeed tends to build up rapidly and care must be taken to ensure that the aircraft does not exceed its Never Exceed Speed (V_{ne}).

During a spiral dive there will be a maximum rate of turn but the ball will be in balance, unlike a spin where there is outward skid. A more detailed comparison between the spiral dive and the fully developed spin is contained in Exercise 11, *Spinning*.

Recovery from a spiral dive is made by closing the throttle, determining the direction of turn, stopping the turn with aileron and rudder, then gently easing out of the dive. As the nose comes up to the horizon power should be added to resume straight and level flight.

Additional Benefits of Exercise 10B

The two hours devoted to Exercise 10B are all dual. While the prime purpose is to avoid stall/spin accidents an important by-product of practising slow flight is that it will enable the student pilot to improve control of the aircraft during an approach to land, particularly when, at a later stage of training, short field landings are being taught.

Pre-flight Briefing

Exercise 10B. Spin Awareness and Avoidance (Incipient Spin and Recovery)

AIM — *To recognize the approach of conditions that could cause a spin and to recover at the incipient stage with minimum height loss.*

AIRMANSHIP — *Height: sufficient for the exercise*
Airframe: flaps as required, brakes off
Security: no loose articles in the cabin, harness and hatches secure
Engine: temperatures and pressures 'in the green'. Watch overheating, carb. heat use. Fuel pump on.
Location: not in controlled airspace, etc., in sight of known landmark
Lookout: clearing turn to check not over towns, airfields or other aircraft
Suitability of weather

AIR EXERCISE (Power off Entry)

Approach to incipient spin	*Spin entry*	*Recovery from incipient spin*
Carb. heat HOT: Power OFF *Raise nose, maintain height* *Low airspeed* *At V_{s1} + 5 kt* *Full rudder in spin direction* *Wheel/stick back*	*Sharp yaw in rudder direction* *Sharp roll in same direction* *Sharp nose drop although wheel/stick is fully back*	*Check:* *Yaw direction* *Power off* *Ailerons neutral* *Recovery Actions:* *Check yaw with opposite rudder* *Ease wheel/stick forward* *Level wings (aileron) after speed has increased* *Ease out of dive and add power*

Effect of power: *Higher nose attitude: Need for rudder (balance): More decisive entry*
Recovery actions: *As before but if nose in diving attitude POWER OFF*

Effect of flap: *Speed decrease more rapid: Less effective rudder and elevators: Lower entry speed*
Recovery actions: *As before but raise flaps to avoid exceeding limit speed*

Exercise 10B. Spin Awareness and Avoidance (Flight at V_{so} and V_{s1} + 10 kt reducing to + 5 kt).

AIM — *To develop confidence and skill in slow flying and to recognize possible spin conditions, taking appropriate actions to avoid entry.*

AIRMANSHIP — *Height: sufficient for the exercise*
Airframe: flaps as required, brakes off
Security: no loose articles in the cabin, harness and hatches secure
Engine: temperatures and pressures 'in the green'. Watch overheating
Location: not in controlled airspace, etc., in sight of known landmark
Lookout: clearing turn to check not over towns, airfields or other aircraft
Suitability of weather

AIR EXERCISE (to be practised at V_{so} and V_{s1} + 10 kt, reducing to + 5 kt)

Setting up slow flight	*Slow flight exercises*	*Return to crusing flight*
LOOKOUT *Reduce power and speed* *At flaps-up V_s + 10 kt* *Constant height* *Trim* *Note visibility ahead* *Feel of controls* *Repeat with various stages of flap* *Repeat in landing configuration*	*Climbing* *LOOKOUT* *Maximum power* *Maintain stalling speed plus 10 kt* *NOTE Climb rate less than normal* *Ball moves to RIGHT* *CORRECT balance with rudder to RIGHT* *Descending* *LOOKOUT* *Reduce power* *Maintain stalling speed plus 10 kt* *NOTE High nose attitude* *High sink rate* *Turning* *LOOKOUT* *Roll on bank (15–20 degrees)* *NOTE Airspeed reducing* *Height loss* *CORRECT with power*	*LOOKOUT* *Raise flap in stages* *Adjust nose to cruise attitude* *Adjust power to cruise RPM* *Re-trim in stages* *Check:* *Altitude* *Airspeed* *Heading* *Balance* *Trim* *Repeat exercises with various stages of flap* *Repeat in landing configuration* *Repeat with full flap/max. power to simulate a 'go round'*

Flight Practice

COCKPIT AND OUTSIDE CHECKS

H A S E L L (Adequate height for recovery is essential while
AIR EXERCISE demonstrating the following exercises).

Recognition of Spin and Recovery at Incipient Stage from various Flight Attitudes and Configurations

Power Off Entry

(*a*) Apply carburettor heat and close the throttle.
(*b*) Trim the aircraft at normal gliding speed.
(*c*) Progressively apply back pressure to the wheel/stick and some 5 kt above stalling speed put on full rudder and move the wheel/stick fully back. The aircraft yaws, rolls and the nose drops.

Recovery at Incipient Stage

(*a*) When the wing has dropped some 45° check further yaw with rudder in opposite direction. Keep the ailerons neutral.
(*b*) Ease the wheel/stick forward until the incipient spin stops.
(*c*) Centralize the rudder.
(*d*) As the airspeed increases level the wings with aileron, raise the nose, apply power and return to straight and level flight.

Entry from a Maximum Power, Low-speed Climb with uncontained Yaw

(*a*) Set up a maximum power climb.
(*b*) Progressively raise the nose to reduce the airspeed.
(*c*) Note the poor rate of climb.
(*d*) The ball is moving away from centre. Make no attempt to correct balance but keep reducing speed with progressive back pressure on the wheel/stick until the aircraft stalls, there is a sudden yaw/roll and the start of a spin.

Recovery at Incipient Stage

(*a*) When the wing has dropped some 45° check further yaw with rudder in opposite direction. Keep the ailerons neutral.
(*b*) Ease the wheel/stick forward until the incipient spin stops.
(*c*) Centralize the rudder.
(*d*) If the nose has dropped into a dive close the throttle, if not leave on power.
(*e*) As the airspeed increases level the wings with aileron, raise the nose and adjust power for straight and level flight.

Recovery from a Climbing Turn

(*a*) Go into a climbing turn at a low airspeed some 5 knots above the stall.
(*b*) Add rudder to speed up the turn and hold off bank to keep the wings at a gentle angle – a misuse of the controls.
(*c*) Progressively allow the airspeed to decrease by applying back pressure to the wheel/stick, simulating a distraction in flight.
(*d*) The aircraft yaws, rolls, stalls and enters a spin.

Recovery at Incipient Stage

(*a*) When the wing has dropped some 45° check further yaw with rudder in opposite direction. Keep the ailerons neutral.
(*b*) Ease the wheel/stick forward until the incipient spin stops.
(*c*) Centralize the rudder.
(*d*) If the nose has dropped into a dive close the throttle, if not leave on power.
(*e*) As the airspeed increases level the wings with aileron, raise the nose and adjust power for straight and level flight.

Entry from a Descending Turn with Power to simulate a Turn onto the Final Approach

(*a*) Set up a descent at an airspeed some 5 kt above the stall, leave on 1500–1600 RPM and set 10° of flap. The aircraft is now in the base leg configuration.
(*b*) Start a turn but prevent the bank from exceeding 5° or 10° with opposite aileron. At the same time add rudder to speed up the rate of heading change – a misuse of controls.
(*c*) Progressively add rudder and back pressure on the wheel/stick until, suddenly the aircraft rolls into a spin.

Recovery at Incipient Stage

(*a*) Immediately the wing drops apply rudder in opposition to yaw. Keep the ailerons neutral.
(*b*) Ease forward the wheel/stick until the incipient spin stops.
(*c*) Centralize the rudder.
(*d*) Raise the flaps before limiting speed is reached.
(*e*) As the airspeed increases level the wings with aileron, raise the nose and adjust power for straight and level flight.

NOTE: *Entry to the spin in this configuration can be very sudden and recovery action must be started immediately the wing goes down. The throttle may have to be closed if a dive develops, otherwise power may be used to assist recovery.*

Flight at V_{so} + 10 kt and V_{s1} + 10 kt, progressively reducing to V_{so} + 5 kt and V_{s1} + 5 kt

Initiation of Slow Flight

(*a*) Reduce power slightly and maintain height by progressively raising the nose.
(*b*) Maintain balanced flight with the rudder in the direction indicated by the ball.
(*c*) Allow the ASI to settle at stalling speed plus 10 kt.
(*d*) Re-trim in stages as the airspeed changes.
(*e*) Adjust power to maintain height.
(*f*) Check trim. Maintain a good LOOKOUT.
(*g*) Now progressively reduce speed to stalling speed plus 5 knots.
(*h*) Note that power may have to be added to maintain height as drag increases at the very low airspeed.
(*i*) Now apply flap in stages. Note that the nose must be lowered to maintain speed.
(*j*) With maximum flap lowered and the airspeed 5 kt above stalling speed a lot of power is required to maintain height. At intervals check engine temperatures and pressures.
(*k*) Check balance and note that considerable rudder pressure is required to keep the ball in the centre.

Return to Cruising Speed

(*a*) Raise part flap, say 10° or 15°, lower the nose slightly and allow the airspeed to increase.
(*b*) Maintain balance with rudder and use the trimmer to assist while altering configuration.
(*c*) As the speed increases, raise the flaps in stages, maintain height and as cruising speed is reached check power, trim and balance.

Climbing (to be demonstrated and practised at V_{so} and V_{s1} plus 10 kt reducing to plus 5 kt)

(*a*) From low speed straight and level flight check that it is clear to climb then open the throttle fully.
(*b*) Maintain the same speed and note that considerable rudder pressure is required to keep the ball in the centre.
(*c*) Trim the aircraft into the climbing attitude and notice that rate of climb is less than that attained at the correct climbing speed. Check engine temperatures and pressures.
(*d*) Relax rudder pressure and the ball will move away from centre, yaw will induce a tendency to roll in the same direction. A slight reduction in airspeed could result in a spin.

(*e*) Add rudder to bring the aircraft into balance. The yaw/roll tendency has now stopped.
(*f*) Practise the exercise with various flap settings.

Descending (to be demonstrated and practised at V_{so} and V_{s1} plus 10 kt reducing to plus 5 kt)

(*a*) From low speed straight and level flight check that it is clear to descend then reduce power to produce a 500 feet per minute rate of descent.
(*b*) Maintain the same speed and notice that less rudder pressure is needed to keep the ball in the centre.
(*c*) Trim the aircraft into the descending attitude.
(*d*) While descending allow the speed to decrease until the stall warning is heard. Notice the aircraft will stall although the aircraft is not in a nose-up. *This lesson is important and it must never be forgotten.*
(*e*) Now return to straight and level flight.
(*f*) Practise the exercise with various flap settings and at various rates of descent.

Turning (to be demonstrated and practised at V_{so} and V_{s1} plus 10 kt reducing to plus 5 kt)

(*a*) From low speed straight and level flight check that it is clear to turn then roll on bank for a medium turn. Less bank is required than for medium turns at normal cruising speeds.
(*b*) There is a tendency for the airspeed to reduce and height to be lost. Prevent this by adding a little power as bank is rolled on.
(*c*) Now turn in the other direction. The amount of balancing rudder has changed. While turning left little or no rudder is required; for turns to the right considerable rudder pressure is needed.
(*d*) Roll out of the turn in the usual manner and as the wings level reduce power to maintain height at the correct speed.
(*e*) Now practise climbing and descending turns at low speed and with various flap settings.

NOTE. *The instruction given in item (c)* assumes that the propeller turns clockwise when seen from behind.

The Overshoot with Un-compensated Trim Changes

(*a*) Set up the aircraft in the approach configuration and reduce power to 1500 RPM. Re-trim the aircraft at the correct engine-assisted approach speed.

(*b*) Imagine that for some reason it is necessary to overshoot. Open the throttle fully.

(*c*) Notice the powerful nose-up trim causing a rapid decrease in airspeed and the need for firm application of rudder to maintain the ball in the centre.

(*d*) Deliberately allow both trim changes to go unchecked.

(*e*) The airspeed is now decreasing and a yaw has provoked roll.

(*f*) When the stall warning sounds check the yaw with rudder and bring the ball into centre.

(*g*) Apply forward pressure to the wheel/stick and re-trim into the correct climbing speed for the flaps-down configuration.

(*h*) Raise the flaps in stages, retrimming at intervals, and establish a flaps-up climb at the correct speed.

Exercise 11
Spinning (non-mandatory)

Background Information

The full spin is a development of the incipient spin described previously in Exercise 10B. While Exercise 11, *Spinning*, is no longer a requirement for the UK Private Pilot's Licence some States include spinning in their training syllabuses. Furthermore, a strong body of informed opinion in flying training regards spinning as of vital importance. Regulations may change but the laws of aerodynamics remain unmoved.

It should be emphasized that, provided the recovery procedure is clearly understood, there is no need for apprehension when approaching this exercise.

Aircraft Suitability

Some aircraft are not certificated for spinning and on no account may these be used for spin demonstration.

Notwithstanding all that is known about aerodynamics and although the designer has at his disposal facilities such as modern computers, predicting the spin behaviour of a newly designed aircraft can, even today, produce surprises when the testing begins. Often a new design intended for spin certification will have to undergo several modifications before its behaviour is regarded as totally satisfactory. Consequently, pilots who deliberately spin aircraft that have not been thoroughly evaluated and pronounced fit for the task are walking into the unknown.

The aircraft manual will state whether or not an aircraft is cleared for deliberate spinning. It will also stipulate these limitations and constraints:

1. Maximum weight at which spinning may be conducted.
2. In the case of four-seat aircraft whether or not the rear seats may be occupied. Usually the two rear seats must remain empty during spinning.
3. Whether or not power may be used for the spin entry.

4. Whether or not aileron may be used to assist entry.

Item 2 is related to the class of Certificate of Airworthiness for the aircraft. Some of the more powerful light trainers have four seats, allowing them to be used as tourers. In this role the aircraft is certificated under the **Normal Category**. When only the front seats are occupied and the aircraft is at lighter weight it may be flown under the **Utility Category**. The significance of this in the spinning context will be explained under *Factors Affecting Spin Behaviour*.

Spin Entry and Recognition

The spin may develop from a number of flight conditions; gliding turns at too low an airspeed, unbalanced flight while at low speed with high power and badly executed aerobatic manoeuvres.

For demonstration purposes the spin is usually entered power off and the sequence of events is as follows:

1. At a speed some 5 kt above stalling full rudder is applied in the required direction of spin.
2. A yaw follows and this induces a roll in the same direction (Further Effects of Rudder, page 33).
3. The down-going wing meets an airflow from below which effectively increases its angle of attack, still further stalling the inner wing and increasing drag which adds to the original yaw.
4. The up-going wing meets an airflow from above which effectively decreases its angle of attack, partly un-stalling the outer wing and reducing drag. A little thought will reveal that this too will add to the original yaw.
5. The nose will now follow the lower wing below the horizon (Further Effects of Rudder, page 33) and after a few turns the aircraft will settle into a state of equilibrium known as **Autorotation**. The aircraft may oscillate slightly in pitch.

These events are illustrated in Fig. 28. During the spin the aircraft is rolling, yawing, skidding towards the outer wing and, relative to the pilot, pitching up. It describes a complex, corkscrew path and although the airspeed is low rate of descent will be very high.

During the spin elevator and rudder effectiveness will be poor. The manoeuvre is not to be confused with a spiral dive where the airspeed is high and increasing, the aircraft is in balance and control effectiveness remains unimpaired.

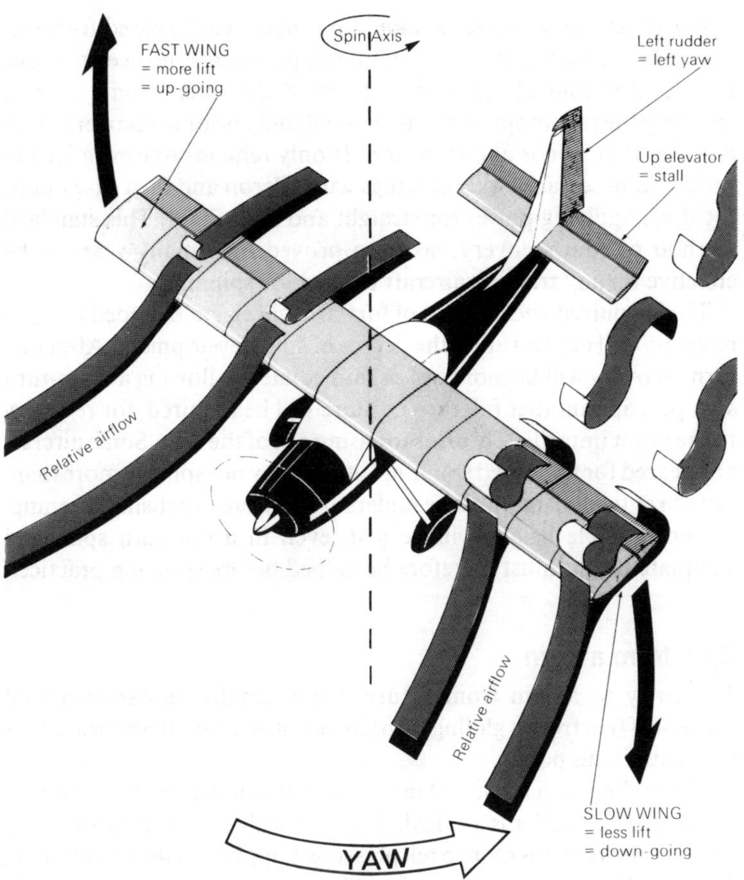

Fig. 28 Aerodynamic forces in a Spin.

Recovery to Level Flight

Since yaw was the original prime cause of the spin it follows that elimination of yaw is an important factor in spin recovery. Moving the wheel/stick forward as a first action would, in a modern aircraft, most likely convert the spin into a spiral dive with an attendant build up of airspeed and considerable loss of height.

To recover from a spin the pilot must ensure that power is off and the ailerons are neutral. Direction of yaw should be confirmed on the turn needle, then *full* rudder in the opposite direction must be applied and held on.

After one or two seconds pause to allow time for the rudder to take initial effect the wheel/stick should progressively be eased forward, un-stalling the aircraft, until spinning stops. There is no need to move the elevator control any further forward than that. Some aircraft speed up the rate of spin rotation immediately prior to recovery. This is a sign that spin is about to stop. It only remains to centralize the rudder, if necessary level the wings with aileron and then ease out of the dive, applying power for straight and level flight. This standard method of spin recovery has been proved over many years to be effective on any training aircraft cleared for spinning.

Time required and amount of forward wheel/stick needed for spin recovery will depend upon the degree of spin development. After one turn recovery will be more or less immediate. Following a three-turn spin perhaps another full turn or more will be required, for recovery to the point immediately prior to easing out of the dive. Some aircraft are cleared for unlimited spinning, others may not spin for more than, say, three turns. This must be understood before practising spinning.

Considerable height will be lost, even in a one-turn spin, and adequate height must therefore be gained before spinning practice.

Spin from a Turn

The entry to a spin from a turn will normally be demonstrated power-off (i.e. from a gliding turn) since not all aircraft are cleared for spin entry with power.

The gliding turn is entered in the normal manner, speed is reduced and excessive rudder is applied. The nose will tend to drop and in an effort to prevent this more back pressure is applied to the wheel/stick. The, combination of yaw and low airspeed will then produce a spin entered at a slightly higher speed than from the straight glide because of 'g' loading.

Recovery is as previously described.

Additional Forces in a Spin

While it is not necessary for pilots to have a deep understanding of all the complex forces that occur in a spin they should at least be aware that such forces exist.

The pitching-up fuselage and the rolling wings each develop gyroscopic properties. The 'wing gyroscope' is disturbed by the pitching-up force to provide a pro-spin reaction while the 'fuselage gyroscope' is displaced by the rolling force to provide an anti-spin reaction.

In practical terms, these gyroscopic forces mean that aircraft with large wings and relatively light fuselages (e.g. light twin-engine designs) will have a fast rate of spin and recovery may be difficult. Aircraft with relatively small wings and long, heavy fuselages will have a slow rate of spin and recovery is usually straightforward.

Less complex and perhaps easier to understand is the effect of engine and tail unit mass while the aircraft is spinning. There is a natural tendency for these to fly outwards due to centrifugal reaction. If this action is predominant there is a risk that a **Flat Spin** may develop (Fig. 29). The danger of such a condition is that airflow over the rudder and elevators no longer approaches from a direction which allows them to have much effect and recovery from a flat spin can prove very difficult or even impossible.

In so far as training aircraft are concerned those certificated for spinning are known to have no flat spin tendencies.

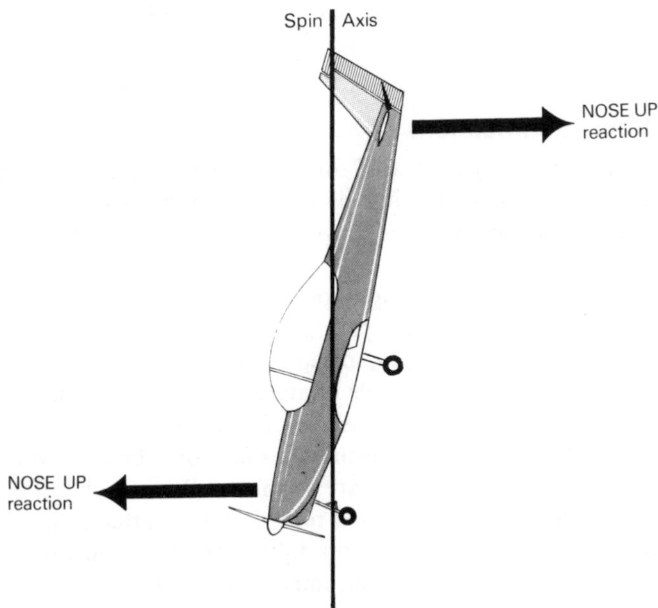

Fig. 29 Rotating masses of the nose and tail surfaces tend to flatten the spin.

Centre of Gravity

The modern light aircraft is less prone to inadvertent spinning than those of pre-war design. Stall/spin resistance has partly been achieved by limiting the effectiveness of the elevators and rudder while flying at low airspeeds. There are, of course, other design features involved but control effectiveness is important in the centre of gravity context.

If the elevators have limited effect at low airspeed it follows that an aircraft with a C of G towards the maximum rear limit will be better able to adopt the high angle of attack required for stall/spin entry than one with a forward C of G. Some aircraft are difficult to spin when loaded with a forward C of G whereas the same aircraft with an aft C of G (e.g. with passengers in the rear seats or heavy baggage in the locker) will prove more prone to spin entry and less easy to recover. Therefore such aircraft may only spin when flown under the Utility Category privileges of their Certificate of Airworthiness.

Although fuel is usually carried at or near the aircraft's centre of gravity, amount of fuel can sometimes affect spin entry behaviour. Section 6 of the aircraft manual will indicate whether or not fuel contents have any material effect on the Centre of Gravity.

Instrument Indications

Some modern trainers cleared for spinning are capable of demonstrating the entry but within half-a-turn or so the airspeed starts to increase, the manoeuvre is no longer a spin and it has become a spiral dive.

In an effort to induce a proper spin there have been cases in the past where pilots have introduced pro-spin aileron (i.e. wheel/stick in the opposite direction to the spin, so depressing the aileron on the down-going wing to increase its angle of attack). At the same time a little power has been left on to increase the effectiveness of the rudder and elevators. These established techniques may only be used when the aircraft manual confirms they are permitted. When such techniques are prohibited there will be good reasons and no attempt should be made to force such aircraft into a fully developed spin which may flatten and prove difficult, if not impossible, to recover.

A comparison of instrument indications for the spin and the spiral dive follows overleaf.

Instrument	Spin	Spiral dive
ASI	Low, fluctuating airspeed	High, increasing speed
VSI	High or maximum descent	High or maximum descent
Altimeter	Large height loss	Large height loss
Turn Needle	Maximum turn rate	Maximum turn rate
Ball	Skid away from spin	In balance
DI	Spinning, perhaps toppled	Spinning, perhaps toppled
AI	Toppled or random	Toppled or random

Toppled Instruments

The direction indicator may be re-set after it has toppled with the setting knob provided. Attitude indicators that are not topple-free require 10 minutes or so to re-erect after any manoeuvre placing them outside their limits.

Pre-flight Briefing

Exercise 11. Spinning (from the Straight Glide and a Gliding Turn)

AIM — *To recognize the developed spin and to recover with a minimum loss of height*

AIRMANSHIP — *Height: sufficient for the manoeuvre*
Airframe: flaps up, brakes off
Security: no loose articles in the cabin, harness and hatches secure
Engine: fuel pump on, carb. heat hot, temperatures and pressures 'in the green'.
Location: not in controlled airspace, etc., in sight of known landmark
Lookout: clearing turn to check not over towns, airfields or other aircraft
Suitability of weather

AIR EXERCISE

Spin entry	*During the spin*	*Spin recovery*
Fuel pump on, carb. heat hot *LOOKOUT* *Close throttle* *Hold up nose to maintain height* *5 kt above stall:* *FULL RUDDER in required direction* *Wheel/Stick FULLY BACK* *Aircraft spins* *Entry from a gliding turn* *From gliding turn hold up nose* *Add excessive rudder* *Prevent nose drop with more up-elevator* *Aircraft spins at higher entry speed (effects of 'g')*	*Low fluctuating airspeed* *Rapid height loss* *Max. turn in spin direction* *Ball shows skid in opposite direction* *Possible oscillation in pitch* *Aircraft rolling, yawing and pitching up*	*Check:* *Power off* *Ailerons neutral* *Yaw direction on turn needle* *Recovery actions:* *FULL rudder opposite to yaw direction. Hold on rudder* *Pause, one/two seconds* *Progressively ease forward Wheel/stick until –* *SPIN STOPS* *Centralize rudder* *Level wings (aileron)* *Ease out of dive* *Add power for S & L*

Flight Practice

Vital Actions 'HASELL'

Height – sufficient for the exercise.
Airframe – brakes off, flaps up.
Security – harness and hatches secure. No loose articles in the cabin.
Engine – fuel pump 'on' carb. heat hot, temperatures and pressures normal.
Location – out of controlled airspace, or in sight of known landmark.
LOOK OUT – carry out an inspection turn: not over towns, airfields or other aircraft.

AIR EXERCISE

Introduction

Before an attempt is made to teach spinning the instructor will acclimatize the student by demonstrating a spin and recovery.

Spin Entry and Recognition

(*a*) Apply carburettor heat, close the throttle and hold the nose just above the horizon.
(*b*) Approximately 5 kt above stalling speed, apply full rudder smoothly in the required direction of spin, at the same time holding the wheel/stick right back.
(*c*) The aircraft is now spinning. (It is possible to count the number of turns by looking out at the horizon and watching for a known landmark each time it comes around.)

Recovery to Level Flight

(*a*) Check: power off, ailerons neutral and direction of yaw on turn needle.
(*b*) Apply and hold full rudder in opposite direction to yaw.
(*c*) Pause, then ease the wheel/stick progressively forward until spinning stops.
(*d*) Centralize the rudder. Level the wings with aileron.
(*e*) Ease out of the dive, add power and resume straight and level flight.

Spin from a Turn

(*a*) Apply carburettor heat, close the throttle and commence a gliding turn at a low airspeed by holding up the nose.
(*b*) Attempt to speed up the rate of turn by additional rudder and hold the nose up by further backward movement of the wheel/stick.
(*c*) The aircraft will begin to spin. Recovery in the usual manner.

(*d*) As flying speed is gained level the wings with ailerons. Ease the wheel/stick back and bring the nose into the straight and level position. Apply cruising power.

Instrument Indications

(*a*) Enter a spin from the glide.
(*b*) Note: the low, fluctuating airspeed,
the high rate of descent,
the maximum rate turn,
the outward skid,
the toppled gyro instruments.
(*c*) Recover from the spin.

Important

1 Power or pro-spin aileron must not be used to assist full spin entry unless this instructional technique is permitted for the aircraft type (see the Owner's Flight/Operator's Manual).

2 While aircraft differ in spin behaviour the degree of forward elevator movement needed to effect recovery will depend upon the state of the spin. A fully developed spin may require full down elevator while only part forward movement will usually be required after two turns of a spin or less.

3 Immediately prior to recovery the rate of spin may increase. This indicates that the correct recovery action has been taken.

Exercise 12

Take-off and Climb to Downwind Position

Background Information

It is preferable to take off with a short ground run (*a*) to avoid unnecessary wear of the tyres and wheel-bearings and (*b*) in order to clear the far boundary of the airfield in safety.

Outside the perimeter of the airfield, buildings and trees must often be over flown immediately after take-off and in consequence an adequate angle of climb in relation to the ground is a further requirement. It is with this consideration in mind that whenever possible the take-off is carried out into wind. One or more ground indications are provided at the aerodrome for the purpose of showing wind direction and these are listed on page 196–7.

How does taking off into wind make it possible to meet the foregoing requirements? It will be remembered that there is a minimum speed below which the aircraft will fail to generate sufficient lift for flight. Light aircraft usually take off some 10 to 15 kt above this speed.

Assuming a take-off speed of 50 kt and remembering that it is airspeed (relative airflow) which generates lift, a 20 kt wind along the runway will provide almost half the required airspeed even when the aeroplane is stationary. In other words it will only be necessary to accelerate to 30 kt **Ground Speed** to attain the 50 kt airflow needed for safe take-off. The effect of a 20 kt headwind is shown in Fig. 30. If an

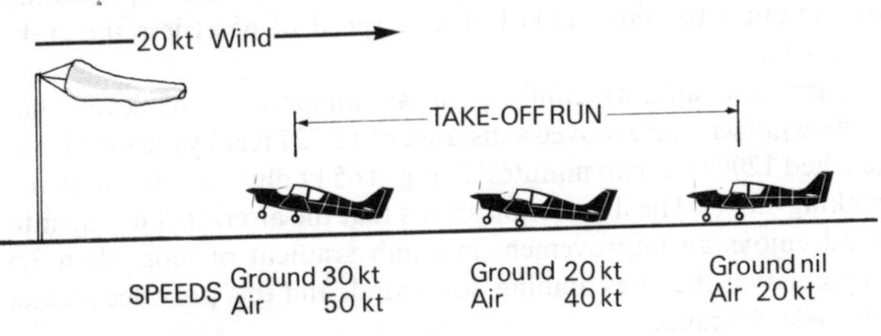

Fig. 30 Taking off into a 20 kt headwind.

attempt was made to take off downwind under the same airfield conditions the ground speed would reach 70 kt before the necessary 50 kt airspeed lifted the aircraft off the ground and Fig. 31 shows that the 20 kt additional ground speed will require a longer run over the ground.

After becoming airborne the aircraft must not be forced into a climb immediately but the airspeed should be allowed to build up naturally to that for best rate of climb.

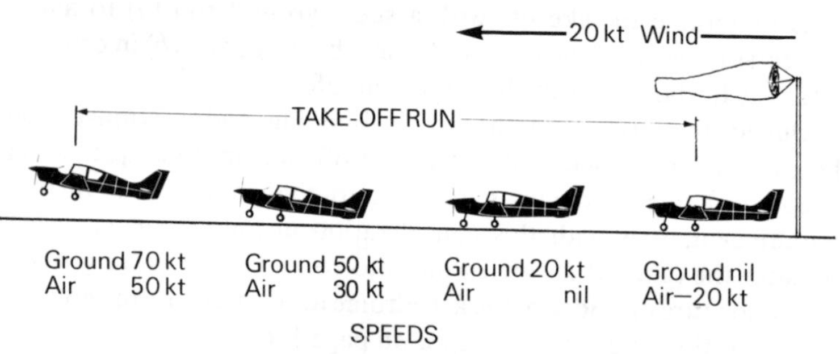

Fig. 31 Taking off with a 20 kt tailwind. The ground roll is greatly increased (compare with Fig. 30).

Effect of Wind on Climb Gradient

Into wind or downwind the rate of climb remains practically the same but **Climb Gradient** (steepness of the climb path relative to level ground) will differ greatly. Figure 32 illustrated the case of an aircraft with a rate of climb of 600 ft/min and a climbing speed of 65 kt. If it takes off into a 20-kt wind, by the time it has reached 1200 feet, distance travelled up the climb path will be 9120 feet, representing two minutes to climb 1200 feet at a speed of 65 kt less the 20-kt headwind.

The same aircraft climbing at an airspeed of 65 knots but downwind will have moved a distance of 17,227 feet by the time it has reached 1200 feet (two minutes' flying at 65 kt plus the 20-kt tailwind making 85 kt). The illustration shows that the aircraft climbing into wind enjoys an improvement in climb gradient of more than 3.5 degrees over the one climbing downwind, and this provides greater obstacle clearance.

At low levels the wind is slowed by friction with the ground,

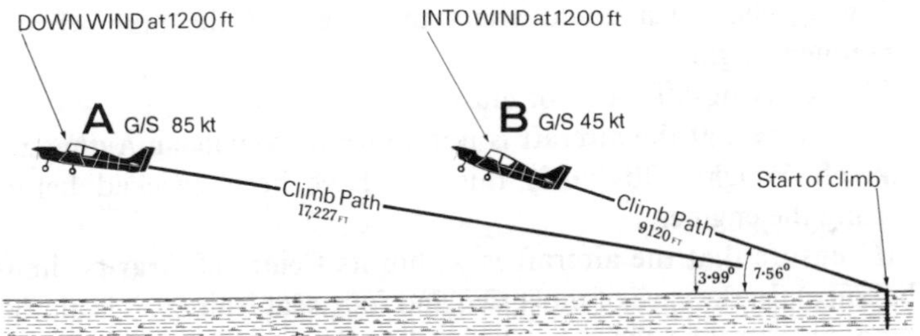

Fig. 32 Climbing after take-off. Aircraft B has reached a height of 1200 ft after flying a distance of 9120 ft. Aircraft A is the one shown in Fig. 31. Being downwind it has had to fly 17,227 ft while climbing to 1200 ft. Note the difference in climb path angle. Aircraft B has good obstacle clearance.

becoming progressively stronger as height is gained. Since the 'into wind' condition entails climbing into an ever-strengthening head-wind the angle of climb will increase for the first few thousand feet, whereas the reverse would be the case when climbing downwind. It should be understood that this effect is only noticeable under conditions of strong wind.

Taking off in a **Crosswind** presents different problems since there will be a tendency for the aeroplane to **Drift** sideways during the procedure. There are times when a partial crosswind cannot be avoided and steps must be taken to counteract the effects of drift in order to safeguard the undercarriage. Crosswind take-offs will be dealt with later in the chapter.

Before flight the engine run-up procedure outlined in Exercise 2 (page 16) must be carried out by the pilot. Usually this is conducted at the **Holding Point** (position near the beginning of the **Downwind end** of the runway). The engine will have warmed up ready for power checks while taxying to the holding point.

Pre-Take-off Vital Actions and Traffic Check

Prior to commencing the take-off it is necessary to complete a number of pre-flight checks, some of them before walking out to the aircraft.

On paper there would appear to be much to do; in practice most of the actions will become routine with experience. However, they are not to be regarded as of little consequence. Every single item in the following check list could affect the safety of the take-off and subsequent flight.

Before taking-off the pilot must:

(*a*) ensure that the aircraft is not above its **Maximum Authorized Take-off Weight**. Obviously this will have been checked before starting the engine.

(*b*) ensure that the aircraft is within its **Centre of Gravity** limits (balance). In the case of most two and four-seat trainers it is almost impossible to load the cabin beyond its limits of balance, but this aspect becomes of importance in some six-seat touring designs and light twin-engined aircraft as well as large transport types. Like (*a*) this check will have been made before starting the engine.

(*c*) determine that the airfield length and conditions will permit the aircraft to take-off safely. A light aircraft operation from a licensed airfield will normally have ample room for the take-off. But when the strip or field is short and the ground is soft (or the grass is long) the Owner's/Flight/Operating Manual must be consulted to find the aircraft's **Field Performance** under those conditions.

Items (*b*) and (*c*) are more fully explained in Chapter 8, of *Flight Briefing for Pilots Vol. 2*.

(*d*) check that the weather is suitable for the planned flight.

(*e*) advise the local Air Traffic Control of the intentions. At busy training airfields this is usually done from the aircraft over the radio.

(*f*) conduct a thorough pre-flight inspection of the aircraft as detailed in Exercise 2, checking that there is adequate fuel for the exercise.

(*g*) start up and taxi to the correct holding point for the runway in use.

(*h*) complete the power checks.

(*i*) complete the pre-take-off Vital Actions.

Pre-Take-off Vital Actions

Before any action is taken which radically alters the existing condition of an aircraft steps must be taken to ensure that all relevant controls, switches, etc. are properly set. By now the student will be familiar with the HASELL check which precedes any manoeuvre that may entail loss of height. Other examples of checks are those required before landing and, in the case of this exercise, immediately prior to take-off. Such checks are known as **Vital Actions**.

Vital actions, when performed in larger and more complex aircraft than light trainers, take the form of printed check lists. Flying such aircraft is usually a two-crew operation and it is the practice for one pilot or crew member to read the check list while the other carries out the actions.

Although training aircraft are, in the main, confined to simple systems and equipment some flying training schools train their students to use check lists as a matter of policy. Unfortunately, those printed in Owner's/Flight/Operating Manuals differ from manufacturer to manufacturer; the order in which the actions are presented is not standard.

A method which, for many years, served aviation well was the standard mnemonic – a word or collection of letters (such as HASELL) which represents the actions to be taken. Provided these mnemonics are thoroughly learned and recited with care they are particularly suitable for light aircraft operation when the pilot is usually the single crew member. The vital actions before take-off that have been in use for many years are remembered by the letters:

T M P F F G H

T: trim for take-off, tighten throttle friction
M: mixture rich, carb. heat cold
P: pitch fully fine ('pitch fixed' in most trainers)
F: Fuel on correct tank, sufficient for flight, fuel pump on
F: flaps up (or as required)
G: gauges and gyros (set DI with compass and check instruments)
H: harness and hatches secure.

No mention is made of the anti-collision beacon in this check list because it would have been switched on before taxying from the parking area.

Use of Flap

On page 36 it was mentioned that the first 15°–25° of flap depression gave a considerable increase in lift and only a moderate rise in drag. Therefore it is common practice for the flap lever to be provided with a take-off position or **Maximum Lift** position. When the flap system is electric or hydraulic take-off setting will be marked on the flap position indicator.

Although the use of flap during take-off has little effect on most light aircraft, the technique is essential to heavier and faster designs providing these advantages over the flaps up take-off:

(*a*) The aircraft becomes airborne at a lower airspeed and therefore leaves the ground after a shorter run. This in turn means that particularly on a rough surface the undercarriage is subjected to less wear and tear.

(*b*) After becoming airborne climb path relative to the horizontal is steeper because although the rate of climb remains unchanged (or may even decrease slightly) forward speed is 5–10 kt lower. Fig. 33 compares the positions, one minute after lift-off, of two identical aircraft, one with flaps up and the other with flaps at the take-off setting. Using flap the steeper climb path increases obstacle clearance and is of particular value when taking-off from a relatively small airfield surrounded by buildings and trees.

It should be realized that no benefit is derived from the climb out with flap unless the correct flaps down climbing speed is used, this usually being some 5–10 kt below the recommended best climbing speed in the flaps up configuration.

Not all aircraft enjoy improved take-off performance when using flap, and before using the technique pilots should consult the aircraft Owner's/Flight/Operating Manual.

ATC Clearance

At airfields which operate an **Air Traffic Service** there will be a control tower staffed by **Air Traffic Controllers**. The purpose of this organization is:

(*a*) to provide airfield, traffic and weather information.
(*b*) to ensure adequate separation of aircraft on the ground and in

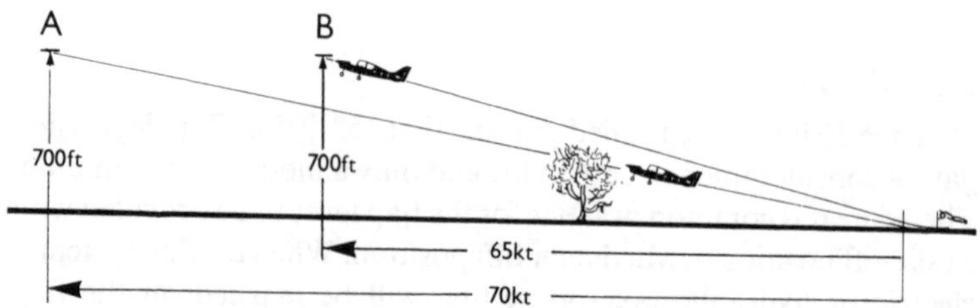

Fig. 33 Effect of flap on the climb after take-off. Illustration shows relative positions of two aircraft when they have climbed to 700 ft. Aircraft B, climbing at 65 kt with recommended take-off flap has a steeper climb path than Aircraft A which is climbing at 70 kt without flap. Not all light aircraft benefit from the use of flap during the climb.

the local **Aerodrome Traffic Zone,** a volume of airspace extending from ground level to a height of 2000 feet. The ATZ extends laterally for a distance of 1½ nm from the aerodrome boundary. Such an airfield is known as a **Controlled Aerodrome.**

The scale of facilities varies from airfield to airfield. At the smaller aerodromes radio communications may be confined to a single frequency operated by an unlicensed controller. Information provided by such a controller will be **Advisory**. Larger or busier airfields or airports will have licensed air traffic controllers and often there will be three separate radio frequencies:

GROUND:	Taxi clearance will be obtained on this frequency before moving off towards the holding point and while taxying back to the parking area after landing.
TOWER:	Permission to take off will be obtained on this frequency. It will also be used while landing.
APPROACH:	After leaving the ATZ the approach controller will handle communications with the pilot until handing over to the next link in the air traffic control chain. When approaching the airfield the pilot will make radio contact on the approach frequency.

Before any attempt is made to taxi onto the runway the pilot will request 'Departure clearance' on the 'Tower' frequency. Use of radiotelephony is explained in Chapter 11, Volume 2 of this series. However, even when the Tower controller has given permission for the aircraft to taxi onto the runway this does not absolve the pilot from checking that no other aircraft is on the approach to land or still on the runway in the path of intended take-off.

Take-off Into Wind

Immediately clearance has been given taxi on to the beginning of the runway without delay, since time wasted by indecision can inconvenience other aircraft waiting to use the runway. To ensure that a minor swing left or right of the take-off path does not involve leaving the runway the aircraft should be positioned on or near the centre line which is usually marked in a manner similar to a motor road.

The controls are given a final 'full and free movement' check. This is to ensure that the controls are free and (when fitted) the Auto-Pilot is disengaged. (A reference point on which to keep straight should be found on the far boundary of a grass airfield.) After a good look behind and ahead to ensure that all is clear the take-off may now begin.

Directional Control

One of the advantages provided by nosewheel aircraft over tailwheel designs is their natural tendency to run straight on the ground without a tendency to swing. This is of particular value to the pilot during the take-off (tailwheel techniques are described in Chapter 20, *Special Techniques*).

During the take-off maximum power will be applied and this will introduce:

1. *Slipstream Effect*: a yawing tendency caused by the helical slipstream moving down the fuselage to impinge on one side of the fin and rudder (described in Exercise 6, page 58)
2. *Torque Effect*: a rolling tendency applied in the opposite direction to propeller rotation which presses one wheel more firmly on the ground than the other. There will be increased rolling friction on this wheel causing an effect similar to applying brake. On aircraft of low power torque effect is of little significance.

The two effects just described will tend to swing the aircraft to the left during take-off, assuming the propeller rotates clockwise when seen from behind. However, the tendency to swing is easily contained by steering the nosewheel through the rudder pedals. In aircraft without nosewheel steering linked to the pedals the inherent stability of nosewheel undercarriages allied to the steering power of the rudder, acting under the influence of full-throttle slipstream, will overcome any tendency to swing.

The take-off is started by opening the throttle smoothly and fully in one movement, maintaining direction by steering with the rudder pedals.

Monitoring Acceleration and Engine Performance

During the early stages of the take-off, immediately after directional control has been assured, the pilot must quickly glance at the engine instruments to confirm that the expected RPM are being obtained, thus confirming that full power is being developed. If, during the take-off, power seems lacking, there are unusual engine noises or misfiring and the aircraft is failing to accelerate normally the throttle must immediately be closed and the take-off abandoned. The ASI should confirm 'airspeed building'.

There can never be any justification for pressing on in the hope that the engine will clear itself and restore full power.

The engine instrument check should not be allowed to divert attention from the important task of keeping straight during take-off. A brief look at the RPM and the engine temperature/pressure instruments will suffice.

Undercarriage Considerations

Immediately directional control is ensured a gentle backward pressure must be applied on the control column to remove weight from the nosewheel. This important technique is sometimes overlooked so that the nosewheel is subjected to considerable unnecessary stress during the take-off run. Furthermore under certain conditions there may occur a development called **Wheel-barrowing**. In effect unless a backward pressure is applied to the wheel/stick during the acceleration prior to lift-off the weight of the aircraft is removed from the mainwheels and, in extreme cases, the mainwheels may leave the ground while the nosewheel remains in contact with the runway. Any crosswind existing at the time will cause the aircraft to weathercock or pivot around the nosewheel like an unstable wheelbarrow. If the situation is allowed to continue unchecked the aircraft will develop an uncontrollable swing leading to total loss of direction which is beyond correction. In the process the undercarriage has been known to collapse.

A wheelbarrow can be corrected by moving back the wheel/stick slightly to restore weight onto the mainwheels.

The Lift-off

When lift-off speed has been achieved a further gentle backward pressure will **Rotate** the aircraft, i.e. raise the nose and increase the angle of attack to the point where lift is in excess of weight. The aircraft will then leave the ground. At this stage the airspeed should be allowed to build up naturally (i.e. without deliberately holding down the aircraft) when the climb can then be established.

Establishing the Climb

The immediate post-lift-off phase allows the aircraft to gain height while accelerating to climbing speed. During this phase the wings must be kept level to maintain direction and, because there is maximum power being applied at a relatively low airspeed, considerable rudder pressure will be required to maintain balanced flight.

Having previously been taught Exercise 7, Climbing, the student should be aware of the technique for setting up a climb. The correct attitude must be adopted relative to the horizon, airspeed must be allowed to settle and then the aircraft is re-trimmed.

Electric Fuel Pump

The electric fuel pump is provided as a back-up to safeguard fuel supply in the event of mechanical pump failure. If the mechanical pump does fail many seconds will be required for the electric pump to restore fuel to the engine. There is a tendency among some pilots to switch off the electric pump immediately after take-off but a failure of the mechanical pump at low level is bound to result in a forced landing at a difficult time.

The electric pump should be left on around the circuit when a landing is to follow, or until the top of the climb when leaving the airfield.

Some high-wing aircraft rely on gravity feed and therefore have no electric pump.

Instrument Indications

During the actual take-off instrument scan is confined to engine power check during the early part of the take-off run and the ASI (to establish lift-off speed). All subsequent references while accelerating and lifting off are visual. Immediately the aircraft is established in the climb the pilot should make the following instrument checks:

ASI	Check for correct climbing speed
VSI	Confirm expected rate of climb
Altimeter	Confirm gain in height
DI	Check that correct heading is being maintained
AI	Confirm bank and pitch with outside visual reference
Turn Needle	No turn

Ball	Centred
RPM	Normal for climbing power
Ts & Ps	All 'in the green'

Visual Traffic Checks

At a busy training airfield there could be a number of aircraft on the circuit and these may be of mixed performance. It is the pilot's responsibility to maintain safe separation from other traffic while flying under **Visual Flight Rules** (as opposed to **Instrument Flight Rules** when this responsibility is assumed by the Air Traffic Control Service).

Modern light aircraft climb in a high, nose-up attitude and this tends to restrict the view ahead. It is therefore important to maintain a careful lookout while climbing away from the airfield.

Depending on local airfield traffic rules it is usual for a climbing turn to commence when the aircraft has reached a height of 500 feet. Pilots must be alert and on guard for aircraft which may turn and cut across their own climbing path. Furthermore, before starting their own climbing turn in preparation for flying around the circuit, a very careful check must be made in the intended direction of turn.

Most circuits are flown in a left-hand direction but there are occasions when a right-hand circuit is in operation. From the left-hand seat, visibility to the right will be more restricted than to the left and a thorough lookout check will have to be made, particularly in high-wing aircraft.

Short Field Take-off

At a later stage in training the student will be shown the special technique to be adopted when it is necessary to fly out of a small airstrip.

When the available take-off run is unknown the pilot will have to pace out the field. Experience shows that while most people imagine one pace to be 3 feet in reality the average person steps little more than 2 feet and this length of pace should be used in calculating available take-off run.

Condition of the grass (long, short, wet), surface (soft, hard) and the existence of an up-hill or down-hill gradient must also be taken into consideration while consulting the Owner's/Flight/Operating

Manual (Section 5, Performance). This will confirm whether or not it is safe to take off under the existing conditions.

Not all aircraft benefit from the use of flap as a means of shortening the take-off run and steepening the climb gradient but when flap is recommended for a short field take-off it is essential that the correct lift-off speed is used. This will be lower than that for a normal, flaps-up take-off. After lift-off the correct speed for steepest climb gradient must be adopted if maximum obstacle clearance is to be achieved.

It will be self-evident that the *full* length of the field must be made available for take-off from restricted airstrips and unorthodox techniques, such as gaining speed, then lowering flap just before lift-off, are not to be used. There is no evidence that this procedure, which diverts attention from the essential task of keeping straight, has any advantages to offer. If the airstrip is so short that the aircraft manual indicates there is insufficient room for a safe departure the pilot has no business attempting a take-off until:

(*a*) wind conditions are more favourable, or
(*b*) take-off weight is reduced by offloading fuel and/or cabin load.
(*c*) expert advice has been obtained.

Performance accidents are usually the result of inexperience, over-confidence and a lack of understanding of the factors involved. These are described in Chapter 8, Volume 2.

Soft Field Take-off

This technique is used when soft ground is likely to cause serious additional rolling drag. In extreme cases this could prevent the aircraft attaining flying speed. The aim is to start the take-off run with the control column almost fully back, thus removing much of the load from the nosewheel. When the elevators become sufficiently active to rotate the aircraft, the wings will be at or near flying speed. In some aircraft the back pressure on the elevators must be relaxed during rotation otherwise the tail may strike the ground.

After lift-off the speed should be allowed to build up before a climb is commenced.

Crosswind Take-off

One of the disadvantages of runways is that choice of landing and take-off direction is restricted. Even a three-runway airfield can only provide six alternatives. On an airfield with a single runway it is possible to have a 90° crosswind. Naturally runways are sited with due regard to the prevailing wind directions in the locality but it is a common occurrence for the wind to be some degrees off the runway.

On some airfields a grass area is made available for light aircraft, these usually being more vulnerable to crosswind conditions than larger multi-engined types. Nevertheless, steps must be taken with all aircraft, big or small, to counteract the drift which will result from a crosswind.

Drift in itself is of no danger to an aeroplane when it is in the air and clear of obstructions. It is the transition from ground to air or air to ground which can be endangered by the presence of drift, subjecting the undercarriage and tyres to considerable side loads.

The aim is to carry out a normal take-off holding the wheels firmly on the ground. In doing this care must be taken not to induce wheelbarrowing. During the early stages of the take-off run the wheel/stick should be moved towards the wind in order to prevent a wing lifting. When ample speed has been attained the aircraft is lifted cleanly off the ground in the certain knowledge that the extra speed will prevent it sinking and touching again with drift. As soon as drift occurs, the nose is turned towards the wind and a climb started with the flight path in line with the runway.

Throughout these explanations reference has been made to the runway. A crosswind may occur on a grass airfield which, for physical reasons, is confined to certain directions only and the same procedures are necessary and applicable.

Crosswind Limits

Although runways are sited with due regard to the most common wind directions in that area it is often the case that crosswind conditions exist and steps must be taken to protect the undercarriage from sideways forces that would result from drift. All aircraft have their published **Crosswind Limits** and these will be listed in the aircraft manual for the type.

The extent of drift depends upon:

(*a*) The wind speed.
(*b*) The angular difference between its direction and the bearing of the runway.
(*c*) The take-off and landing speeds of the aircraft, fast aircraft being less affected by any given crosswind conditions than those with low take-off and landing speeds.

It is usual for a simple diagram to be included in the aircraft manual which will enable a pilot to see whether or not the crosswind conditions are within limits. When it is stated that a particular aircraft type has a crosswind limit of 15 kt this is assumed to be at 90° to the runway. When, for example, there is only a 10° difference between the runway and the wind direction a stronger wind could be tolerated and the diagram in the aircraft manual is a simple device for calculating that component of the wind which is effectively at 90° to the runway.

Climb and Circuit to the Downwind Position

Having taken off, climbed to 500 feet and made a climbing turn through 90° the aircraft will then be on the **Crosswind Leg**. It will continue climbing before levelling off at circuit height (800–1000 feet according to local airfield traffic rules). At a suitable distance to one side of the runway a second 90° turn will place the aircraft on the **Downwind Leg**. The Circuit is described in the Exercise 13, Circuit, Approach and Landing.

Exercise 12. Take-off and Climb to Downwind Position.

AIM — *To take off down the centre of the runway at the correct speed and to establish a climb in preparation for a circuit.*

AIRMANSHIP
ATC Liaison
Vital actions: TMPFFGH or Check list
Circuit traffic
Suitability of weather

AIR EXERCISE

Take-off into wind

Holding point position
Power checks
Vital actions
Taxi clearance over radio
LOOKOUT
Full length of runway
Line-up on centreline
Full and free control movement check
Full power
Keep straight
Power check (RPM, Ts & Ps)
Weight off nosewheel (elevator)
Rotate at correct speed
Lift-off. Speed build-up
Establish climb attitude
Trim
LOOKOUT
Climb to 500 ft then turn

Short field take-off

Establish field length
Assess surface
Consult aircraft manual
Take-off flap if recommended
Use full length of field
Take-off as before
Lift-off at correct short field speed
Climb at correct max. climb gradient speed
When all obstacles cleared:
Flaps up
Re-trim

Soft field technique
Take-off with wheel/stick held back
Be prepared for sudden rotate
Relax back pressure during rotate
Establish climb
Re-trim

Crosswind take-off

Preliminaries as before
Check direction of wind
Wheel/stick held into wind
Full power
Keep straight
Power check (RPM, Ts & Ps)
Hold aircraft on runway
Rotate cleanly at higher than usual speed
When aircraft leaves ground watch for drift
Turn nose away from drift
Track along runway centre
Note heading and hold
Re-trim
LOOKOUT
Climb to 500 ft then turn
Continue circuit

Flight Practice

COCKPIT CHECKS

Vital actions

T M P F F G H

T: Trim for take-off.
Tighten throttle friction.

M: Mixture rich. Carb. heat COLD.

P: Pitch fine. ('pitch fixed' in training aircraft)

F: Fuel on and sufficient for flight. Electric pump ON, fuel pressure normal.

F: Flaps up or take-off position.

G: Gauges and Gyros checked.

H: Harness and hatches secure.

(*Note*: Although light trainers have fixed-pitch propellers the student pilot should recite this part of the vital actions with the words 'pitch fixed' since it prepares the way for the time when more advanced aircraft may be flown.)

OUTSIDE CHECKS

(*a*) Position the aircraft so that aeroplanes taking off, flying around the circuit and landing can be seen.

(*b*) Obtain clearance from the Tower and ensure no aircraft is on final approach before taxying on to the take-off area/runway.

AIR EXERCISE

Take-off Into Wind

(*a*) Line up on a distant point on which to keep straight. Check controls for full and free movement. Look behind and ahead. When clear to go, open the throttle sufficiently to move the aircraft at a brisk taxi pace.

(*b*) When sure the aircraft is running straight open the throttle smoothly and fully. Check the engine instruments.

(*c*) As the aircraft gathers speed ease back the wheel/stick to relieve the load on the nosewheel. Keep straight.

(*d*) Keep the aircraft running on its mainwheels. At the correct speed ease back the wheel/stick and lift the aircraft off the ground.

Establishing the Climb

(*a*) Keep the wings level and allow the airspeed to build up naturally while the aircraft continues to gain height.

(*b*) At the correct speed establish the climbing attitude and re-trim.
(*c*) Maintain a good lookout and climb ahead to 500 feet.

Instrument Indications

During take-off run

(*a*) Immediately directional control is assured check RPM for full power and confirm all temperatures and pressures are normal.

After lift-off

(*a*) Check for correct airspeed, heading and balance. Attitude indicator should confirm visual reference.
(*b*) Note rate of climb on VSI and height gain on altimeter.

Short Field Take-off

(*a*) Having selected take-off flap during vital actions (or as recommended in the aircraft's Owner's/Flight/Operating Manual), line up the aircraft in the usual way. Check the controls for full and free movement. Look behind and ahead.
(*b*) With the brakes on, open up to full power, check the engine instruments, then release the brakes.
(*c*) Lift-off at the lowest safe speed (according to type).
(*d*) Keep the wings level and allow the aeroplane to accelerate to the flaps down climbing speed, then go into the climb. Find a new reference point.
(*e*) Clear all obstacles by a safe margin then raise the flaps (if applicable) and increase the airspeed to the flaps-up climbing speed.

Soft Field Take-off

(*a*) Having carried out all the preliminaries, ensured that there is sufficient take-off run for the aircraft and, if recommended in the Owner's/Flight/Operating Manual, set the flaps, line up in the usual way. Check the controls for full and free movement. Look behind and ahead.
(*b*) Hold the wheel/stick fully back then open the throttle to maximum power. Check the engine instruments.
(*c*) Be prepared for the elevators suddenly to become effective.
(*d*) As the aircraft rotates relax some of the back pressure on the wheel/stick to protect the tail bumper. Lift-off occurs with rotation.
(*e*) Keep the wings level and allow the aircraft to accelerate to the flaps-down climbing speed (if flaps have been applied).

(*f*) Find a new reference point on which to keep straight, clear all obstacles by a safe margin, then raise the flaps (if applicable) and increase the airspeed to the flaps-up climbing speed.

Crosswind Take-off

(*a*) Line the aircraft up along the required take-off run, select a point on which to keep straight and note from which side the wind is coming. Check controls for full and free movement. Look behind and ahead.

(*b*) Hold the wheel/stick towards the wind, open the throttle smoothly and fully and keep straight. Check engine instruments.

(*c*) Deliberately hold the aeroplane down and concentrate on keeping straight.

(*d*) When well above the usual take-off speed lift the aeroplane cleanly off the ground.

(*e*) Counteract the drift by turning the nose towards the wind.

(*f*) Establish the aircraft's track in relation to the take-off run and climb away as usual.

Exercise 12E
Emergencies (take-off)

Background Information

With the general improvement in engine reliability that has come over many years of development a power failure during any phase of flight is, these days, very rare. The few instances that occur during or shortly after take-off are usually the result of inadequate pre-flight inspection or incorrect cockpit procedure.

Although engine failure is unlikely in modern aviation pilot training includes various emergency drills which have been devised to develop an ability to handle the situation automatically. Provided these emergency drills are understood and practised at intervals, and provided pilots resist all temptations to act against well-defined principles that have been proved correct over many decades, in the great majority of power failures during or after take-off there is no reason why the pilot or passengers should suffer injury. Properly handled, the emergency need not even result in serious damage to the aircraft.

Captain's/Instructor's Brief before Take-off

In airline flying it is standard flight deck procedure for the captain to brief the first officer on the take-off which may be handled from either seat. There will be a clear understanding about who will handle each function (for example, the captain may look after the nosewheel steering while the first officer will handle the power and flying controls). There will also be a statement from the captain detailing the actions to be taken in the event of power failure during the take-off.

The pilot under training at a flying club or school may one day fly more complex aircraft, consequently, flying training often contains practices that should stand the student in good stead when more advanced aircraft are encountered in the future, even if they are nothing more complex than a single-engine tourer with a retractable undercarriage.

The captain's/instructor's brief before take-off will relate to:

(*a*) Abandoned take-off
(*b*) Engine failure after take-off

Abandoned Take-off

There could be a number of reasons for abandoning a take-off. Someone may have taxied onto the runway without permission or a car might suddenly have appeared unexpectedly. Alternatively, a full or partial engine failure may occur.

If, during the initial instrument scan it is evident that the engine is not delivering its normal full power RPM, or if there are obvious signs of engine malfunction in the form of vibration, unusual noise, smoke, etc., the take-off must immediately be abandoned. On a large airfield, with adequate runway lengths for light aircraft operations, the procedure is confined to closing the throttle and vacating the runway as quickly as possible to avoid disrupting other traffic.

When there is a smell of burning or smoke the fire risk must be dealt with as a priority. This will entail closing the throttle, operating the idle cut-off (mixture control in the fully back position) and turning off the fuel, ignition and master switch.

Small Airfields

Flying training is sometimes conducted from airfields that offer adequate room for normal take-offs and landings in light aircraft but obviously there is less runway/grass strip length available for emergencies such as an abandoned take-off.

In cases when there is a risk of hitting an object on the ground, fuel, ignition and the master switch must immediately be turned off, the idle cut-off should be operated and, if necessary, evasive action must be taken by steering away from the hazard, even if there is a risk of damaging the undercarriage.

When the take-off is being abandoned while on a hard runway of modest length stopping distance can often be reduced by deliberately steering onto the adjacent grass and adding to rolling friction. However when the grass is wet braking efficiency will be better on the runway and the pilot should have made his decision on whether or not to leave the runway *before* starting the take-off.

Engine Failure After Take-off

In the early days of flying engine failure after take-off was an everyday occurrence. It soon became clear that pilots who attempted to turn back towards the airfield in the hope of making a downwind landing almost invariably involved themselves in a serious crash, following the inevitable spin that resulted from tight turns at low speeds as a desperate attempt was being made to reach the field.

To this day, notwithstanding many years of hard evidence which, time and again, proves that the only safe action to take following engine failure is to make the best possible landing more or less ahead, pilots continue to try the impossible – turn back.

Risks of Turning Back after Engine Failure

Experiments have revealed that when an unexpected loss of power occurs the average pilot takes four seconds to react. During that time, remembering the high nose attitude of a climbing aircraft, airspeed will rapidly decrease unless the nose is immediately lowered. Fig. 34 is drawn to scale, except for the aircraft which is shown some thirty times larger. It will be seen that a Rate 1 turn would place the aircraft almost one mile to the side of the runway, entailing a turn through 180° + 45° = 225° before heading back towards the airfield. Although radius of turn would be smaller if a Rate 2 or Rate 3 were adopted, stalling speed increases accordingly and the pilot would have to adopt a higher gliding speed, thus increasing the rate of descent. The following figures for a typical light aircraft show the amount of time needed to turn through 180°, the number of extra degrees required to head back to the airfield and the total time required to complete the turn.

Rate of turn	*Time for 180° turn (sec)*	*Extra degrees required*	*Total time (sec)*
1	60	45	75
2	30	30	35
3	15	10	16
4	7.5	7	8

When these times are translated into height loss it can be calculated that, allowing for the four second reaction time, in an engine failure at 300 ft above runway level a pilot attempting a Rate 1 turn towards the airfield would roll out of the turn, still some distance from the runway

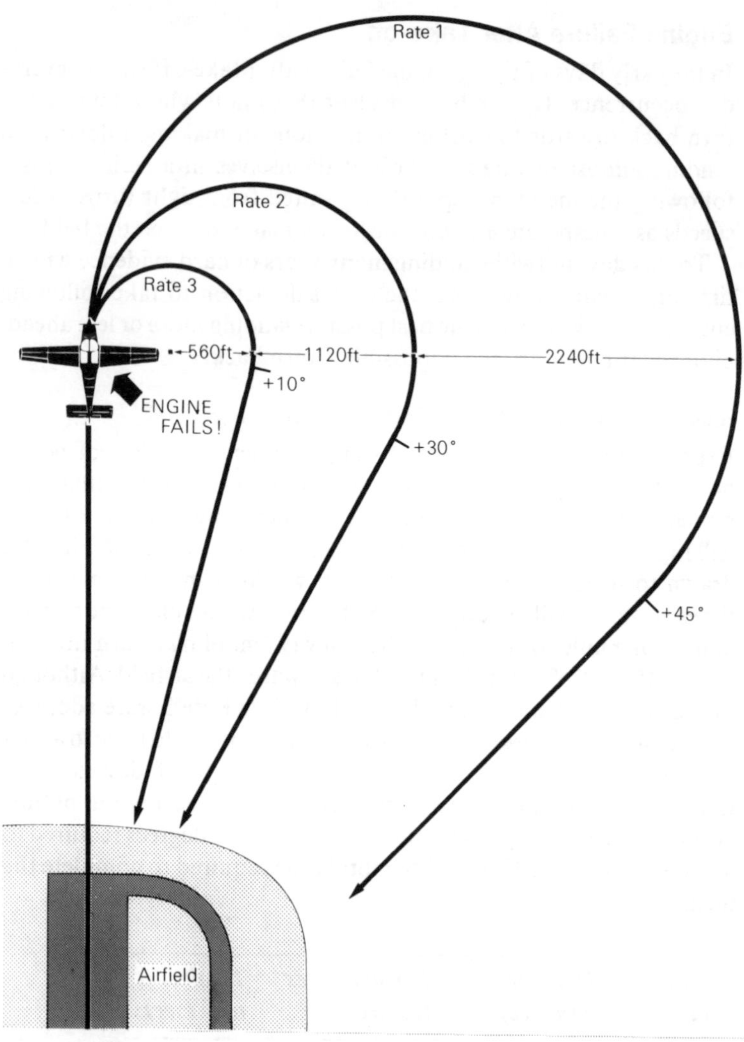

Fig. 34 Engine Failure after Take-off and the risks of turning back. The illustration, which is drawn to scale shows the distances involved and the text explains why it is often impossible to reach the airfield in safety after turning back.

but more than 1000 feet *below* ground level. A Rate 3 turn, even allowing for the higher gliding speed and consequent higher descent rate, would terminate more than 30 feet below ground.

There is the additional hazard, even when sufficient height is available for a return to the airfield, namely the risk of landing against traffic taking off, not to mention the crosswind problems of approaching the field at an angle. The oft-repeated DO NOT TURN BACK is as valid today as it was at the dawn of flight.

Dealing with an Engine Failure after Take-off

When the engine fails at, say 300–400 feet the nose should immediately be lowered to attain gliding speed. QFE will have been set before take-off so providing the pilot with accurate height information above ground level. The QNH setting would entail mental arithmetic and a knowledge of the airfield elevation at a critical time. The area ahead should be scanned through some sixty degrees left and right of the nose. It is rare indeed that no open space is available for a forced landing.

While gliding towards the selected landing area turn off the fuel, operate the idle cut-off, if there is time put out a Mayday call, then devote the attention to reaching the field. The master switch will have to be left on until after flap has been lowered unless these are operated mechanically.

If possible, aim to overshoot slightly, then when it is certain the field can be reached apply full flap and make the best possible approach, avoiding obstacles by making turns.

Shortly before touchdown the door(s) should be unlatched and slightly opened, a precaution against jamming in the event of a heavy landing. The landing should be made by adopting a tail-down attitude so that touchdown speed is as low as possible.

When there is a moderate wind landing speed will be low and impact with fences or other objects will cause little damage.

Engine failure after take-off should be practised at six-monthly intervals in the company of a flying instructor who, most likely, spends much of his time teaching this and other emergencies. Professional pilots, civil and military, are constantly re-checked throughout their career. The private pilot should not regard such revisional training as anything other than good insurance for the future.

Pre-flight Briefing

Exercise 12E. Emergencies (Take-off)

AIM — *To deal safely with engine failure during take-off while on the ground or during the initial climb.*

AIRMANSHIP — *Captain's brief*
Liaison with ATC
Other traffic
Suitability of weather

AIR EXERCISE

Abandoned take-off	*Engine failure after take-off*
Captain's Brief: *Division of responsibility* *Actions to be taken* *On runway/off runway* *ATC approval of exercise* *Importance of QFE setting* *Start take-off* *At 45 knots:* *POWER FAILURE WITH SMOKE* *Actions:* *Close throttle* } *Idle cut-off* } *Fuel off* } *Ignition off* } *Master switch off* } *SIMULATE* *Brake to a standstill* *Avoid obstacles* *Turn onto grass (if necessary)* *Leave the aircraft*	*Preliminaries as for abandoned take-off* *Warn ATC of intentions* *Other traffic considerations* *At 400 feet:* *POWER FAILURE* *Actions:* *Lower nose to gliding attitude* *Look left and right through 60° and find landing area* *Head towards area* *Re-trim* *Close throttle* } *Idle cut-off* } *Fuel off* } *Ignition off* } *SIMULATE* *Aim to overshoot slightly* *MAYDAY call if time* *Lower flap when sure of field* *Master switch off* *Unlatch doors (SIMULATE)* *Overshoot and climb away*

Flight Practice

COCKPIT CHECKS
(Vital Actions)

T M P F F G H

OUTSIDE CHECKS

Airfield traffic suitable for the exercise
Avoid noise sensitive areas during demonstration

AIR EXERCISE

Abandoned Take-Off

While accelerating down the runway it is decided to abandon the take-off.

(*a*) Close the throttle.
(*b*) Keep straight and apply the brakes.
(*c*) If there is evidence of fire or when there is a risk of over-running the take-off area:
Operate the idle cut-off
Turn off fuel and ignition
Turn off the battery master switch.
(*d*) When there is a risk of hitting objects take firm avoiding action, if necessary turning onto the grass to increase rolling friction.
(*e*) Advise ATC of abandonment at first opportunity.

Engine Failure After Take-off

In the unlikely event of engine failure shortly after take-off:
(*a*) Depress the nose to the gliding attitude immediately.
(*b*) Close the throttle.
(*c*) Look through an arc 60° left and right of centre and select the best landing area.
(*d*) Put the aeroplane into the best open space available, using flap to lose height as necessary. Switch off petrol and ignition when committed to the landing and unfasten the door latch.
(*e*) Aim to land in a low-speed, tail-down attitude.
NEVER ATTEMPT TO TURN BACK TO THE FIELD

Exercise 13
Circuit, Approach and Landing

Background Information

Within the basic exercises required for a Private Pilot's Licence the take-off and particularly the landing usually present the student with most difficulty. In consequence during the flying course several hours will be spent on the circuit practising these two exercises.

The circuit finds its main application during training since it enables pilots under instruction or those on solo practice to take off and land a number of times within a relatively short period of time. The circuit is a particularly valuable training exercise because it brings into play many of the earlier exercises – climbing, straight and level flight to accurate limits, climbing, medium and descending turns and, of course, the take-off which formed the subject of the previous exercise.

Establishing the Circuit

The sides of a circuit are known as **Legs**. After take-off the initial climb is usually continued up to 500 feet before making a climbing turn through 90° onto the **Crosswind Leg**. On some airfields local traffic rules may stipulate a straight climb up to circuit height (usually 1000 feet but there are exceptions to this rule). At this height above airfield level the aircraft will continue flying crosswind until it is about the width of the airfield to one side of the runway. It was at this stage that the previous exercise terminated. The remainder of the circuit will now be described.

The Downwind Leg

At a busy training airfield there will be a number of aircraft on the circuit, some flying faster than others, and it is essential that a careful lookout is maintained at all times. In the early stages of circuit training some instructors teach their students to establish a proper circuit pattern by selecting turning points on the ground. It is a

technique that works well. When the correct position has been reached another 90° change of heading, this time a medium turn, will place the aircraft on the **Downwind Leg**.
Early on the downwind leg the pilot should check:

1. HEIGHT: Correct for local circuit rules
2. SPEED: Correct for the circuit
3. RPM: Suitable for the circuit (about 65 per cent power)
4. TRIM.

A common fault of even quite experienced pilots is to converge with the runway centreline. In low-wing aircraft there is an easy method of preventing this and at the same time determining that the downwind leg is being flown at the correct distance from the intended landing path. In an ideally positioned downwind leg the wingtip should trace down the runway. This is illustrated in Fig. 35.

In high-wing aircraft it is sometimes possible to relate another part of the structure to the runway but, in any aircraft, the direction indicator can be used to ensure that the downwind leg is being flown on a reciprocal heading to the take-off and landing direction. The importance of not converging with the runway will become clear later in this chapter.

Before take-off it is necessary to complete the vital actions. Likewise, an aircraft must be set up correctly for the landing and, during the downwind leg the following vital actions must be taken by the pilot. A check list may be used but when the training policy is to use mnemonics a long-standing one in common use is remembered by the letters:

B U M P F

B: BRAKES: off
U: UNDERCARRIAGE: down and locked ('fixed' on light trainers)
M: MIXTURE: rich. Check for carburettor ice
P: PITCH: fine ('fixed' on light trainers)
F: FUEL: on the correct tank, sufficient for an overshoot, fuel pump on, fuel pressure normal.

Like the pre-take-off vital actions this is a standard list suitable for more advanced aeroplanes and it should be recited in full even when all the items do not apply.

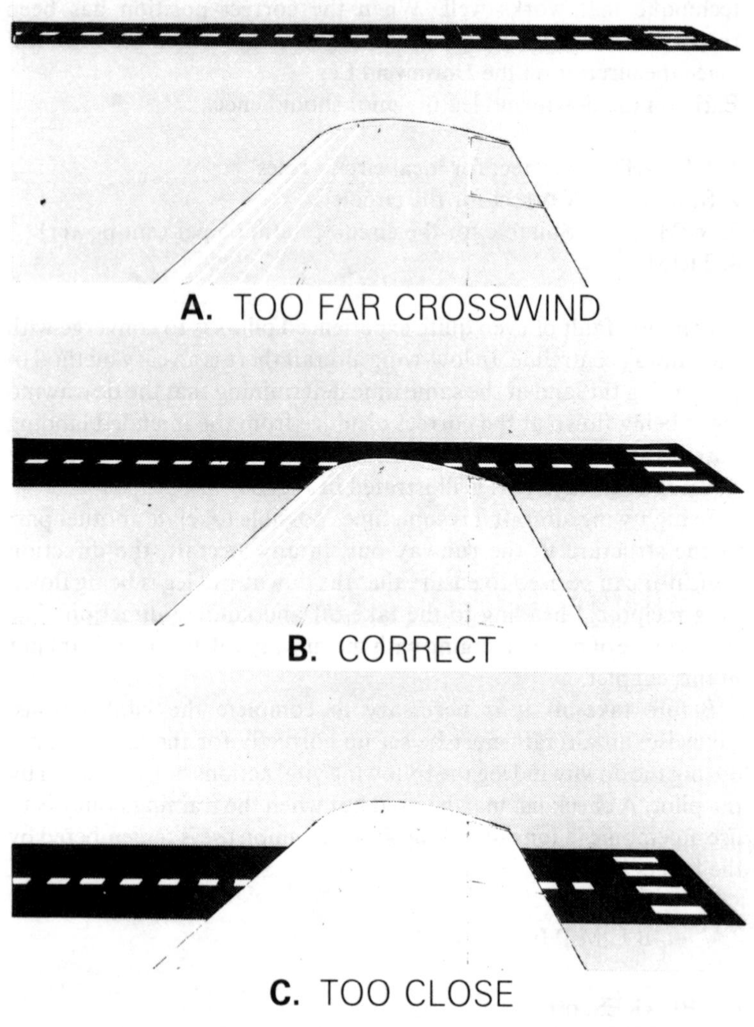

Fig. 35 Monitoring the downwind leg by using a wingtip (low wing aircraft only).

Establishing the Base Leg

When the downwind boundary of the airfield appears to lie at some 45° behind the pilot another turn is made through 90°, bringing the aeroplane on to the **Base Leg**. On the base leg, it is possible to assess the wind strength by the amount of drift away from the airfield and in some cases to correct this it will be necessary to turn slightly towards the field. At this stage preparation for the approach will commence.

The standard method used for positioning an aircraft ready for landing is known as the **Engine Assisted Approach**. It is no more than a powered descent, controlled to accurate limits in terms of descent path and lateral position relative to the **Extended Runway Centreline** (i.e. an imaginary line extending back from runway centre to depict an ideal approach path). Soon after the turn from downwind to base leg it will be necessary to place the aircraft in the **Initial Approach** configuration. Power will be reduced (1500 RPM would be a typical setting), speed is decreased to recommended **Final Approach** speed plus 5–10 kt (according to aircraft type), part flap is lowered (typically half flap) and the aircraft is accurately trimmed.

Turning onto Finals

While on the base leg power should be adjusted so that the aircraft is allowed to descend to approximately 600 feet. The **Turn onto Finals** (or **Final Turn**) is then made through 90°, rolling out on the extended runway centreline at a height of not less than 400 feet. The aircraft is now on the initial approach.

The circuit so far described is illustrated in Fig. 36. It is a well established fact that, in so far as pilots of limited experience are concerned, a good landing must be preceded by a good approach, a good approach is founded on a good base leg, a properly positioned base leg will only follow a good downwind leg and a good downwind leg will not be attained if the turn from the crosswind leg was in the wrong place.

Although experienced pilots learn to make good landings even when for some reason the approach is not ideal, accurate circuit flying is of prime importance to the student pilot or the inexperienced PPL.

Effect of Wind on Circuit Pattern, Drift and Angle of Descent

A distorted circuit pattern can be caused by inaccurate flying. It can also be the result of wind drifting the aircraft from the intended legs.

Even when the wind is blowing down the runway speeds in excess of 15 kt or so can induce drift on the crosswind leg and the base leg. If no compensating action is taken by the pilot drift on the crosswind leg will move the turning point for establishing the downwind leg to a position where it will be shortened and there may be insufficient time

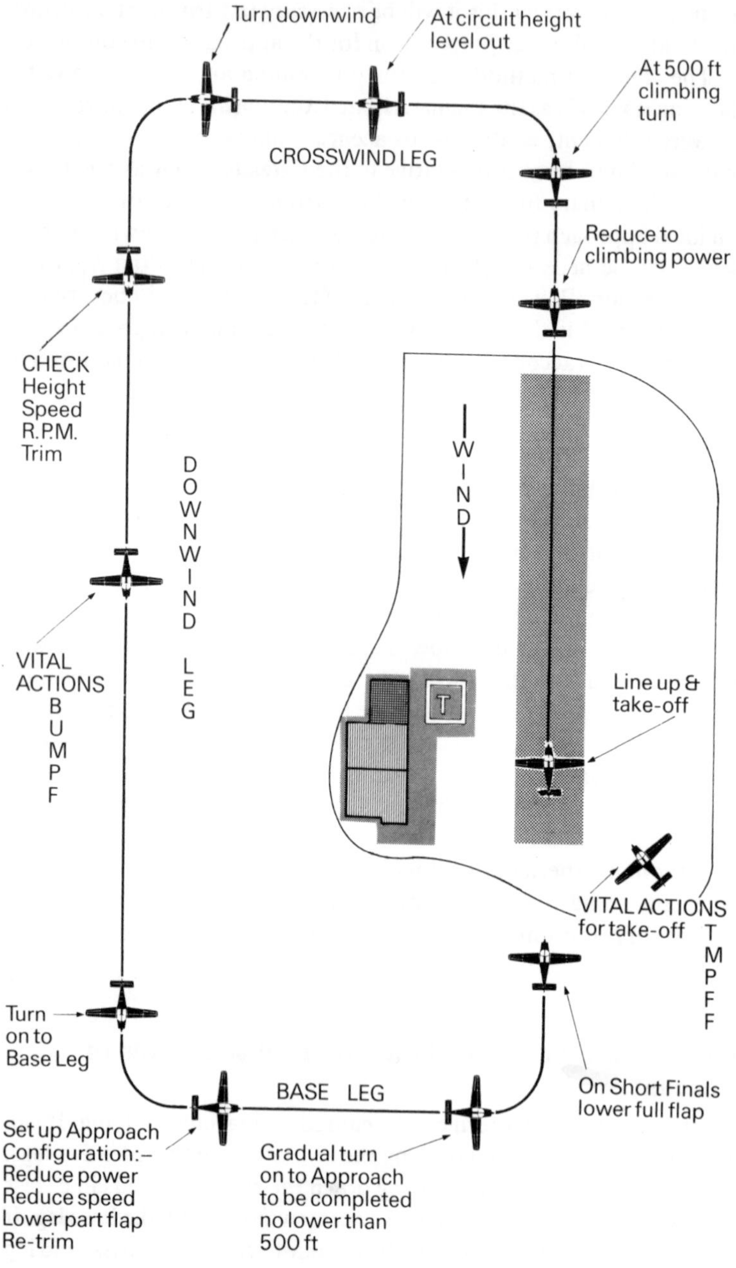

Fig. 36 The circuit should be firmly imprinted on the mind of the student pilot.

for the pilot to complete the vital actions in time for the turn onto base leg. Likewise a strong wind will tend to drift the base leg away from the **Runway Threshold** (the end of the runway over which the aircraft will fly for landing).

In the presence of crosswind conditions during the take-off the following faults may arise (according to direction of drift, left or right):

1. Drift to the left or right during the climb out following take-off.

2. Lengthening or shortening of the crosswind leg.

3. Because of (2) possible misplaced start to the downwind leg. A tendency to drift towards or away from the correct downwind leg which may be checked using the method shown in Fig. 35.

4. Lengthening or shortening of the base leg.

5. Drift to left or right during the approach.

Earlier in this chapter it was stressed that the downwind leg must not be allowed to converge with the runway. If this is allowed to happen the base leg will be shortened to the point where there will be insufficient time for the pilot to establish the aircraft at the correct speed and rate of descent before turning onto the final approach.

Whatever the wind conditions all the foregoing circuit faults may be corrected by intelligent interpretation of the wind, an understanding of the direction of drift to be expected and a determined attitude towards arriving over the chosen turning points.

The Final Turn

A common error among pilots is to fly up to the extended runway centreline before starting the final turn. Such a technique can only result in flying through the centreline and having to turn back towards it. The final turn should be gradual and of large radius so that, as the wings are rolled level, the aircraft is positioned over the extended runway centreline. Figure 37 shows the correct and incorrect way of making the final turn.

Angle of Descent

The incline down which a landing aircraft descends towards the runway threshold is surprisingly flat, 3° being ideal. Such an approach is adopted for normal airfield traffic although STOL (short

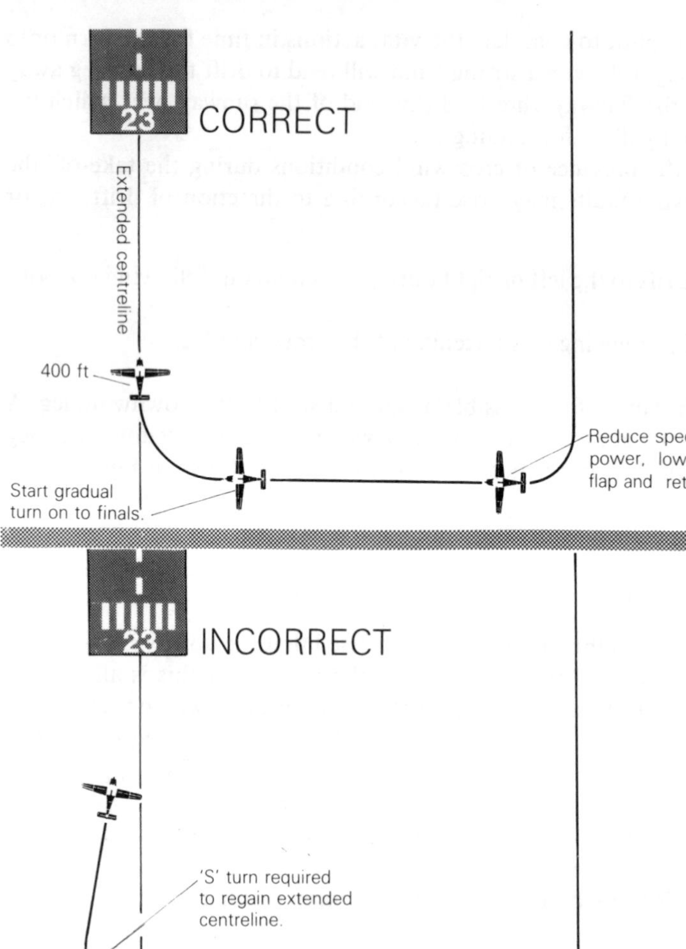

Fig. 37 The base leg and turning onto 'finals', correctly (top) and incorrectly.

take-off and landing) aircraft operating into and out of primitive airstrips often land off a much steeper approach so that trees and other obstacles may be cleared in safety.

Reverting to the engine-assisted descent (Exercise 8. *Descending*) it will be remembered that rate/angle of descent may be controlled to fine limits by the pilot, using various power and flap settings. This technique is used during the approach.

Use of Flaps/Attitude/Power

To obtain maximum control over an aircraft's rate of descent it must be flown at a speed lower than that for best gliding performance. With power off, a relatively high rate of sink is required so that rate of descent may be adjusted with the throttle, from a throttle-closed, high sink rate to power-on, level flight.

The descending path of the aircraft towards the runway threshold is known as the **Glidepath**, something of a misnomer because use of power during this phase of flight is important. Glidepath angle is measured from the horizontal and, as previously mentioned, an ideal glidepath would be 3°. Glidepath angle is affected by the following factors:

1. *Speed and Rate of Descent*: For any given rate of descent a high speed will produce a flat glidepath and a low speed will result in a steep glidepath.
   For any given speed a high rate of descent will cause a steep glidepath whereas a reduced descent rate will cause a flatter glidepath.
2. *Wind*: For any given airspeed/descent rate combination a strong wind will reduce groundspeed and therefore steepen the glidepath. When there is no wind the same airspeed/descent rate combination will produce the flattest glidepath.

The variables described are more straightforward than they seem because control of the aircraft on the approach may be resolved into visual pictures. On the approach the aircraft will be trimmed at the correct engine-assisted approach speed and rate of descent will be controlled with the throttle - more power/less descent: less power/more descent. Handling technique will later be described in greater detail but at this stage it should be understood that, on the approach, the two controls, throttle and elevator, are closely inter-related.

Use of Flaps

After the aircraft has become established on the initial approach, and the pilot has made such minor corrections as are necessary to maintain the extended runway centreline and airspeed/rate of descent, the Final Approach phase will begin. This makes the aircraft ready for arrival over the threshold so that a landing can be made, giving the pilot a maximum of runway length for the subsequent landing roll.

Full flap should be lowered and the aircraft should be re-trimmed into the correct final approach speed.

In recent years there has developed a practice of teaching student pilots to land with only part flap. One of the reasons for this is that some trainers exhibit large trim changes calling for excessive control effort when maximum power is applied for **Overshoot** or **Missed Approach Action** (these terms refer to a situation where it is necessary to abandon the landing for some reason and climb up for another circuit). This reluctance to use full flap had its side effects and a breed of pilots emerged who were reluctant to use full flap because this was regarded as some extreme form of control. From it followed an increasing number of minor landing accidents – over-running the runway and ending the landing in the far hedge, damaged nosewheel struts caused by pilots trying to land at high speed and then stop the aircraft before running out of airfield, and so forth.

While there may be a case for using part flap during certain phases of training this should be regarded as a matter of convenience, not correct handling. Even when there is a strong wind full flap should be used for the landing, although lowering of the final stages of flap may be delayed until the aircraft is nearer the runway threshold. Only when full flap has been lowered will the aircraft make the best transition from air to ground.

There is an exception to the full flap requirement – **Crosswind Landings**. As the term suggests these are landings made when the wind is not in line with the runway direction. Crosswind landings will be explained later in the chapter but most aircraft manuals recommend pilots to use the minimum flap that conditions will allow while landing in a crosswind. This is because when the flaps are lowered airflow over the rudder and elevators is, in some aircraft, disturbed and control effectiveness is reduced.

Control of Airspeed, Maintaining Glidepath, Selection of Touchdown Point

An important skill the student pilot will have to develop is that of maintaining the recommended approach speed while power adjustments are made to achieve the ideal glidepath. Use of power/attitude was described in the previous section. It will, by now, be obvious that proper use of the trimmer will relieve the pilot of some workload in so far as a more steady airspeed will be maintained. However it will be necessary to adjust the trim whenever power changes are made in response to glidepath adjustments.

The previously mentioned, perhaps a little confusing, speed, rate of descent, wind variables that affect the glidepath are relatively simple to control provided the student can train both eyes and perception to react according to visual pictures while on the approach.

Visual References

Figure 38 shows the runway as it will be seen by the pilot during the approach. Drawing A is how it should appear while on the glidepath. B depicts an **Undershoot** situation, where the runway threshold has moved up the windscreen and flattened in perspective. If the approach continues the aircraft will arrive in a field some distance short of the aerodrome. In this case, power must immediately be added and the nose raised slightly to regain picture A. Speed must remain constant during these adjustments.

Drawing C shows an **Overshoot**; the threshold has dropped down the windscreen and the runway is beginning to stand on end. If the pilot takes no corrective action the aircraft will, at best, touch down too far along the runway (with a risk of not stopping within its available length) or it may fly into the far boundary.In this case power must immediately be reduced and the nose lowered slightly to regain situation A.

Using the visual reference technique the aircraft can accurately be made to arrive over the chosen touchdown point which will have been selected during the final approach. Normally this will be the runway threshold.

It is essential to keep the wings level while flying the approach, otherwise there will be a tendency to depart the runway centreline.

Wind Gradient

Wind near the surface is slowed by friction and the effect of this on a landing aircraft is that it will, under **Wind Gradient** conditions, be descending into a wind of rapidly decreasing speed. Rate of speed decrease will depend upon the nature of the surface.

Not all approach areas are prone to wind gradient but, in mild form, it always exists although the normal margin between approach and stalling speeds is adequate to cater for *gradual* wind strength reductions of this kind. The dangers of wind gradient are mainly those that follow an increase in rate of descent – risk of sinking heavily on to the ground.

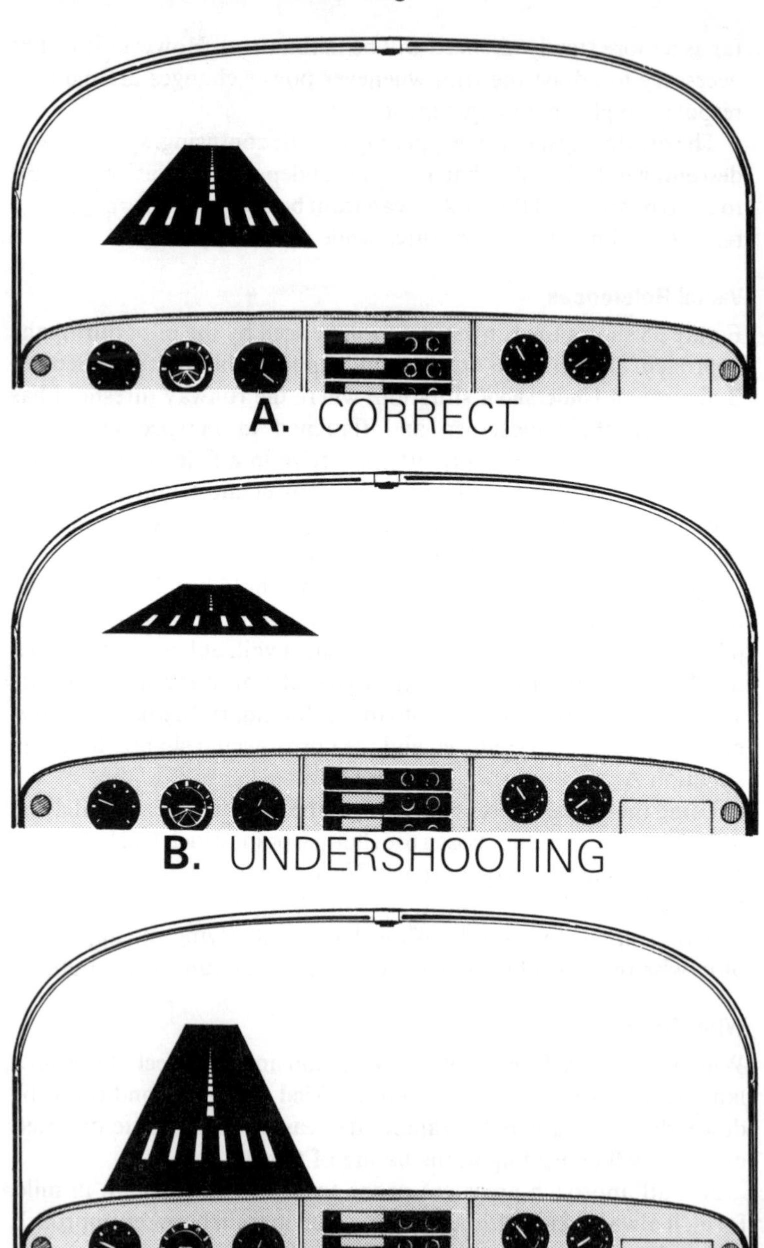

Fig. 38 Visual cues on the approach provided by aspect of the runway and the position of its threshold relative to the glareshield/windscreen.

The Landing

At the end of the approach is the landing, an accomplishment that presents something of a problem to many student pilots, yet without doubt one of the most satisfying experiences in flying is a perfect landing.

The descent must first be checked by easing back the wheel/stick and bringing the aeroplane into the level attitude. The height at which **Round Out** (sometimes referred to as the **Flare**) is effected is the subject of much controversy amongst experienced pilots. Some talk in terms of commencing the round out at the height of a double-decker bus whereas others refer to so many feet above the ground. The actual point is best demonstrated by the flying instructor. Executed properly the flare places the aeroplane in the level attitude with its wheels a few feet above the ground. The throttle is then closed. Naturally the aeroplane will lose speed as it glides parallel to the ground and if the angle of attack were held constant it would drop on to its wheels because of the decreased lift. As the aeroplane slows down and begins to sink the wheel/stick is brought back sufficiently to reduce the rate of sink so that when contact with the ground is made the wheels touch gently.

The geometry of a nosewheel undercarriage is so arranged that the mainwheels are behind the centre of gravity, consequently, a pitch forward will occur, reducing the angle of attack and thus ensuring that the aircraft remains firmly on the ground. It is however important to maintain a little backward pressure on the control column so relieving the nosewheel of undue strain and avoiding the possibility of wheelbarrowing (page 163). As soon as directional control is assured and provided the nosewheel is on the ground, brake may be applied to bring the aircraft to a halt. The aircraft is then taxied off the landing area and the after landing checks are completed.

The landing sequence just described is illustrated in Fig. 39.

During a landing the aircraft must be positioned on the runway centreline thus ensuring space left and right of the landing roll to allow for a minor swing should one develop.

Landing Difficulties

A number of teaching methods have been evolved to assist student pilots who are experiencing difficulty while being taught to land. The following information should forewarn the reader of possible hurdles that may be encountered during this part of the PPL course.

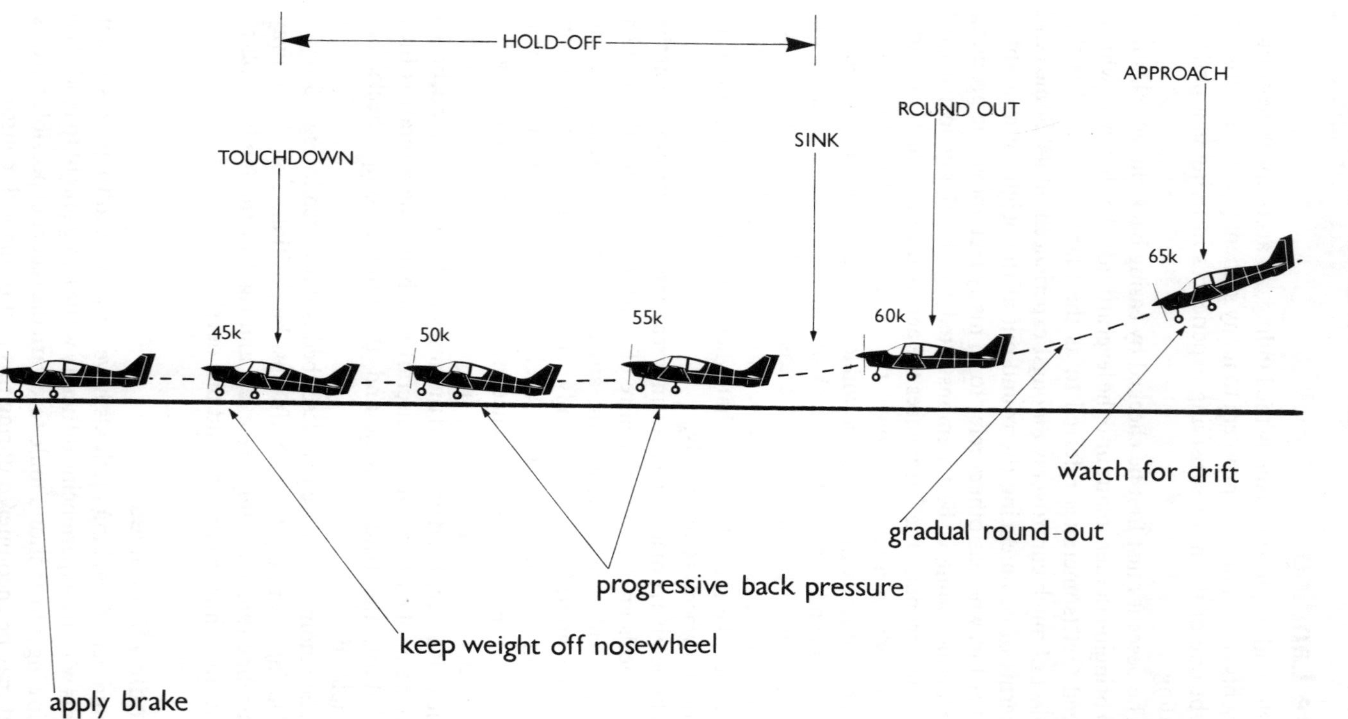

Fig. 39 Typical landing sequence for a nosewheel aircraft.

The Round Out

It is better to aim for a gradual round out rather than descend to within a few feet of the ground and then be faced with a sudden levelling out. Figure 40 shows both methods and alternative 'A' is a much better proposition for the student pilot.

The actual height at which to initiate the check is difficult to define since quoting a number of feet from the ground conveys little when viewed from the aeroplane. 'When the ground seems to expand around the aeroplane' is one description, while another which may help is 'when the ground rushes up to meet the aircraft so that something must be done to stop the descent'. Only practical experience can really convey the true picture.

Guard against bringing back the wheel/stick too quickly, otherwise the aeroplane will climb away from the ground, the angle of attack being too great for level flight at that speed. An aeroplane is said to **Balloon** when this occurs, its speed decreasing rapidly in extreme nose-up cases so that the hand must be kept on the throttle ready to put on power and gradually lower the aeroplane to the ground. On these occasions power should be left on until the actual touch-down when the throttle must be closed immediately. When

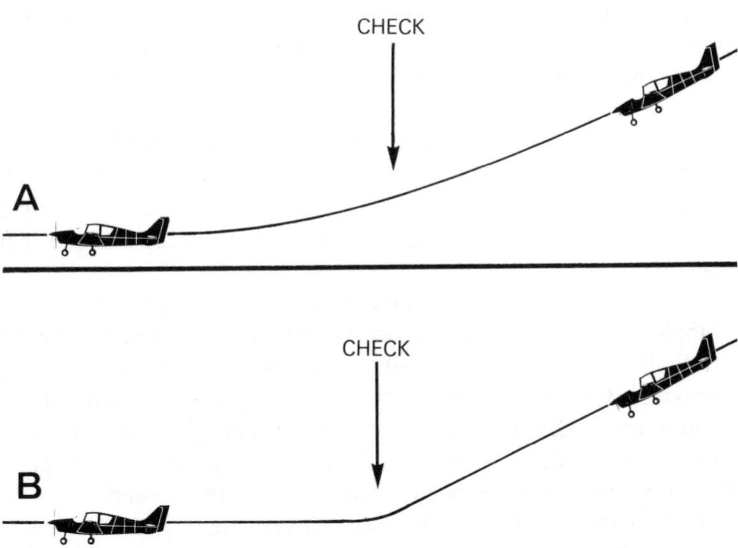

Fig. 40 Checking the descent prior to landing. The roundout made by Aircraft A is easier to judge.

there is insufficient landing area ahead, it may be necessary to open the throttle fully and 'go round again' for another attempt at the landing, and the flying instructor will advise his student according to the size of the airfield and whatever other factors may apply. The balloon and its correction is shown in Fig. 41, but during early training the student is encouraged to overshoot and try another landing.

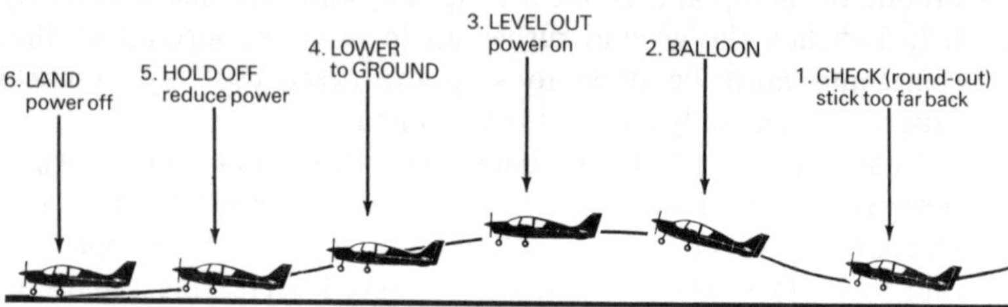

Fig. 41 The 'balloon' and recovery.

The Hold-off

The relative ease with which a modern nosewheel undercarriage aircraft may be landed has, over a period of time, bred a generation of pilots who have lost the art of touching down correctly. It is, of course, possible to continue the approach to within a few feet of the ground, then bring back the control column, level the aircraft and wait for the wheels to make contact with the ground. Such an arrival (it cannot be called a landing in the true sense) must cause the aircraft to alight at a relatively high speed, hence the ever increasing number of broken nosewheel struts and the propeller/engine damage that follows.

The aim when landing a nosewheel undercarriage aircraft is to complete the round-out, reduce further sink to a minimum by gradually increasing the angle of attack as the speed decreases (i.e. by applying back pressure on the control column) when the main wheels should make gentle contact with the ground. This is known as **Holding-off**. Properly executed, the aircraft should be in a slightly tail-down attitude with the nosewheel clear of the ground.

One of the more difficult skills a student pilot must acquire is appreciation of where the ground is relative to the wheels while landing. If this cannot be expressed in words they can, at least, explain how to develop the necessary ability. Simplistic as it may seem, sitting in the aircraft for some minutes while it is on the ground will help develop a mental picture of how the picture should look in the final stages of landing.

During the hold-off on no account fix the gaze in one position; the eyes should look along the left of the nose, moving back to just in front of the aircraft and forward to a point some 30 metres ahead. A valuable training technique while learning to recognize the proximity of the ground is for the instructor to fly the length of the runway with the wheels a few feet above. Speed should be near that for an approach and the student should lightly hold the wheel/stick and say 'rising' – 'sinking' – 'rising' – 'sinking' as the instructor makes the aircraft gently undulate relative to the ground (Fig. 42).

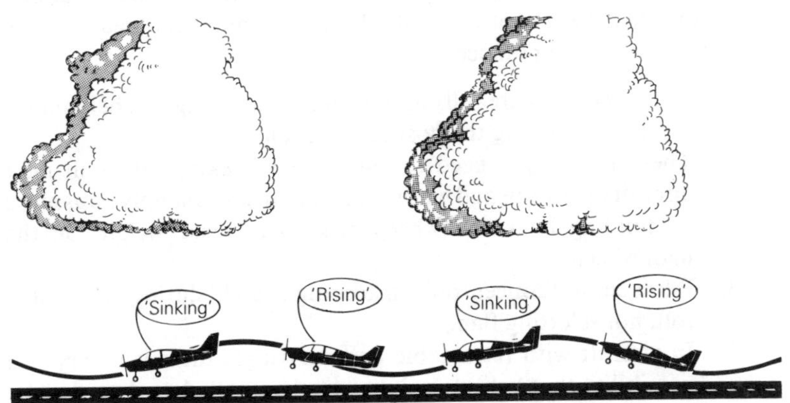

Fig. 42 An exercise to develop recognition of height above the surface during the hold-off.

Recovering from a Bounce

The recovery from a bounce is a question of degree. A gentle one requires no action other than a second touch-down a few yards ahead. A heavy touch-down can, of course, produce a large bounce and here the same correction as that used for ballooning applies. In the early stages of training full power must immediately be applied, fore and aft level maintained and 'go around' action taken so that another circuit, approach and landing can be made. With experience a bounce can be converted into a good landing by using power to minimize the rate of sink.

During the correction for both the bounce and the balloon, the aircraft *must* be kept straight and the wings *must* be level.

With a view to teaching the hold-off technique the instructor can deliberately approach at a high speed so that the resultant lengthy float allows the student time to think.

The student will soon find that there can be times when every landing is a good one. Then a period may follow when nothing seems to go right. One must not be discouraged at this; there are so many variables during the landing that two are hardly ever the same. Initially the aim must be for safety; polish can come later.

Raising the Flaps

A practice that has been canvassed in recent years is raising the flaps after touch-down. Supporters of this technique claim that it allows the brakes to be used more firmly. There are a number of reasons why this is not a good practice:

1. When the flaps are fully applied maximum drag will help retard the aircraft during the post-landing roll.
2. There is rarely a need for maximum braking effort in light aircraft operation and wheel traction can more simply be assured by holding back the wheel/stick to increase weight on the mainwheels.
3. Pilots should concentrate on keeping straight during the landing roll, not selecting flap.
4. In aircraft with retractable undercarriages there is a very real possibility that 'wheels up' may inadvertently be selected while aiming for the flap lever at a time when the aircraft is rolling fast down the runway. Although such aircraft have a safety device, which prevents retraction of the undercarriage when the wheels are supporting the weight of the aircraft, runways are not billiard tables and if the wrong selector has been moved it only requires the effects of a slight bump followed by a gentle hop for the retraction to start. There have been a number of such accidents all over the world.

The correct time and place for raising the flaps is after the runway has been vacated and the aircraft has been stopped for the after-landing checks.

Finding the Landing Direction

Information concerning take-off and landing direction will, at most airfields, be passed to the pilot over the radio. In addition there will be a **Signals Area** near the control tower (ground signals are shown on page 348).

Some airfields provide a grass area for light aircraft although these are less common now that ground handling and ability to cope with stronger winds have made modern designs more independent of conditions.

Runways are marked with two numbers which give the heading to the nearest 10°.

Wind direction, which may differ from the runway heading by many degrees, will be indicated by one or more of the following methods:

(*a*) The Wind Sock – a tubular fabric drogue which swings in the wind from the top of a pole. Its appearance gives an indication of wind strength as well as direction.

(*b*) The 'T' – a large sign painted white and displayed in the Signals Area. Sometimes arranged to swing with the wind but usually set in the correct direction by the control staff.

(*c*) Smoke. When the wind is too light to provide a wind sock indication, a smoke generator is sometimes used on grass, non-runway airfields to give a take-off or landing direction. This method can also be used to determine the position for aligning the 'T'.

Crosswind Landing, Wing-down and Crab Methods

Two methods of dealing with crosswind landings are available to the pilot. In each case the aim is to prevent drift, particularly at the instant of touchdown when any sideways movement of the aircraft relative to the surface could damage the undercarriage and tyres. In some aircraft there could be a risk of causing a wingtip to strike the ground.

The two techniques used for dealing with crosswinds are:

1. The **Wing-down** method.
2. The **Crab** method.

In each case corrective action begins on the approach as soon as drift is detected.

Wing-down Method (Fig. 43)

The wing-down method is based on the fact that, in a sideslip, an aircraft drifts towards the lowered wing. If this drift can be made to balance that caused by the crosswind the aircraft will have no sideways movement relative to the runway.

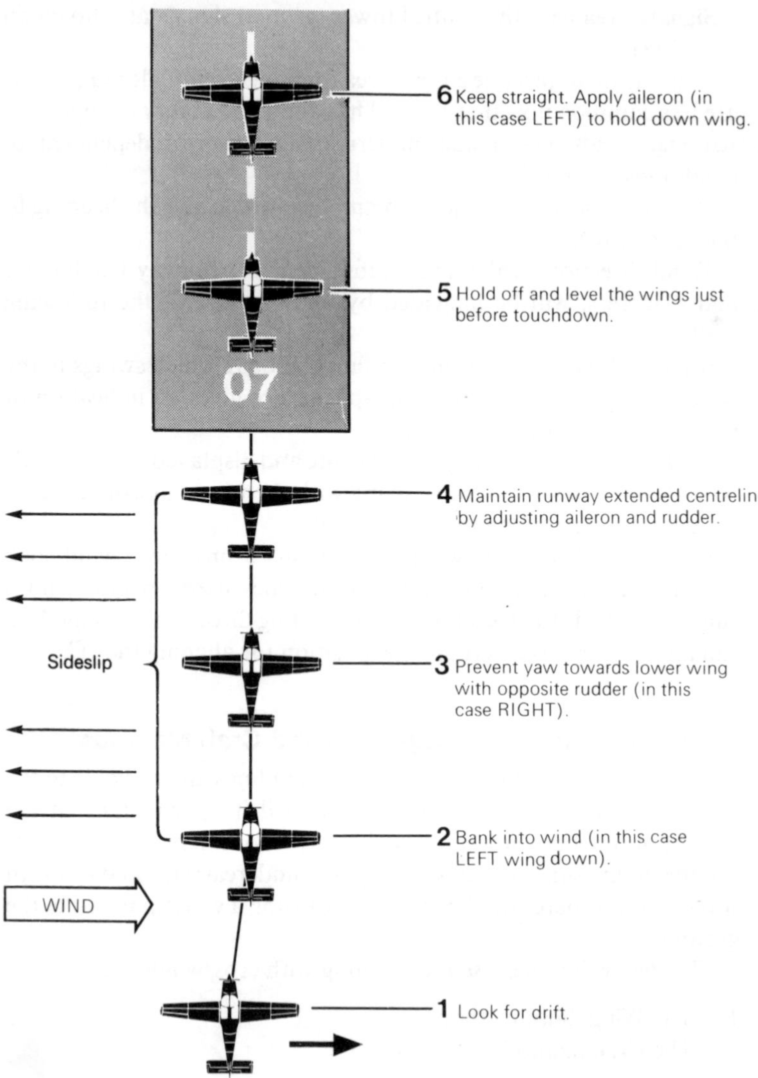
07
6 Keep straight. Apply aileron (in this case LEFT) to hold down wing.
5 Hold off and level the wings just before touchdown.
4 Maintain runway extended centreline by adjusting aileron and rudder.
Sideslip
3 Prevent yaw towards lower wing with opposite rudder (in this case RIGHT).
2 Bank into wind (in this case LEFT wing down).
WIND
1 Look for drift.

Fig. 43 Crosswind Approach and Landing, Wing-down method.

Imagine a wind from the left is drifting the aircraft to the right and that this becomes apparent after turning onto finals. The left wing should be lowered (i.e. the aircraft is banked into wind) and yaw must then be prevented by opposite (in this case right) rudder.

It may be necessary to adjust the angle of bank in response to deviations from the extended runway centreline. For example, if the aircraft moves to the left (i.e. towards the low wing) too much sideslip has been applied for the prevailing crosswind conditions. Bank and opposite rudder should be reduced. If, on the other hand, a drift to the right persists the sideslip is insufficient to counter the crosswind. More bank and opposite rudder will have to be applied.

During the sideslip aircraft heading should agree with the runway direction.

As the threshold is approached and the roundout is made wind strength will decrease and sideslip angle will have to be adjusted to balance drift. If drift is pronounced bank will have to be maintained, even if this entails landing on one wheel. In less severe crosswind conditions the wings can be levelled immediately before touch-down but not too soon, otherwise drift will commence.

After landing the wheel/stick must be held towards the wind to help prevent a wing from lifting.

Crab Method (Fig. 44)

The crab method entails turning towards the wind as a means of countering drift.

When drift is recognized on the approach, again to the right in this example, this means that the wind is coming from the left. The aircraft is then turned towards the wind (i.e. left in this case) by a number of degrees that will depend on the strength of the crosswind. Often there will be a 5 to 10 degree drift angle unless there is a strong crosswind.

On the approach a close watch is made to ensure that the extended runway centreline is being maintained and corrections are made by adjusting the drift angle accordingly. If the aircraft continues to drift to the right insufficient drift angle has been allowed for the crosswind and a further turn to the left will have to be made, placing the nose a few more degrees to the left of runway centreline. Conversely, if the aircraft is seen to be drifting to the left of runway centre too much drift correction has been applied and a turn to the right through several degrees will be necessary.

The allowance for drift, or as it is often known, **'Crab'** angle must be adjusted throughout the approach so that the extended runway centreline is maintained. Crab angle is held over the threshold and during the hold-off. Immediately before touchdown a flat turn to the right is made, using rudder to yaw the aircraft into line with the runway. A little opposite aileron may be needed to keep the wings

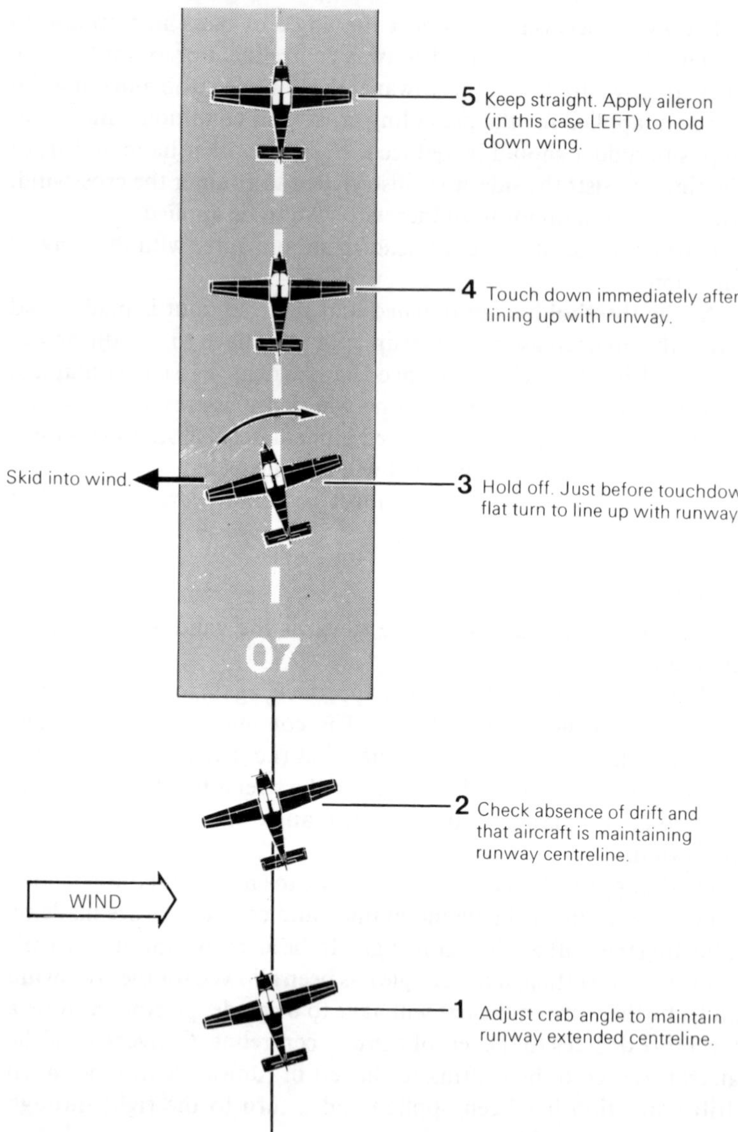
5 Keep straight. Apply aileron (in this case LEFT) to hold down wing.
4 Touch down immediately after lining up with runway.
Skid into wind.
3 Hold off. Just before touchdown flat turn to line up with runway.
07
2 Check absence of drift and that aircraft is maintaining runway centreline.
WIND
1 Adjust crab angle to maintain runway extended centreline.

Fig. 44 Crosswind Approach and Landing, Crabbing method.

level but a flat turn is essential since the object is to cause a skid into wind for the purpose of countering drift during the landing.

Why turn at the last moment? It would, of course, be more simple to touch down in the position shown at aircraft number 3 in Fig. 44 but, after landing, it would start to run off the left-hand edge of the runway.

During the after-landing roll the wheel/stick must be held to the left, preventing the left wing from lifting in the crosswind.

Advantages and Disadvantages

Each of the two methods described has its supporters and detractors and some pilots even use a combination of both systems while landing in a crosswind. This is not recommended on grounds of unnecessary complication. The wing-down method is particularly suited to light aircraft but either method is effective. In each case the problem is to judge exactly when to remove the correction (bank or drift angle) immediately prior to touch-down.

Crosswind Limits

Crosswind limits for the aircraft type were explained on page 167 (Exercise 12, Crosswind Take-off section). These limits are also important while landing. ATC will give the wind strength and direction along with the runway in use and it is the pilot's responsibility to ensure that these conditions are within:

(*a*) The crosswind limits of the aircraft.
(*b*) The capabilities of the pilot.

Air traffic controllers who spend most of their time dealing with larger aircraft are not always aware of the problems facing light aircraft in strong crosswind conditions. Often there is an alternative runway offering less crosswind and the pilot should request a landing on this, or the grass area if one is available.

Glide Approach

In the days when most light aircraft were biplanes it was almost invariably the practice to land from a glide approach. The technique has given way to the engine assisted approach previously described, mainly because it is easier to fly accurately and the touch-down point can be controlled to fine limits. However the glide approach and landing is a valuable exercise since it forms the final stage of a later exercise, *Forced Landing Without Power* which is described on page 247.

Since power is not used in a glide approach, and indeed will not be available if an engine failure has occurred, the obvious situation that must be avoided is an undershoot. Without power nothing can be done to gain the airfield if the aircraft is too low on the approach.

Excess height, on the other hand, is an asset provided an overshoot is not allowed to develop to the point where the aircraft is too high to land on the airfield. To make the task of judging gliding distance as easy as possible it is advisable to turn onto base leg immediately the runway threshold is alongside the pilot's side windows. This will ensure a short final approach. The aim should be to position the aircraft on the approach in a position that would land it approximately one third of the way into the airfield. When it is certain the landing area can be reached flap is lowered in stages to bring back the touch-down point so that a round out can be made over or near the threshold.

A glide approach is somewhat steeper than an engine assisted approach and the round out will entail a larger than usual change in attitude. The landing that follows is conducted in the same manner as one following an engine assisted approach.

Short Landings

Not all aerodromes provide airport-size runways and it is sometimes necessary to land at an airfield where a normal engine assisted approach would entail over-running the limits allowed by the landing area. In some parts of the world such airfields are the rule rather than the exception but the ability to make short landings when the occasion demands is of value to the pilot, extending skill and increasing the scope of a PPL.

In essence, a short landing is a standard engine assisted landing flown at the lowest speed consistent with safety. In Exercise 10B a section was devoted to slow flight and this is a useful foundation for the type of approach demanded of a short landing.

At low speed the nose is inclined upwards at an angle from the descent path and, in consequence, engine power will make a small but useful contribution to lift. This may better be understood by imagining an aircraft with sufficient power to climb with the nose pointing vertically upwards. In such a climb engine power would be providing all the lift needed to oppose weight.

The effect of this small contribution to lift is a reduction in stalling speed and that, in turn, will allow a lower than normal approach speed to be flown in safety. The aim during a short landing is to make a standard engine assisted approach, but just before short finals there is a further reduction in airspeed (to be demonstrated by the instructor), then power is adjusted to provide a descending approach.

To help establish the aircraft in a steady, low speed approach at a lower than usual airspeed the pilot must avoid crowding the final part of a short field circuit. Consequently it is sound practice to extend the downwind leg a little before turning onto the base leg, thus ensuring a longer final approach in which to settle the aircraft on the glidepath.

An older approach technique was known as the **Creeper**. Figure 45 compares this with the descending approach and clearly the latter offers a number of advantages. Visibility on the approach is better than that provided by the somewhat nose-up creeper method. Also when there is a fence or hedge to cross before reaching the threshold touch-down point will be closer to the downwind end of the airfield because of the descending angle of glidepath.

On no account may the throttle be closed until after the round out when the wheels are near the surface. At the low speeds used for short landings the removal of power will cause a high sink rate which could result in a heavy landing.

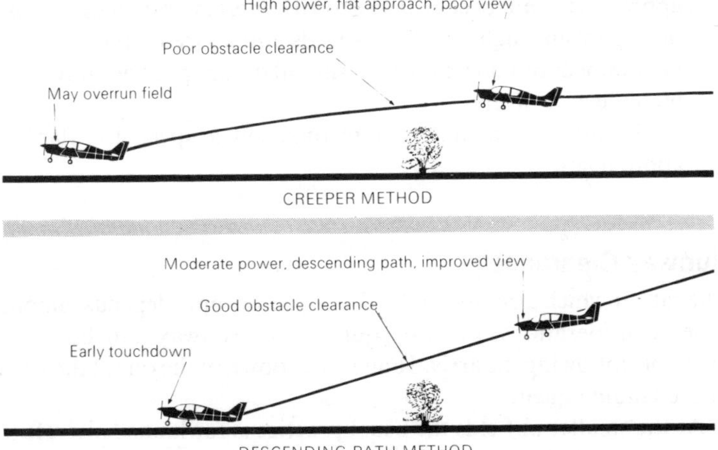

Fig. 45 Short field Approach. The descending path method (lower picture) results in a shorter landing when there is an obstacle to cross. It also provides a better view ahead than the creeper-type approach.

Flapless Landing

In the unlikely event of failure to lower flap due to an electric or mechanical fault a perfectly satisfactory landing is possible but there are a number of factors to be borne in mind:

1. Without flap the aircraft will approach in a more nose-up attitude and forward visibility may, in some types, be impaired. The problem can to some extent be dealt with by avoiding excessive power during the approach.
2. An aircraft without flap will glide more efficiently and, even with the throttle fully closed, rate of sink may not be sufficient to correct an overshoot situation. The sideslip will have to be used in place of flap but in order to maintain the extended runway centreline it will first be necessary to turn the aircraft slightly so that the nose is to the right of runway centre. Remember that in a sideslip the flight path lies somewhere between the nose and the lowered wing and this flight path must be aligned with the runway.
3. In trying to improve the view ahead it is tempting to allow the gliding speed to increase but this can only store trouble for a later stage. Excess speed will entail a lengthy **Float**, i.e. prolonged hold-off due to excessive speed, and this could be a problem when runway length is not over generous. During a flapless approach the airspeed will be 5 to 10 knots faster than usual; loss of flap means higher stalling speeds and a safe margin must be maintained. But on no account should the airspeed be allowed to increase excessively.

   The aircraft manual will recommend a speed for flapless approaches.

Runway Clearance

The rate at which aircraft can land at an aerodrome depends, among other considerations, on how quickly the runway can be made available following the arrival and touch-down of the one at the head of the landing queue.

At the smaller airfields the usual practice is for landed aircraft to taxi to the end of the runway or landing area before turning off and taxying towards the parking area. At some of the larger airfields with longer runways there may be several **Turn-off Points** or **Intersections** and ATC will instruct the pilot on where the runway should be vacated. Some of the major airports have **Fast turn-off points**, taxiways that branch from the runway at a small angle and continue in a gentle curve towards the airport terminal. These enable an aircraft to leave the runway while continuing the final part of the landing roll at relatively high taxying speeds.

After Landing Checks

When the runway has been cleared the aircraft is taxied a short distance to ensure adequate separation from landing aircraft and those taking off. It is then stopped and the **After Landing Checks** are completed, their purpose being to ensure that the aircraft is in the proper condition for taxying. The following actions are taken:

1. *Flaps Up*: These should be raised to protect them from the possibility of damage caused by stones or other debris while taxying.
2. *Fuel Pump Off*: When an electric fuel pump is fitted it will have been turned on during the pre-landing vital actions. After clearing the runway it should be switched off since it is not required for taxying.
3. *Carburettor Heat*: If icing conditions required the use of carburettor heat during the approach it must be returned to the COLD position before taxying. In the HOT setting the air filter is by-passed and there is a risk of the engine ingesting abrasive dust particles.
4. *Electrics*: Switch off all non-essential electric load.

Bad Weather Circuit Procedure

The student under training will not be exposed to circuit flying in poor weather until later in the PPL course. Airline pilots accept landing in low cloud and reduced visibility as routine; they are flying into suitable airports and the aircraft are equipped with modern approach aids, even autoland. Furthermore, both the captain and his first officer will have gained an Instrument Rating, a qualification demanding the ability to fly accurately on instruments in response to radio navigation commands.

Although holders of the PPL may gain an Instrument Rating, or the IMC Rating, these qualifications require additional training. Holders of the basic PPL will therefore be confined to flying under suitable weather conditions. Nevertheless there are times when the visibility is poor and a lowering cloud base might demand circuits at a lower than usual height.

Often a low cloud base is accompanied by good visibility and in these conditions no special problems arise other than the need to remain clear of cloud and in sight of the surface. When there are tall structures in the local area the position of these must be known and the aircraft routed accordingly so as to avoid risk of flying into, for example, a television mast.

Liaison with ATC is of vital importance while flying a circuit in poor weather, particularly reduced visibility where location of the airfield may be difficult. In some cases it may not be possible to keep in sight of the airfield throughout the circuit and the following steps should be taken:

1. The aircraft should be flown at **Low Safe Cruising Speed**. Part flap (usually 10–15°) will be lowered to add drag and require additional power. This will increase slipstream over the tail surfaces and improve rudder and elevator response. Use of flap will also lower the nose for any given speed and thus improve forward visibility.
   By reducing speed the pilot will have more time to think and to recognize local ground features.
2. The direction indicator should be used to fly accurate headings around the circuit.
3. A tighter than usual circuit should be flown in an effort to maintain visual contact with the airfield.
4. If, because of reduced visibility, the runway comes into view to one side of the nose an angled approach is perfectly acceptable so long as this does not entail heading changes with high angles of bank or prejudice the approach/landing.
5. At a later stage the instructor will demonstrate how the watch can be used to time circuit legs and assist in flying an accurate circuit even when the airfield cannot be seen.

If, during a bad weather circuit, the pilot becomes lost and is unable to locate the airfield ATC must be advised without delay. Air Traffic Controllers are there to assist in such cases. Above all, keep calm and aim to arrive over the airfield with the assistance of ATC. From that position it is possible to line up with the runway and fly a tight circuit using the direction indicator to maintain accurate circuit legs.

Missed Approach/Going Around Again

There are a number of reasons why it may be necessary to execute **Missed Approach** or **Go Around** action. The aircraft could be too high for a landing, it may be so out of alignment with the runway that unacceptable last minute heading changes are required, another aircraft may not have cleared the runway quickly enough or there could have been a bad bounce.

Techniques for handling a bounce or balloon were described in pages 193 to 196. Missed approach action or, as it is sometimes called, **Going around** is handled as follows:

1. The decision to go round again will often be taken by ATC. If there is any doubt about separation between aircraft on the approach one of them, usually the following aeroplane, will be instructed to 'go around'.
   If the 'go around' decision has been made by the pilot it should be taken early rather than late and ATC must be advised.
2. Full power must be applied and the pilot should be prepared for a nose-up trim change which, in some aircraft, can require considerable forward pressure on the wheel/stick to prevent the nose from adopting a dangerously high attitude. The elevator trim must be used to relieve control load.
3. The flaps should be left down until a minimum of 300 feet. Then they should be raised in stages, adjusting the nose attitude and retrimming in similar steps to maintain the correct climbing speed until the flaps are fully raised.
4. While the aircraft is at low speed and maximum power the wings must remain level and the ball must be kept in the centre with rudder to ensure balanced flight. The risks of high power/low speed/out of balance flight were explained in Exercise 10B.

When the aircraft is in the normal climbing configuration the circuit may be continued in the usual manner.

Soft Field Landings

Generally when a grass airfield is known to be soft, possibly following prolonged, heavy rain, it is best to avoid using it. However if a landing is essential under these conditions the **Soft Field Landing** technique will be used.

The approach is made as though a short landing is intended. In fact the short field landing procedure applies until the touch-down which should be made in a low-speed tail-down attitude. Soft fields present these hazards to light aircraft:

1. There is a risk after landing of the wheels sinking into the ground.

It will then be impossible to move the aircraft under its own power and it may even be necessary to employ a tractor although usually the task of moving an aircraft stuck in soft ground can be handled by a number of people manhandling as a team.

2. The nosewheel may become embedded in the soft ground. In a bad case there is a danger of the nosewheel strut being bent or even broken. The damage that follows can be considerable since it may entail a damaged propeller, a shock loaded crankshaft, damaged engine cowlings and a distorted engine bay firewall.

To avoid these risks it is important to keep the weight off the nosewheel during the landing roll. It is also essential to keep the aircraft moving until the parking area is reached.

After touchdown a little power should be left on to keep the aircraft rolling and the wheel/stick must be held fully back to assist the nosewheel by providing a down-load on the tail.

On soft ground there should be no need to use the brakes since there will already be excessive rolling friction. The aim should be to keep the aircraft rolling until hard ground is reached.

Exercise 13. Circuit, Approach and Landing (Engine Assisted)

AIM — *To fly an accurate circuit positioning the aircraft for an engine assisted approach and landing.*

AIRMANSHIP — *Lookout*
Circuit shape
Downwind checks: BUMPF
ATC liaison
Suitability of weather

AIR EXERCISE

Establishing the circuit	*Approach*	*Landing*
Downwind turn at correct position *Check:* *Parallel to runway* *Correct circuit height* *Correct airspeed* *LOOKOUT* *'Downwind' radio call* *Pre-landing Vital Actions* *Brakes* *off* *Undercarriage down and locked (fixed)* *Mixture* *rich. Check carb. heat* *Pitch* *fine (fixed)* *Fuel* *on correct tank* *sufficient for circuit* *fuel pump on* *fuel pressure normal* *Runway threshold 45° behind pilot* *Base leg turn* *Reduce power and speed* *Part flap* *Re-trim* *Turn onto finals (wide radius turn)* *'Finals' radio call*	*LOOKOUT for others on longer approach* *Check for drift* *Maintain glidepath –* *Power/attitude adjustments* *Check:* *Airspeed* *Runway centreline* *Glidepath* *Trim* *On short finals:* *Full flap* *Adjust airspeed* *Re-trim* *Check:* *Runway threshold* *Runway centreline* *Airspeed* *Wings level*	*Over runway threshold:* *Check for drift and correct* *Watch ground along left of nose* *At correct height:* *Round out* *Close throttle* *Keep wings level* *Hold off:* *Keep eyes moving, back and forth* *Keep in the air with Wheel/stick back pressure* *Slight tail down attitude* *Mainwheels touch* *Allow nosewheel to make contact* *Brake as required* *Hold back wheel/stick* *Clear runway* *After landing checks*

Pre-flight Briefing

Exercise 13. Circuit, Approach and Landing (Crosswind, Wing down and Crab methods).

AIM *To fly an accurate circuit and land maintaining directional control in conditions of crosswind.*

AIRMANSHIP *As for Engine Assisted Approach and Landing*
Crosswind limits

AIR EXERCISE (Drift to RIGHT)

Crosswind circuit	*Wing down method*	*Crab method*
Ensure crosswind leg is correct length *Drift compensation downwind* *Base leg corrections for wind* *Avoid flying through Extended Runway centreline* *LOOKOUT*	*When drift appears:* *Lower LEFT wing* *Apply RIGHT rudder* *Check:* *Airspeed* *Glidepath* *Runway centreline* *Adjust bank/sideslip to maintain centreline* *Over threshold:* *Round out* *Close throttle* *Keep straight* *Just before touchdown:* *Level wings* *OR* *Land on one wheel*	*When drift appears:* *Turn left* *Maintain crab angle* *Check:* *Airspeed* *Glidepath* *Runway centreline* *Adjust crab angle to maintain centreline* *Over threshold:* *Round out* *Close throttle* *Hold crab angle* *Just before touchdown:* *Flat turn RIGHT* *Keep wings level* *Land*

IMPORTANT-IMPORTANT-IMPORTANT
Wing down method: *Do not level wings before ready to touch down. In strong drift land on one wheel.*
Crab method: *Do not kick off crab angle until ready to touch down.*
DRIFT CAN START AGAIN IF THESE ACTIONS ARE TOO EARLY.

Exercise 13. Circuit, Approach and Landing (Glide Approach, Short Landing and Soft Field Landing)

AIM — *Glide Approach: to make an accurate approach from the base leg and land without the use of power.*
Short Landing: to approach at a low speed and land in safety using a minimum of airfield length.
Soft Field: to land on a soft surface and taxi the aircraft to the parking area.

AIRMANSHIP — *As for Engine Assisted Approach and Landing*
ATC liaison
Integration with other traffic on normal approach

AIR EXERCISE

Glide approach and landing	*Short landing*	*Soft field landing*
Early turn onto base leg	*Late turn onto base leg*	*Base leg and approach as for a short landing*
Carb. heat – close throttle	*Reduce to normal powered approach speed*	*Over threshold:*
Trim at best gliding speed	*Part flap*	*Round out*
Assess wind strength (drift amount)	*Re-trim*	*Power off*
Turn finals	*Turn finals*	*Hold off longer and adopt tail down attitude*
Aim to land third of way down field	*Full flap*	*Land*
Part flap	*Trim at low speed*	*Hold back wheel/stick*
Re-trim	*Check for drift and correct*	*Add power to keep aicraft moving*
Assess glidepath (visual refs.)	*Assess glidepath (visual refs.)*	*ON NO ACCOUNT STOP*
Check for drift and correct	*Aim for early touch-down point*	*Taxi to hard/parking area*
Add flap as required	*Check airspeed*	*Avoid using brakes*
WHEN CERTAIN TO MAKE FIELD:	*Round out THEN close throttle*	
FULL flap	*Lower nosewheel*	
Check speed	*Brake to a standstill*	
Anticipate round out		
Land in normal manner		

Flight Practice

(The circuit is made up of exercises which have already been covered in previous chapters and only the approach and landing are detailed here.)

OUTSIDE CHECKS

(*a*) Altitude: 800 or 1000 feet (according to airfield rules).
(*b*) Location: on correct circuit for runway in use.
(*c*) Position: on downwind leg with runway threshold just under the trailing edge of the wing.

COCKPIT CHECKS

Pre-landing Vital Actions

'B U M P F'

B: Brakes off.
U: Undercarriage down and locked ('fixed' on most trainers).
M: Mixture rich. Carb. heat, check for carb. ice.
P: Pitch fine, or in pre-landing position (pitch 'fixed' on most training aircraft).
F: Fuel sufficient for 'overshoot', electric pump 'on' and fuel pressure normal.

AIR EXERCISE

Powered Approach and Landing

(*a*) Before turning on to the base leg continue flying downwind until the airfield boundary is well behind the trailing edge of the wing, say, forty-five degrees behind the pilot.
(*b*) On the base leg reduce power immediately and (i) reduce speed to that required for a powered descent, (ii) lower about half flap, (iii) open the throttle to control the rate of descent, and (iv) re-trim. Now slightly loosen the throttle friction.
(*c*) Look around, check for other aircraft that may be on a long approach and when clear to line up with the landing area turn on to the approach.
(*d*) Advise ATC 'on finals'. Control the rate of descent with the throttle and the airspeed with the elevators. If the runway begins to stand on end and the threshold moves down the windscreen, you are overshooting and must decrease the power and lower the nose slightly. If undershooting, indicated by a flattening runway with the threshold moving up the windscreen, open the throttle and raise the nose slightly, maintaining the airspeed constant throughout.

Look out for and correct drift by turning away from it.

(*e*) On short finals and when committed to the landing, lower full flap.

(*f*) Aim to cross the boundary and touch down on a predetermined spot. Check the airspeed.

(*g*) Watch the ground as it comes up along the left of the nose. Ease the wheel/stick back and check the descent.

(*h*) Keep straight and progressively close the throttle. Prevent the aeroplane from sinking rapidly by progressive back pressure on the wheel/stick until in a slightly tail down attitude the mainwheels make gentle contact with the ground.

(*i*) Keep the weight off the nosewheel by maintaining backward pressure on the wheel/stick. Maintain direction on the rudder pedals then, as speed decreases and the nosewheel lowers to the ground, bring the aircraft to a halt with gentle application of brake.

(*j*) Clear the runway and complete the after-landing checks.

Crosswind Landing

1. Wing Down Method

(*a*) Turn the aeroplane on to the approach in the usual way. Keep a sharp lookout for aircraft on normal approach.

(*b*) When drift appears lower a wing in the opposite direction – into wind – and keep the aeroplane heading in the landing direction with opposite rudder to bank. The aeroplane is now sideslipping into the wind and counteracting drift. Do not lower full flap.

(*c*) Maintain the intended landing path by adjusting the angle of bank as necessary while keeping straight with rudder.

(*d*) Proceed down to the hold-off and level the wings just before touch-down. Concentrate upon keeping straight and during the landing run hold the wheel/stick over towards the wind.

(*e*) Bring the aeroplane to a halt, still holding the wheel/stick hard over to prevent the wing rising and correct any tendency to swing into wind.

(*f*) In a strong crosswind allow the aircraft to land on one wheel.

2. Crab Method

(*a*) Turn the aeroplane on to the approach in the usual way. Keep a sharp lookout for aircraft on normal approach.

(*b*) When drift appears, turn the nose in the opposite direction – into wind. Do not lower full flap.

(*c*) Maintain the intended landing path by adjusting the position of the nose so that drift is counteracted right down to the hold-off.

(*d*) Shortly before the aircraft touches down, yaw the aircraft into line with the landing path with rudder while keeping the wings level with aileron. Land in the normal way.
(*e*) Bring the aeroplane to a halt, if necessary holding the wheel/stick hard over towards the wind to prevent the wing from rising. Correct any tendency to swing into wind using rudder.

Glide Approach and Landing

(*a*) Immediately the runway threshold passes under the trailing edge of the wing, turn through 90° on to the base leg and, when sure the field can be reached close the throttle, reduce to gliding speed and re-trim.
(*b*) Assess the wind strength by the amount of drift experienced and should this be great turn slightly towards the field.
(*c*) Lower flap as required.
(*d*) Make sure no other aeroplane is ahead or below and at about 500 feet execute a gliding turn on to the approach. Resume gliding speed and the aeroplane should have a straight-in approach from approximately 400 ft. Dependent on the wind strength lower the flaps still further. Call 'turning finals' on the radio.
(*e*) Lookout for drift and correct by turning in the opposite direction to it.
(*f*) With the hand on the throttle, watch the ground as it comes up along the left of the nose. At the correct height move the wheel/stick back and check the descent, thus making the aeroplane glide a few feet above the ground.
(*g*) Keep straight and prevent a rapid sink with progressive backward pressure on the wheel/stick until a slightly tail down attitude the main wheels make gentle contact with the ground.
(*h*) Keep the weight off the nosewheel by maintaining backward pressure on the wheel/stick. Maintain direction on the rudder pedals then, as speed decreases and the nosewheel lowers to the ground if necessary slow the aircraft with gentle application of brake.
(*i*) Clear the runway and complete the after-landing checks.

Short Landing

(*a*) When on the approach throttle back slightly, reduce to as low a speed as possible consistent with safety and open the throttle again to give steady rate of descent. Re-trim. The glidepath should be a descending one adjusted with throttle. Glance at the airspeed indicator from time to time and control its reading with the elevators. Be prepared to add power should the rate of descent increase due to wind gradient near the ground. Look out for drift and correct in the usual way.

(*b*) Cross low over the boundary and check the descent with elevator. The aeroplane will now be in a nose-up attitude and when near the ground the throttle should be closed, allowing it to sink to the ground.

(*c*) Hold the wheel/stick right back and bring the aircraft to a halt with careful use of brake.

Flapless Landing

(*a*) Establish an approach at the correct flaps up speed. Use a minimum of power to avoid a high nose attitude and loss of foward visibility.

(*b*) At the correct height check the descent in the usual way. Note the prolonged hold-off and the tendency to float.

(*c*) Be prepared for a longer than usual ground run.

Bad Weather Circuit Procedure (Visual Manoeuvring)

(*a*) When joining an airfield circuit in reduced visibility obtain the assistance of ATC and position over the runway in the landing direction at the correct height. Fly at low safe cruising speed and add power for the turns.

(*b*) Check the direction indicator and, if necessary, re-set.

(*c*) At the upwind end of the runway turn onto the crosswind leg, using the DI to establish headings.

(*d*) Aim to make a smaller circuit than usual, keeping the airfield in sight. When the visibility will not allow this rely on the DI and use the watch to time the legs.

(*e*) Complete the circuit, maintaining a close liaison with ATC.

(*f*) If the runway threshold appears at an angle to the aircraft continue the approach unless this entails last minute large alterations of heading.

Missed Approach/Going Around

(*a*) Open the throttle smoothly and fully and prevent the nose from moving up sharply. Be prepared for a change of trim necessitating a heavy load on the stick on some aircraft. Keep straight and maintain balanced flight.

(*b*) If the flaps are down, trim to correct climbing speed for this condition. Do not raise the flaps below 300 feet and then only in stages. Re-trim after each reduction in flap angle.

Soft Field Landing

(*a*) Set up an approach as for a short landing.

(*b*) Over the threshold, check the descent, close the throttle but do not allow the aircraft to touch down immediately.

(*c*) Continue the hold off until the aircraft touches down in a tail-down attitude.

(*d*) Keep the wheel/stick fully back and leave on some power to help lighten the load on the nosewheel. At all costs keep the aircraft moving towards the parking area.

(*e*) Avoid using the brakes until the aeroplane is on firmer ground.

Exercise 14
First Solo and Solo Consolidation

Background Information

All the previous exercises have been leading to the point when the instructor climbs out of the aircraft and sends the student around the circuit alone for the first time. Clearly before this step is taken the flying instructor will be satisfied that his student can not only take off, fly around the circuit and land, but also that he can deal with such eventualities as

1. Recovery from a misjudged landing.
2. 'Going around' procedure.
3. Engine failure after take-off.

A number of successful landings is usually required by the instructor during the pre-solo check, and safety rather than perfection is expected of the student. An indifferent landing sensibly corrected is a sure indication that the student is fully aware of the events taking place throughout the manoeuvre.

Most trainee pilots feel ready for solo before their instructors are prepared to send them and as their standard of flying improves the event is eagerly awaited by both parties. Students are often told to expect the aircraft to take off after a shorter run once the instructor is out of the cockpit, but in practice the difference is not easy for an inexperienced pilot to appreciate and in any case it is so slight that it is of no consequence.

The take-off, circuit and landing should be flown as if the flying instructor were in his usual place and first solo pilots are usually so occupied with their own thoughts that there is little time to reflect on the vacant seat in the aircraft. For the first solo most instructors require the student to complete one circuit and landing only, however, there should be no hesitation to go round again if, for any reason, the approach is unsatisfactory.

It is particularly important that the student should maintain a good lookout since there will be no additional pair of eyes to warn of the proximity of other aircraft.

It is generally accepted that students learn most about flying while on their own although it is equally true to say that long periods of solo without supervision are often the cause of bad habits being formed. For this reason after first solo the student's flying will be punctuated with dual checks when any handling faults will be brought to the attention and corrected.

When time comes for the instructor to climb out of the aircraft he will stow his safety harness so that it cannot interfere with the controls. It is then for the student to enjoy this never-to-be-repeated occasion. For there can only be one first solo.

Weather, Traffic and Fuel Checks

For the first solo the instructor will choose a day when the visibility is good, the cloudbase is well above circuit height and there are no strong winds. Ideally the wind will be down the runway but, by first solo stage, the student should be able to cope with modest crosswind conditions.

At an active training airfield it may be difficult to find a time of day when there are not a number of other aircraft on circuit flying. While a busy circuit can present problems to a first solo pilot there are advantages to having been trained in such an environment. Pilots who learn to fly at a busy airfield rapidly learn the importance of maintaining a good lookout.

The instructor will warn ATC that he is about to send his student on the first solo and particular consideration will be given to the fledgling pilot; when aircraft become 'bunched' on the approach it is usually the practice to give the first solo right of way. Other pilots will be told to go round again.

Very occasionally there are cases of pilots on their first solo having to make a number of circuits. Perhaps there is heavy traffic and ATC have been unable to give priority, an emergency may have occurred on the runway (an aircraft with a burst tyre or a damaged undercarriage), or the first landing could have been a bad one. To cater for unexpected delay and the need to circle the airfield there should be plenty of fuel in the aircraft before the student is allowed to take off on the first solo.

Consolidation Period

Following the first solo it is common practice to consolidate all that has been learned during the course. Early solo flights are punctuated with frequent checks by the flying instructor until the student is ready to leave the circuit on his own for the purpose of practising some of the upper air exercises. Before he is allowed to do this he will be taught:

1. The correct leaving and circuit joining procedures.
2. The locality of any restricted airspace within the training area.
3. Turns on to a particular heading and how to maintain that heading.
4. The ability to obtain a Magnetic Bearing (QDM) in the event of requiring navigational assistance while returning to the airfield.
5. Altimeter setting procedures.

With this consolidation period should come improvements in skill and increased confidence when undertaking the remaining exercises required to complete the PPL course.

Exercise 15
Advanced Turning

Background Information

The steep turn is a development of the medium turn which was explained in Exercise 9. Exactly the same principles apply, but there are additional factors to be considered.

During any turn lift must be increased to provide a turning force in addition to its usual function – opposing weight. While the increase is small at gentle angles of bank, twice normal lift is required in order to maintain height during a turn with a 60° angle of bank. Beyond 60° the demand for more lift increases rapidly and at just over 84° ten times more lift than usual is needed to avoid loss of height. The aeroplane would then be turning at a very high rate (Fig. 46).

A force causes an equal and opposite reaction. For example, when a rifle is fired, there is a kick back in the opposite direction to the line of fire. During turns the reaction known as **Centrifugal Reaction**, often incorrectly referred to as 'Centrifugal Force'. Like the kick experienced when firing a rifle, it is the reaction to the force turning the aeroplane **(Centripetal Force)** and, when it is permitted to fly on a straight path, centrifugal reaction disappears. Nevertheless this reaction is important to pilot and aircraft alike since both body and machine are subjected to an increased loading. In a correctly executed turn at 84¼° angle of bank a 140 lb (63.5 kg) man would exert a force in his seat of 1400 lb (635 kg)!

Putting this into other terms centrifugal reaction results from pilot and aeroplane resisting the turn while attempting to obey the law of nature which demands that an object must move on a straight path unless diverted by some exterior force. When, in this case a turning force is applied, there is an equal and opposite reaction and during turns such a reaction takes the form of loading.

On page 117 of Exercise 10A, Stalling it was mentioned that one of the factors affecting stalling speed is loading. Figure 47 shows the increased loading at different angles of bank together with the attendant increase in stalling speed. A graph showing the increase in loading with bank is on page 345.

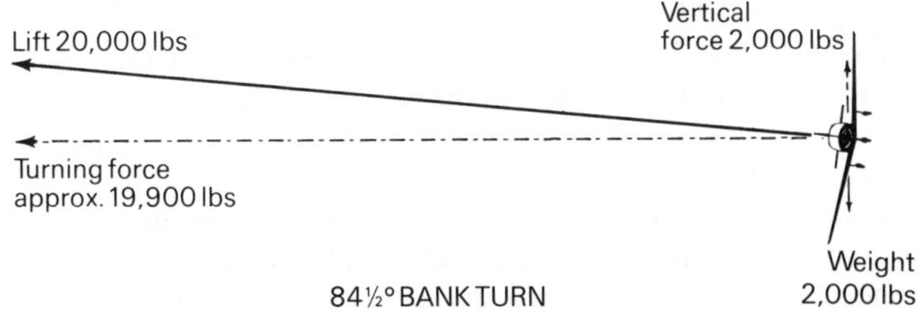

Fig. 46 Increased loading in a Steep Turn. At the bank angle shown ten times normal lift is required to maintain height and, because of their limited engine power, such a turn cannot be sustained without loss of height in a light aircraft.

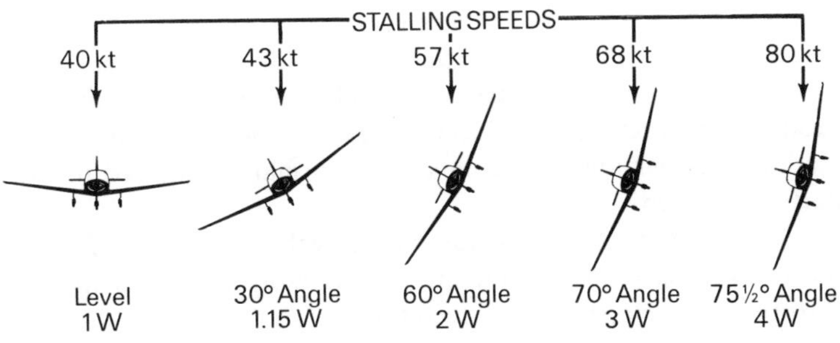

Fig. 47 Stalling speed increases with angle of bank.

The wings will stall at the same angle of attack as in straight flight but, because of loading, speed associated with that angle will be higher and a steep turn is an excellent means of demonstrating the high-speed stall referred to on page 124. During any manoeuvre which increases loading, such as a steep turn, to produce the extra lift required speed will be higher than normal for any given angle of attack. Conversely, if speed is not increased, the angle of attack must be greater to produce the extra lift necessary under loading conditions. Since an increase in angle of attack means an increase in drag, more power will be required to overcome the drag in a steep turn. In practice the steeper the angle of bank, the more power required. So far as light aeroplanes are concerned 65° bank is usually the maximum attainable without loss of height, this limit being dictated by the engine power available to overcome the high drag. In high-speed aircraft the limiting factor is usually the pilot who will 'black out' when the loading is beyond his physical capabilities.

Radius of turn is dependent upon angle of bank and the airspeed. Assuming a constant angle of bank, the higher the speed the larger the radius. This is explained in Fig. 48.

Because of the momentum of an aeroplane it is possible to exceed the maximum angle of bank for level turns for brief periods. Such a manoeuvre is not a correct turn and cannot be sustained without loss of height.

Since there are a number of variable factors involved the student is apt to become confused with the principles relating to steep turns and the following summary should clarify the foregoing text:

(*a*) The steep turn is a developed medium turn.

(*b*) At a constant angle of bank an aeroplane will turn on a smaller radius at lower airspeeds and vice versa.

(*c*) At a particular speed, to tighten the radius of turn centripetal force (turning force) must be increased by obtaining more lift. This produces a reaction called centrifugal reaction. Because of the loading a larger angle of attack than normal is required for any particular speed.

(*d*) To overcome the extra drag caused by the larger angle of attack, more thrust is required and this means more power so that the throttle must be opened in a steep turn.

(*e*) The steeper banked the turn the greater the loading and the more power required. In a light aeroplane engine power is the limiting factor and in a sustained turn it is rarely possible to exceed a 65° angle of bank.

(*f*) The tightest turn (smallest radius) will be at the steepest bank possible for a particular aeroplane and when flown at its lowest

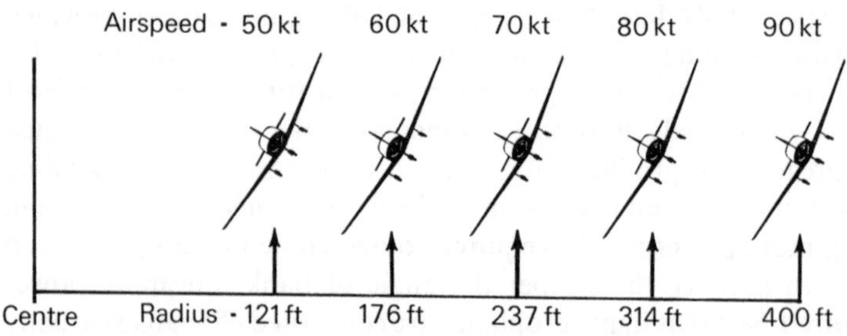

Fig. 48 Relationship of radius of turn to airspeed at a constant angle of bank, in this case 61 degrees.

speed, since for a fixed angle of bank the slower the speed the smaller the radius of turn.

(*g*) The stalling speed is higher in any turn and it increases rapidly as the angle steepens beyond 60°.

Safety Checks

In the early stages of training there is a common tendency to lose height in steep turns, in extreme cases leading to a spiral dive.

In addition to the usual careful lookout all around, ending with a check in the direction of turn, it is important to check below the aircraft before entering a steep turn in case there is a height loss. When the steep turn has been mastered it will itself prove of value as a means of checking below the aeroplane prior to stalling, spinning or any manoeuvre which entails a loss of height.

Entry into the Steep Turn, and Steep Descending Turns

In practice a Rate 3–4 is aimed at during steep turns and the approximate angle of bank applicable to the aeroplane should be related to the horizon.

The position of the nose in relation to the horizon can be deceptive in aeroplanes with side-by-side seating because when turning to the right the student is sitting above the centre-line of the aircraft, the reverse being the case when turning to the left. Experience will show how the aeroplane should look in relation to the horizon under these circumstances.

When a level steep turn is to be made it is entered in exactly the same manner as a medium turn. As the bank angle reaches 35° or so it is allowed to become steeper. At the same time power is added and the wheel/stick is brought back sufficiently to prevent the nose dropping into a spiral dive. The ailerons are centralized when the required bank angle is reached.

In the early stages of training the student may experience difficulty in applying the correct amount of back pressure. This is dealt with in the next section.

Steep Descending Turn Entry

The aircraft will normally be in a descent before starting a steep descending turn. Entry is the same as for any other descending turn but the bank angle is allowed to steepen. However, to cater for the increased loading (which in turn, will increase the stalling speed) the pilot must either add a little power or lower the nose slightly to increase speed. Either option may be used according to the descent rate required.

Use of Power and Controls to maintain Constant Height and Rate of Turn

It was previously mentioned that extra power (not full power) is needed for a steep turn and the instructor will demonstrate this, usually in terms of throttle movement rather than a specific number of extra RPM.

The correct angle of bank can usually be established with reference to a part of the aircraft structure or interior trim but the usual difficulty experienced by pilots under training is how to assess the amount of back pressure to be applied on the wheel/stick. Too little back pressure will result in the nose dropping below the horizon while too much is bound to cause a climb.

When the nose has dropped slightly below the horizon it may be raised by back pressure on the wheel/stick, but if the situation is allowed to develop, a spiral dive will follow and this requires a different recovery technique which will be described later.

Correct use of aileron to maintain bank angle, elevator to hold the nose on the horizon, rudder for balance and power to ensure a steady height demands of a pilot good co-ordination. Consequently, the steep turn is an excellent test of pilot skill. This skill may be further developed by practising steep turns around a reference point on the ground and rolling from steep turns in one direction to the other while maintaining a constant height. During a steep turn downward visibility is greatly enhanced.

Rolling out of the Turn

There is a tendency to gain height during the roll out from a steep turn. Having first checked that it is clear to stop the turn bank is taken off and, as the wings roll through about 35 degrees towards the level position, back pressure on the wheel/stick is relaxed and power is reduced to cruising RPM.

Approach to the Stall/Incipient Spin

This topic has already been dealt with in Exercise 10A, *Stalling* but in the steep turn context the important factor to remember is that, because of the 'g' loading that occurs, stalling will take place at a higher than usual speed. Furthermore, if there is an out of balance condition present at the stall, spin entry is probable. Spin entry under high 'g' can be very rapid in some types of aircraft which are known to **Flick** in these circumstances.

Some light aeroplanes are difficult to high-speed stall during a steep turn, but if the manoeuvre is attempted at normal cruising power or slightly less and the radius is tightened by determined backward movement of the wheel/stick, a stall will occur. Some aircraft will 'buffet' immediately before the stall which will occur at a higher than normal speed for the reason already explained. One wing may stall before the other on certain aircraft and the usual recovery procedure should be adopted to prevent a spin developing.

Spiral Dive and Recovery

If, during the entry to a steep turn, the nose has been allowed to drop and early corrective action has not been taken, a spiral dive will rapidly develop. A stage will be reached where it is not possible to raise the nose with the elevators.

The angle of bank must be decreased immediately when the elevators will resume effective control of nose position relative to the horizon. Prevention rather than cure is the secret of success and any tendency for the nose to drop during a steep turn should be corrected immediately by applying further back pressure on the wheel/stick. This can require some effort in certain types of aircraft.

Maximum Rate Turns

The **Maximum Rate Turn** finds its main application as a means of taking evasive action. It is a somewhat extreme manoeuvre, certainly not one to be enacted while carrying passengers who are not used to flying in a light aircraft.

If, for any reason, a maximum rate turn is required bank must be applied decisively with firm use of the ailerons. At the same time the throttle is *fully* opened and the wheel/stick is brought back as the bank is allowed to adopt a steep angle. Back pressure is applied until the first signs of buffet and then relaxed very slightly. These actions are accompanied by balancing rudder.

To roll out of the turn, bank is taken off smartly, the wheel/stick is allowed to position the nose on the horizon and power is reduced to cruising RPM. In light aircraft of low power the maximum rate turn cannot be sustained through 360° without loss of height.

Instrument Indications

During a steep turn a good lookout must be maintained although this will be difficult in high-wing aircraft since the area on the inside of the turn will be blocked by the lower mainplane.

The attitude indicator will register angle of bank and it will also show the position of the nose as being somewhat above the horizon. The direction indicator will indicate a rapid change of direction and the turn needle/turn co-ordinator will show a high rate of turn. The magnetic compass will give unusable readings as it moves erratically.

Of particular importance are the readings given by the VSI and the altimeter. The VSI can alert the pilot to incorrect use of elevator and this will be confirmed by a steady gain or loss of height shown on the altimeter when an incorrect nose position has been adopted. Airspeed will be somewhat lower than normal cruising.

Exercise 15. Advanced Turning (Steep turns and Steep Descending Turns)

AIM — *To turn on a tight radius with a steep angle of bank in level or descending flight.*

AIRMANSHIP — *Height: sufficient for the exercise*
Airframe: flaps up
Security: no loose articles in the cabin
Engine: temperatures and pressures 'in the green'. Engine handling in steep turns
Location: outside controlled airspace etc., in sight of known landmark
Lookout: clear to turn and clear BELOW aircraft
Suitability of weather

AIR EXERCISE (Steep Turn)

Entering the turn	*During the turn*	*Roll out*
LOOKOUT, particularly sideways and below *Enter turn* *At Medium-Turn attitude:* *Continue adding bank* *Increase power* *Hold up nose*	*LOOKOUT* *Check:* *Bank angle* *Constant height* *Lower than usual airspeed* *Balance*	*LOOKOUT* *Take off bank* *At Medium-Turn attitude:* *Relax back pressure on wheel/stick* *Continue roll out* *Reduce power*

AIR EXERCISE (Steep Descending Turn)

Entering the turn	*During the turn*	*Roll out*
LOOKOUT, sideways and below *Add power or increase speed* *Roll into turn* *Allow to steepen* *Hold correct bank angle*	*LOOKOUT* *Check:* *Bank angle* *Airspeed* *Balance* *Rate of descent*	*LOOKOUT* *Take off bank* *Level wings* *Reduce power or speed* *Resume straight descent*

Pre-flight Briefing

Exercise 15. Advanced Turning (Approach to the Stall: Spiral Dive: Maximum Rate Turn).

AIM — *To avoid and, if necessary, recover from a stall, spin or spiral dive situation and to turn at maximum rate.*
AIRMANSHIP — *As for Steep Turns*

AIR EXERCISE

Approach to stall/incip. spin	*Spiral dive and recovery*	*Maximum rate turn*
LOOKOUT *Reduce power slightly* *Enter steep turn* *Do not correct out of balance* *Increase back pressure on wheel/stick* *AIRCRAFT REACTION* *Buffet (some aircraft types)* *Stall warning* *Tendency to FLICK into spin* *RECOVERY ACTION* *Relax back pressure on wheel/stick* *Rudder in opposition to yaw (ball)* *Add power* *When airspeed increases:* *Level wings with aileron* *IF SPIN IS ENTERED:* *Close throttle* *Centralize ailerons* *Take spin recovery action*	*LOOKOUT. Check altimeter* *Enter steep turn* *Allow nose to drop* *Prevent engine overspeed* *Try and raise nose with wheel/stick. IMPOSSIBLE.* *AIRCRAFT REACTION* *Rapid airspeed increase* *Rapid height loss* *RECOVERY ACTION* *Level wings (aileron)* *Close throttle* *Gently ease out of dive* *As nose meets horizon:* *Open throttle to cruise RPM* *Set up straight and level flight* *Check height loss*	*LOOKOUT* *Roll smartly into very steep bank* *FULL power* *Decisive back pressure on wheel/stick* *At Buffet/Stall warning:* *Relax back pressure slightly* *Note airspeed* *ROLL-OUT* *LOOKOUT* *Take off bank* *Relax back pressure on wheel/stick* *As wings roll level:* *Reduce to cruise RPM* *Check balance and trim*

Flight Practice

COCKPIT CHECKS

(*a*) Trim for level flight.
(*b*) No unsecured items in cockpit.

OUTSIDE CHECKS

Altitude: sufficient for manoeuvre.
Location: not over town or airfield, or in controlled airspace.
Position: check in relation to known landmark.

AIR EXERCISE

Steep Level Turns

(*a*) Look around and enter the turn in the usual manner.
(*b*) Allow the angle of bank to increase and, as it becomes steeper than for a medium turn, open the throttle further.
(*c*) Prevent the nose from dropping below the horizon by backward pressure on the wheel/stick. This will also increase the rate of turn. Maintain the required angle of bank with the ailerons and balance with rudder. Keep a good lookout throughout the manoeuvre and notice the excellent downward vision because of the steep angle of bank.
(*d*) To resume straight flight, look out ahead and, if clear of cloud or other aircraft, roll out of the turn in the usual way using the aileron control in co-ordination with sufficient rudder to prevent skid.
(*e*) As the wings become level move the wheel/stick forward to keep the nose in the correct position relative to the horizon and reduce power to cruising RPM.

Steep Descending Turns

(*a*) From a descent add a little power or increase the airspeed some 5 kt beyond that for a normal descending turn. Check that it is clear to turn.
(*b*) Go into the descending turn in the usual way but hold the aeroplane in a steeper angle of bank. Maintain the higher speed with the elevators.
(*c*) Come out of the turn in the usual manner and return to normal descent speed when the wings are level.

Approach to the Stall/Incipient Spin

(*a*) Go into a steep turn without increasing the power.

(*b*) Tighten the radius of turn by backward movement of the wheel/stick.
(*c*) Shortly before the stall, buffeting may be felt. Notice the stall is at a higher speed than usual when power is on and the wings are level.
(*d*) To recover, ease the wheel/stick forward and increase the power. The ailerons will again become effective and capable of correcting any alteration in angle of bank caused by one wing stalling before the other.

Maximum Rate Turns

(*a*) After making sure that it is clear to turn, apply bank smartly in the required direction, at the same time opening the throttle fully. Control balance with rudder.
(*b*) Steepen the angle of bank still further and increase the rate of turn by moving the wheel/stick back until just before the stall. The aeroplane is now turning at its maximum rate. Maintain a good lookout throughout the turn.
(*c*) Roll out of the turn quickly by taking off bank in co-ordination with sufficient rudder to prevent skid.
(*d*) Reduce to cruising power and move the wheel/stick forward to keep the nose in the correct position relative to the horizon as the wings become level.

Spiral Dive and Recovery (to be demonstrated at a safe height)

(*a*) Go into a steep turn without adding power. Note the height.
(*b*) Allow the nose to drop below the horizon while holding on the bank.
(*c*) Prevent the RPM from exceeding the red line limit.
(*d*) Note the instrument indications:
Rapid increase in airspeed.
High descent rate.
Maximum rate of turn.
Aircraft in balance.
(*e*) Note that the elevators are unable to raise the nose.
(*f*) To recover:
Close the throttle.
Roll the wings level.
Ease gently out of the dive.
Open the throttle as the nose comes up to the horizon.
(*g*) Now check the very considerable height loss.

Instrument Indications (during a Steep Turn)

(*a*) Go into a steep turn and when the aircraft is settled note the instrument indications:

Airspeed lower than for cruising flight
Turn needle showing Rate 4
Attitude indicator giving angle of bank and nose position relative to the horizon
Direction indicator showing rapid heading change
VSI and altimeter indicating steady height
Ball in the centre.

Exercise 16
Operation at Minimum Level

Background Information

Often the need to fly low arises from such deteriorating weather conditions as a lowering cloudbase and/or poor visibility. There are, however, commercial requirements which involve low flying, typical examples being aerial photography and crop spraying. Additionally it is sometimes necessary to leave or enter controlled airspace at low level.

Low flying for its own sake is a fascinating exercise but its very attraction can be a source of considerable danger to the inexperienced pilot and the cause of annoyance to those on the ground. Because the sensation of flying near the ground is very different from that experienced normally, low flying must be considered as a separate exercise to be taught by a flying instructor before it is attempted by the pilot under training. Aspects of training peculiar to operation at minimum level may be regarded under two headings:

(*a*) Aircraft handling at low level.
(*b*) Navigation at low level.

While no minimum height is laid down for flights over open country, under para 5. Sec. 2 (General) Rules of the Air and Air Traffic Control Regulations, it is an offence to fly within 500 feet of buildings, people, etc. To prevent accidental infringement of this regulation, for all practical purposes 500 feet above ground level should be considered the minimum height to fly over unfamiliar open country. When a pilot has been forced by weather or other cause to descend below this limit, he is advised to report the circumstances to the controlling authority on landing. Whatever instructions are issued by ATC it remains the captain's responsibility to ensure that flight over a congested area is at sufficient height to clear the area in the event of engine failure. Certain considerations are common to all forms of low flying and these should be understood before the various exercises are explained.

Checks before Descent to Minimum Level

1. Whenever possible all cockpit checks should be completed prior to the descent so that full attention may be directed outside the aircraft once it is near the ground.
   These are listed in the Pre-flight Briefing and Flight Practice sections of this exercise.
2. When the area chosen for the exercise includes farmland, animals must be avoided since they are usually afraid of low flying aircraft.
3. During operation at minimum level prior study of the intended route and all obstructions is of special importance. Fold the map so that it is easy to handle leaving plenty of area either side of track to cater for possible deviations left or right of required track.
4. The Navigation Log should be prepared as for a normal cross country. When computing fuel required for the flight it should be remembered that at low levels the engine burns more gallons per hour thus reducing the range of the aircraft.

Visual Impressions of Low-level Flight

1. At low levels, map reading is complicated by the speed at which ground features approach and disappear from view. Low-level map reading demands a technique quite dissimilar to that used during normal cross country flying when ground features remain in view for many minutes unless the visibility is poor.
2. High-tension cables are not easy to see and a sharp lookout must be kept for pylons and other obstacles.
3. Because of the proximity of the ground there is obviously no margin for error and since the height of the ground may vary the altimeter is of limited use in determining immediate terrain clearance. For this reason it is important to learn to estimate height above the ground when flying low. Form the habit of looking at the horizon then comparing this impression with the ground just ahead of the aircraft. It is impossible to assess height above the ground by looking directly below.
4. Whereas the ground may appear quite flat when viewed from several thousand feet, every feature and undulation of the terrain will become apparent during low flying, and the pilot must watch for ground rising at a steeper rate than the maximum climbing path of the aircraft. In bad weather avoid flying through a valley

in mountainous country since this could lead to a 'blind alley'.

5. Turbulence is often associated with flight near the ground, partially because of the disturbance which results from the flow of wind over such ground features as trees, hills, ridges, buildings, etc. In summer the effect of uneven ground-heating adds to this turbulence and the effect on the aircraft is twofold:

   (*a*) variations in height
   (*b*) airspeed fluctuations.

   It is therefore necessary to make allowance for possible down draughts when flying over undulating ground and the hand must be kept on the throttle throughout all low flying. For the same reasons a safe airspeed must be maintained when turbulent conditions prevail.

6. The aircraft has inertia and violent changes of attitude occasioned by an evasive manoeuvre may cause an aircraft with a fairly high wing loading to 'mush' (sink) and allowance must be made when pulling up over an obstacle. In the extreme case sudden evasive action may provoke a high-speed stall. With these thoughts in mind low flying is normally practised some 500–600 ft above ground level. Tactical exercises at tree-top height are flown by Service pilots and crop spraying is performed at similar or lower levels. In each case considerable experience is required.

7. When low flying over smooth water appreciation of height often becomes difficult and level snow over featureless countryside will present the same problem. Height above water, smooth or otherwise, is difficult to judge in conditions of poor visibility and when the sun is discernible through the haze, inexperienced pilots may become disorientated.

8. Low flying will seriously affect the range of VHF radio equipment. Requests for navigational assistance or other communications may not be clearly received unless the aircraft is relatively close to the transmitting station.

Low Safe Cruising and the effect of Flap

In several previous exercises mention has been made of low safe cruising speeds. The purpose of this mode of flight is to allow more time for:

(*a*) Recognition of ground features, possibly in poor visibility or while operating at low levels where ground features remain in

view for only brief periods.
(*b*) Complying with ATC instructions.

Low speed means adopting a higher angle of attack and this, in turn, entails a reduction in forward visibility. Furthermore, during low speed flight airflow over all control surfaces is reduced and there is bound to be a corresponding reduction in their effectiveness. Both problems are resolved by lowering 10–15° of flap. The effect of this is:

1. For any particular speed there is a lower nose attitude and this will materially improve forward visibility.
2. With flap there will be an increase in drag requiring additional power. More slipstream will be generated and this will increase airflow over the elevators and rudder, improving their effectiveness.
3. With flap set to the 10–15° position there is a small but useful reduction in stalling speed.

Notwithstanding (3) above, when the aircraft is being flown at its low safe cruising speed there will be a tendency for the airspeed further to reduce during turns, consequently a little power must be added before rolling on bank and taken off as the wings level.

Effect of Wind, Drift and apparent Slip/Skid

When flying near the ground, wind effect is clearly defined. Not only does drift become most apparent but the effect of the wind on ground speed can be appreciated.

The pilot's first impression during the exercise will be that of speed over the ground, the sensation becoming more apparent the lower the aircraft is flown. When flying downwind ground speed will be high and the pilot must resist the temptation to reduce power particularly when flying at 'low safe cruising speed'. Remember that *airspeed* keeps the aircraft flying and not ground speed.

When flying into wind the reduction in ground speed is most marked.

The effects of drift will by now be well known and allowance should be made when flying crosswind in close proximity to high obstructions. Drift during turn is the cause of certain illusions. When turning crosswind after flying into wind drift will create the impression that the aircraft is slipping in (Fig. 49).

Conversely when turning crosswind after a downwind flight path the aircraft will appear to skid outwards (Fig. 50). Both impressions may cause the pilot to make unnecessary and in some cases positively

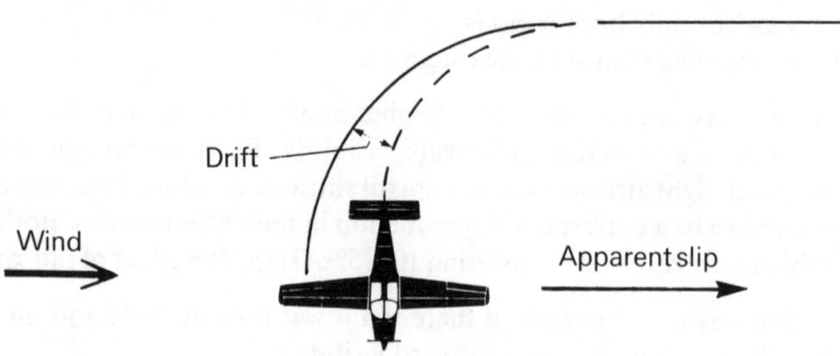

Fig. 49 Apparent slip during a turn downwind at low level.

dangerous rudder corrections and a quick glance at the turn and slip indicator will correct any false impressions while turning with drift near the ground.

During this part of the exercise the effects of speed, drift and the illusion of slip or skid should be noted, when the apparent inaccuracies may be checked with the instruments. Resist any tendency to make corrections which are not necessary, particularly the urge to reduce power. When flying at low level the aircraft must be correctly trimmed and while accurate flying is important this must not be at the expense of lookout.

Effect of Turbulence on Flight at Minimum Level

The obvious effects of turbulence are those felt by the occupants of an aircraft in the form of bumpiness. To some extent the severity of these

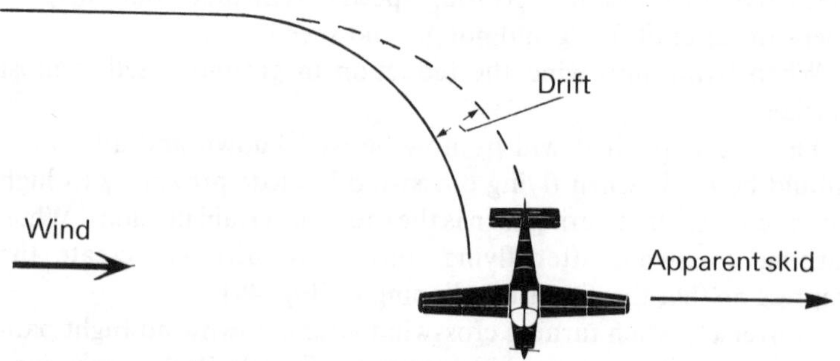

Fig. 50 Apparent skid during a turn into wind at low level.

disturbances will be reduced by flying at low safe cruising speed and it is often possible to anticipate the presence of turbulence by observing the nature of the terrain.

Turbulence is usually to be expected downwind of large structures, hills/ridges and woods. Provided attempts are not made to adopt too low an airspeed, offering a dangerously small margin above stalling speed, such turbulence can be dealt with by prompt application of extra power immediately the aircraft sinks.

Of more importance are the effects of **Wind Sheer** (sudden changes in strength and/or direction). These changes can occur in the vertical plane, for example, a reversal of wind direction while descending.

Horizontal wind sheer is the result of sudden alterations in wind direction. In each case an aircraft exposed to, say, a 25 kt headwind component may suddenly fly into an air mass providing a 5 kt tailwind, a net reduction in wind strength of 30 kt. This would entail a considerable loss of lift until:

(*a*) The angle of attack was increased by the pilot.
(*b*) The aircraft is able to accelerate and restore the 30 kt airspeed lost by the sudden change of wind.

By now the importance of keeping the hand on the throttle, ready for instant action, will be clear.

Low Flying due to Low Cloudbase, Poor Visibility or Precipitation

Unlike the professional pilot flying according to clearly defined company rules, the private pilot must base his decision whether or not to fly on a knowledge of his own limitations and guidance on the subject provided in the Air Information Publication. Particularly during operation at minimum level, self-discipline is all-important.

Changes in weather can upset the best flight plan. Actual weather reports are, in isolation, insufficient and should be studied before the flight in conjunction with 'forecasts' for the relevant area. Avoid smoke by flying to the windward side of large towns. Particularly when looking into sun, smoke can seriously reduce low-level visibility.

When using specified entry/exit lanes to airfields within control zones the weather must be such as to allow visual contact with the ground. Before the flight all obstructions and danger/prohibited areas must be studied on an up to date map.

Determining Minimum Height

(*See* 'Altimeter Settings', Vol. 2)
Regional QNH (as opposed to airfield QNH) is the altimeter setting to use during operation at minimum level since altimeter readings are then related to msl (mean sea level) and it only remains to add 500 feet to ground level below the aircraft (or the height of obstructions to be overflown) to ensure legal en route clearance. Built up areas must be avoided.

When joining traffic at an airfield QFE should be set on the altimeter thus relating it to airfield level. A minimum 500 feet should then be maintained around the circuit.

Low Flying due to Low Cloudbase

Since light aircraft are equipped with comprehensive radio aids, the completion of a journey below a lowering cloudbase is usually avoided in favour of flying over the cloud layer and effecting a radio let down at the destination. However, only pilots with an instrument qualification are able to adopt this procedure.

On occasions when no radio facilities are available or in the event of a radio failure, low flight below cloud may be necessary. In addition to the various considerations already outlined in this chapter, there are other factors which should be understood.

It is important to anticipate any high ground ahead and if necessary the flight plan should be altered to re-route the aircraft over flat terrain.

Under no circumstances should cloud be entered while low flying and a height should be chosen giving the optimum ground and cloud clearance. When the visibility is normal the higher the flight the greater the range of vision ahead of the aircraft but endeavour to remain at least 200 feet below cloud.

To conform with the Rules of the Air always fly to the right of such line features as railways when these are being followed. The feature will then remain in view from the first pilot's seat. Other aircraft which may be following the same line in the opposite direction will adopt this procedure so ensuring that both aircraft pass on opposite sides of the line feature. Be constantly on the lookout for other aircraft and be prepared for an emergency calling for extra power.

As an aid to map reading it is usually preferable to fly at low safe cruising speed although when a reasonable amount of ceiling is available and the visibility is good normal cruising speed may be used.

Low Flying in Poor Visibility

If in bad visibility the pilot is forced to fly low the speed of the aircraft should be reduced to low safe cruising. The opening of windows or clear vision panels may improve visibility.

Whenever low flying is occasioned by poor weather once a certain minimum visibility or cloud base has been reached, no attempt should be made to fly on into steadily deteriorating conditions. The pilot should either (*a*) divert to another area where conditions are known to be better or (*b*) return to the home base. The point when it is decided to abandon the original flight plan will depend upon the pilot's capabilities, the terrain ahead and to some extent the type of aircraft. When the weather conditions have deteriorated to the extent that it is neither possible to divert nor turn back a Forced Landing with Power will be the only alternative (Exercise 17b) advising ATC.

Since low flying in poor visibility is very tiring it should, whenever possible, be avoided. When conditions make such a flight unavoidable adjust cruising height to determine the best level for visibility, always remembering not to descend below the legal limit and not to climb into controlled airspace.

Low Flying in Precipitation

Rain, hail or snow will reduce visibility and in aircraft without windscreen wipers it may be necessary to open the clear vision panels provided. The effectiveness of wipers may be improved by reducing airspeed. When precipitation is of a localized nature it should be avoided by making small alterations in heading, air traffic and other considerations (such as terrain and obstacles) permitting.

Fuel Consumption

At low levels fuel consumption will be at its highest, particularly since drag will have been added while flying at low safe cruising speed. A combination of high gallons per hour and lower than normal speed for the power setting used can only add up to greatly reduced air miles per gallon.

When flying at minimum level, either as an exercise or through force of circumstances, high fuel consumption/reduced range must not be overlooked.

Navigation in Poor Visibility

The main problems while navigating at low level in poor visibility are:

1. At low level VHF communications will be lost and VDF, Radar scan and VOR/DME reception will cease unless the stations are within line of sight distance of the aircraft.
2. Such ground features as can be seen will remain in view for relatively few seconds.
3. Major features on either side of the aeroplane, some of them important as aids to map reading, may not be seen due to reduced visibility.

As a matter of principle, it is rarely good practice to descend below 1000–1500 feet in poor visibility unless forced to do so by a low cloudbase. **Slant Range** (i.e. distance from aircraft to a ground feature ahead) will obviously be greater than at 600 feet or so but the assistance of radio aids is of greater value than the ability to see well ahead. Furthermore obstacle clearance while flying at 1000–1500 feet is of less concern to the pilot than at low level.

If low flying in poor visibility cannot be avoided navigation will almost invariably be confined to map reading until near the destination when radar or VDF assistance may be forthcoming. To assist in map reading these steps should be taken:

1. Settle the aircraft at low safe cruising speed.
2. Take advantage of convenient line features – motorways, main roads, railways, etc., even if this means altering the route. Adopt the usual Rules-of-the-Air practice of flying to the right of line features.
3. Avoid significant obstructions (TV masts, high ground, etc.) by a wide margin.
4. Anticipate good features that are shown on the map (lakes, rivers, road/rail crossings, etc.) and use them as 'stepping stones' to guide the aircraft on its way to the destination.
5. Maintain a listening watch on the appropriate radio frequency and establish contact as soon as the aircraft has flown into range.

Low-level Circuit Procedure, Approach and Landing

When the destination airfield is within controlled airspace access may be confined to specific **Entry/Exit lanes**. The horizontal and vertical extent of these lanes must be clearly understood before the flight and a good lookout must be maintained since other aircraft may be using the lane.

All instructions from ATC must be complied with and if for any reason this cannot be done, stay well clear of the circuit and notify control accordingly.

At major airports the circuit joining procedure in poor weather will usually entail positioning by radar. At this stage of training the student pilot will confine his low flying to **Visual Meteorological Conditions** (VMC).

The circuit will be flown allowing adequate clearance from cloudbase but not below 500 feet agl (above ground level). All the steps outlined under 'Bad Weather Circuit Procedure', page 205 in Exercise 13, apply and full use must be made of the direction indicator, watch and, when the airfield is well known, ground features on which to establish turning points.

A normal engine assisted approach and landing will be made and when the airfield is equipped with **High Intensity Lighting** this will materially aid the pilot in lining up with the runway.

Pre-flight Briefing

Exercise 16. Operation at Minimum Level

AIM To fly the aircraft safely when, for operational reasons, it is necessary to conduct the flight at low level.

AIRMANSHIP *Planning*
Pre-descent checks:
Minimum safe height
Aircraft management for Low Safe Cruising/Reduced range
ATC Liaison
Suitability of weather

AIR EXERCISE

Visual impressions (low flight)	*Effects of wind*	*Low cloud poor vis./precipitation*
High speed *Undulations of terrain* *Loss of map reading 'picture'* *Inconspicuous power/telephone lines* *Fleeting ground features* *Difficult map reading* *Value of low safe cruising* *Improved forward visibility* *Better elevator/rudder response* *More time to map read* *Calmer environment*	*Flying downwind:* *High speed impression* *Resist temptation to throttle back* *Flying into wind:* *Low speed impression* *Resist temptation to open throttle* *Turning crosswind from into wind:* *Illusion of Slipping In (Confirm Ball Central)* *Turning crosswind from downwind:* *Illusion of Skidding Out (Confirm Ball Central)*	*In low cloud conditions* *Minimum terrain clearance* *Clear of cloudbase* *Option of turning back or Forced Landing With Power* *Avoid rising ground* *NEVER enter cloud* *Poor visibility* *If 1000–1500 ft not possible:* *Minimum terrain clearance* *Avoid smoke from towns* *Aim for line features* *ATC liaison when in range* *In precipitation* *Clear vision panel* *As for poor visibility*

Flight Practice

PRE-DESCENT CHECKS

(*a*) ***Height:*** Minimum safe height for the area Check altimeter setting

(*b*) ***Airframe:*** Flaps set to recommended Low Safe Cruising position

(*c*) ***Security:*** No loose articles in cabin (to distract pilot in turbulence)
Harness tight (in case of turbulence)
Hatches secure

(*d*) ***Engine:*** Fuel on fullest tank
Mixture rich. Carb. heat as required
Fuel pump on
Temperatures and pressures 'in the green'

(*e*) ***Location:*** As directed by ATC or, during practice, away from built up areas, airfields and animals

(*f*) ***Lookout:*** Be alert for power/telephone cables and undulating terrain.

AIR EXERCISE

Visual Impressions of Low Level Flight

(*a*) The aircraft is now at 600 feet above ground level. Note the impression of speed.

(*b*) Study a ground feature. It remains in view for a very short time.

(*c*) The area looked flat from 2000 feet. Now the terrain can be seen to undulate.

(*d*) At normal cruising speed map reading is difficult.

Low Safe Cruising and Effect of Flap

(*a*) Re-trim at the new speed. Notice the reduced forward visibility and the ineffective elevators and rudder.

(*b*) Lower 10–15° of flap (according to aircraft type). Maintain speed by lowering the nose and adding a little power. The nose is now lower, forward visibility is improved and so is rudder and elevator response.

(*c*) Ground features remain in view a little longer and map reading is easier.

(*d*) To turn at low safe cruising speed a little power must first be added. After rolling out of the turn power may be reduced to the previous setting.

Effect of Wind, Drift and Apparent Slip/Skid

Note. The following exercise is best demonstrated when there is a moderately fresh wind.

(*a*) At low safe cruising speed fly crosswind. Notice both drift and ground speed. Maintain a constant lookout for obstructions and other aircraft.

(*b*) Add power for all turns when flying at low safe cruising speed. Now turn into wind. Note the decrease in ground speed.

(*c*) Turn through 180° and as drift occurs notice that the aircraft appears to be slipping in although the ball is in balance. Maintain a constant height above the ground.

(*d*) The aircraft is now downwind and the ground speed is appreciably higher than before although the airspeed is unchanged. The temptation to compensate for higher ground speed by reducing power must be resisted.

(*e*) Turn upwind and as drift occurs the aircraft appears to skid outwards although the ball is in balance.

(*f*) Now become accustomed to turning in both directions at various angles of bank. Allow for drift and inertia when turning near obstacles.

Effect of Turbulence on Flight at Minimum Level

(*a*) The aircraft is now flying downwind of the wood/hill (find a suitable feature) at normal cruising speed. Turbulence is more pronounced than usual and the importance of a tight harness will be obvious.

(*b*) Now set up the aircraft at low safe cruising speed. Note that turbulence has less effect.

(*c*) Be prepared for the aircraft to sink by immediately increasing power. The hand must remain on the throttle at all times.

(*d*) With experience turbulence may be minimized by avoiding areas likely to produce disturbances.

Low Flying due to Low Cloudbase, in Poor Visibility or Precipitation

Low Flying due to Low Cloudbase

(*a*) Imagine the cloud has lowered to 700 feet above ground level. Descend until the aircraft is well clear of cloud but remain at a safe height above the ground.

(*b*) Reduce to low safe cruising speed and lower flap.

(*c*) Maintain a constant lookout for other aircraft, particularly when flying along a line feature. Be prepared to take evasive action by

keeping the hand on the throttle throughout.
(*d*) Re-route the flight when high ground lies ahead.

Low Flying in Poor Visibility

(*a*) Imagine the visibility has deteriorated and for operational reasons it is necessary to descend.
(*b*) Reduce to low safe cruising speed and lower flap. Increase power for all turns.
(*c*) Maintain a constant lookout for other aircraft particularly when flying along a line feature. Make full use of the clear vision panel or open windows to improve visibility. Be prepared to take evasive action by keeping the hand on the throttle throughout.
(*d*) Make frequent but brief reference to the instruments to confirm attitude.
(*e*) Maintain a listening watch and contact ATC when the aircraft is within VHF range.
(*f*) Avoid smoke from towns. Should the visibility deteriorate, abandon the flight and either divert towards better weather or return to base.

Low Flying in Precipitation

(*a*) Fly at a height appropriate to the conditions.
(*b*) Switch on the pitot heater and make full use of the demister (if fitted).
(*c*) Carburettor heat as required.
(*d*) Open the clear vision panel to improve visibility.
(*e*) Make alterations in heading to avoid heavy precipitation.

Navigation in Poor Visibility

(It is assumed that the usual preparations as explained in Exercise 18 have been completed.)
(*a*) Fly at the correct height and speed according to prevailing conditions.
(*b*) Note the limited field of vision and that landmarks remain in view for only brief periods. Main pinpoints should be anticipated.
(*c*) Make full use of line features, flying to the right of them.
(*d*) Note the limited range of radio equipment.
(*e*) Be sure the minimum legal clearance from property is maintained and throughout the flight keep a good lookout for obstructions.
(*f*) Make frequent brief checks on the engine and flight instruments and remember that at low levels the range of the aircraft will be less than normal.

(*g*) Make early corrections to any deviation from the planned route. At low levels accurate navigation is vital.

(*h*) Should the weather ahead deteriorate make an early decision to divert or return to base.

Low-Level Circuit Procedure, Approach and Landing

(*a*) Make early contact with ATC and remain clear of the airfield until permission to join has been obtained.

(*b*) Keep a good lookout for other aircraft.

(*c*) Check altimeter setting and minimum circuit height.

(*d*) Join the circuit as instructed by ATC and report 'down-wind' and 'finals' at the appropriate time.

(*e*) Use the direction indicator and, when possible, known ground features to establish the circuit legs. In some conditions the watch may be of assistance in timing legs.

(*f*) Make a normal engine assisted approach, making use of the runway lighting in the final stages.

(*g*) If unable to comply with *any* instructions advise ATC immediately.

Exercise 17A

Forced Landing Without Power

Background Information

These days the chances of engine failure occurring in the air are remote; nevertheless the pilot must be competent to deal with the situation should it arise and take immediate action to select a suitable landing area, plan the circuit and approach and complete a successful landing.

The majority of light aircraft land at low speed and can therefore be placed in a medium-sized field with little or no risk of injury or damage. Because of the low speeds involved when the correct procedure is adopted, injury to the occupants of the aircraft is even unlikely when damage to the machine does occur.

By far the most common cause of forced landings without power is lack of fuel. Fuel gauges are not always as accurate as they might be, but provided the inaccuracy is understood and allowance is made when reading the instrument, there should never be any doubt about the fuel situation. In any case a visual check on fuel contents is a vital part of the pre-flight inspection. Indeed, a thorough pre-flight inspection and power check is the pilot's best insurance against a forced landing. As a double check the pilot should be fully conversant with fuel consumption per hour and the endurance of the particular aircraft.

If circumstances are such that the pilot finds himself on a cross-country away from aerodromes and with little fuel, a **Forced Landing With Power** (Ex. 17b is explained in the next chapter) should always be conducted in preference to this exercise. Without power time cannot be spent searching for a good field and once chosen the field cannot be thoroughly inspected nor the approach controlled so accurately. Furthermore in the event of a bad approach, without power it will not be possible to go round again.

Immediate Actions after Engine Failure

In the absence of engine power height is an obvious advantage since height means time in the air and greater gliding distance. With these thoughts in mind cross country flights should seldom be routed below 2000 feet, preferably higher.

Imagine that while on a cross country flight the engine fails at a height of 2000 feet. Convert speed into distance by maintaining the level attitude of the nose then assume best gliding speed. The wind direction should be determined using smoke or, if none is visible, the take-off direction should be used. A good field ahead or to one side of the aircraft would be fortuitous but more likely one will have to be found. Unless a suitable landing area is immediately apparent a turn downwind should be made since this will have the effect of stretching the glide and offering more opportunities for field selection.

Attaining Best Gliding Attitude/Speed and Trim

Many a forced landing without power has ended with the aircraft failing to make the selected field because height had been dissipated throughout the glide by adopting the wrong speed.

Immediately the engine fails the following drill must be enacted:

1. Maintain the level attitude and aim for best gliding speed.
2. At best gliding speed allow the nose to adopt the correct attitude.
3. Make minor adjustments to the attitude until the correct gliding speed is indicated, then re-trim.

There are some who advise that when the engine fails the nose should be held up to 'convert speed into height'. This finds its roots in wartime operations when heavy military aircraft which, flying fast, had sufficient inertia to gain a considerable amount of height when an engine failed. The technique does not work on light aircraft. Speeds are too low and by attempting to climb without power airspeed invariably reduces to a low figure and more height is lost than gained while trying to attain the proper gliding speed.

The aircraft must be settled into best gliding speed without delay when the engine fails, using the technique recommended in this manual. This is vital to the success of a forced landing without power.

Field Selection

Qualities required for an ideal forced landing field are:

(*a*) It should be large.
(*b*) Clear approaches are important.
(*c*) It should be level.

(*d*) Good surface is an advantage.
(*e*) Near a road and habitation so that help may be obtained after landing.

The choice of a field presents certain problems:

(*a*) While it is relatively easy to select one large enough for the landing and subsequent take-off, it is difficult to assess the surface.
(*b*) In an emergency a ploughed field can be used provided a landing is made with the furrows, although in general these should be avoided because of the soft nature of the ground.
(*c*) Colour is one indication and a large field similar in appearance to the home airfield should be sought. (See '*Surface Recognition of a Forced Landing Field*' on page 261)
(*d*) If a field can be found near a main road and habitation, so much the better for obvious reasons.

Knowing or Assessing Wind Velocity

A forecast wind velocity for the route will have been obtained before the start of the cross country flight and this will provide the pilot with a degree of guidance. However local winds, particularly at low levels, can be distorted to the point where they bear little resemblance to the general pattern.

Wind direction may ideally be determined by smoke, when there is any in the forced landing area. Tall crops often display wind streaks and in broken cloud their shadows on the ground will give a good approximation of both wind speed and its direction. Claims that animals face downwind (or into wind) while feeding are not always reliable. If there are no indications, use the take-off direction.

An opportunity to assess wind strength more accurately will be presented when the aircraft has turned onto base leg. This will later be described.

Circuit Plan and Selection of 1000 ft Point

A normal circuit to the left or right should be planned, according to the aircraft's position in relation to the field and a marker point should be selected at the end of the downwind leg from which to turn on to the base leg. **The marker point is the keystone of the operation and the pilot must aim to arrive over it at 1000 feet**. Incorrect selection is a common cause of misjudged forced landings and Fig. 51 shows that it

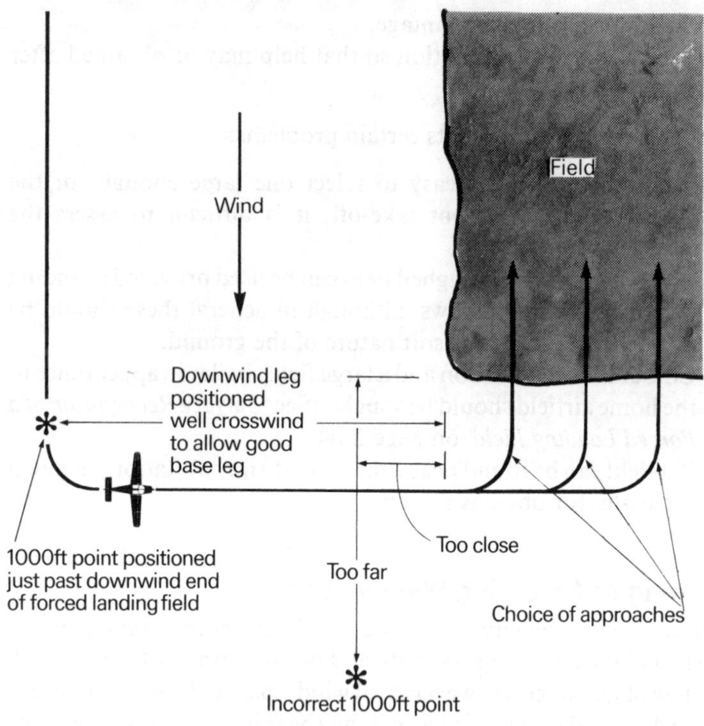

Fig. 51 Locating the '1000 ft point'. Correctly positioned it offers a choice of approach paths.

should be a full **Airfield** width from the field. Although a normal, airfield-type circuit is ideal there may be circumstances where unorthodox patterns will have to be adopted if the 1000 ft point is to be reached. For the first demonstration the instructor will probably base the procedure on a normal circuit but, at a later stage, the student will be shown how to arrive over the 1000 ft point when insufficient height or other considerations make it necessary to take 'short cuts'.

Attempts to cure a severe overshoot situation by making a 360° turn usually culminate in an incurable undershoot. There are better ways of dealing with an overshoot and these will later be explained but, as a matter of general principle, it is bad practice to turn away from the selected forced landing field.

Examination for Cause of Failure/Restart Action

Having selected a landing area and planned the circuit the pilot is now free to investigate the cause of engine failure by checking the following possibilities while gliding towards the 1000 ft area.

(*a*) Switches: They may have been turned 'off' accidentally.
(*b*) Fuel: Change to another tank (if fitted). Primer: check it is in and fully locked. An unlocked primer can seriously disturb the mixture. Check mixture control.
(*c*) Mechanical fuel pump failure: Switch on the electric pump.
(*d*) Carburettor icing: Select carburettor heat.

If no remedial action can be taken, the normal pre-landing vital actions should be completed but the fuel and ignition must be 'off' with the mixture control in the idle cut-off position. When practising the exercise this part of the check is simulated, but in the case of a real engine failure both fuel and ignition must be 'off' to prevent the engine from temporarily restarting only to fail again. Such engine behaviour would be typical of fuel starvation and may prompt the pilot to climb away, only to find himself in a worse situation than before. Idle cut-off is selected as a precaution against fire.

R/T Distress Call

Now that the pilot is committed to a forced landing without power, time permitting, a distress call should be made using the MAYDAY procedure described in Chapter 11 of Volume 2. Obviously it is in the interest of everyone to be aware of the situation. ATC will be in a position to alert the police who, in turn, will be able to render assistance should this be required.

In making the call the best possible description of the location should be included in the message, e.g. 'forced landing in field near large wood five miles south of Newtown', etc.

The prime objective is to land safely in the chosen field and a MAYDAY call should not be made if it is likely to distract the pilot from the main task. However, if a MAYDAY call is made it should be transmitted while the aeroplane is still high enough to be in VHF range.

Approach and Landing

Having disposed of the restart checks and the pre-landing vital actions the pilot must concentrate on arriving over the 1000 ft point, bearing in mind that the altimeter is on QNH and the height of the terrain (as shown on the map) will have to be assessed or subtracted from the altimeter reading.

On arrival over the 1000 ft marker point a gliding turn on to the base leg is effected and at this stage a deliberate 'overshoot' should be planned aiming to touch down about one third of the way into the field. There are several reasons for this:

1. Excessive height can always be lost by using full flap or sideslipping, whereas without engine power nothing can be done to correct for insufficient height.

2. In a real forced landing the glide will be steeper without the engine firing, the propeller windmilling and causing drag.

3. If an error must occur, running into the far boundary at taxying speed is preferable to flying into the downwind hedge or fence.

Correcting for Wind Conditions

While gliding on the base leg the strength of the wind can be assessed by the amount of drift away from the field. When the wind is strong the aeroplane must be prevented from drifting away from the field by turning slightly towards the boundary. Conversely, when the aircraft is obviously too high, a turn away from the field may be made. These adjustments are shown in Fig. 52. In any event the base leg must be positioned close to the boundary of the field and Fig. 51 will show how this precaution allows the pilot several alternative turning points on to the approach according to height. A base leg some distance from the field would entail a long, straight, gliding approach and this is very difficult to judge accurately.

It is, of course, possible to overshoot excessively, particularly when the wind is lighter than anticipated. When the aeroplane is so high that height must be lost before turning on to the approach, one of the following methods may be used:

1. A sideslip away from the field before turning in.

2. In extreme cases the pilot can glide past the turning point and then turn back across wind before turning in. The two methods are shown in Fig. 53.

All turns must be made towards the field. As previously explained, turning away places it out of view and there is the attendant risk of undershooting because of the considerable height lost in a prolonged

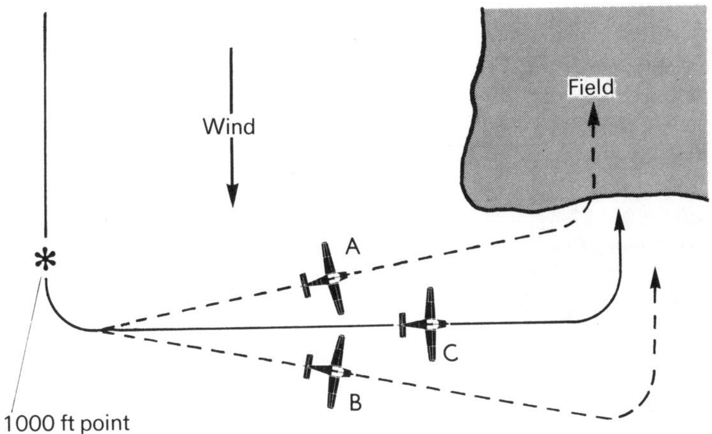

Fig. 52 Alternative method to that shown in Fig. 51 for adjusting the final approach.

gliding turn. Once a field has been selected the pilot should not change his mind. A change of plan during the last stages of a forced landing is rarely warranted and usually produces the wrong results.

Throughout the forced landing circuit pilots should develop a talent for assessing the aspect of the field relative to the aircraft. The ability to distinguish between too flat, too steep and the correct aspect of the chosen field will assist in guarding against gross misjudgement of the final approach.

When the final turn on to the approach has been made and it is certain the field can be reached, surplus height is lost by lowering full flap. This will bring back the touch-down point towards the downwind boundary of the field (Fig. 54).

Pre-crash Vital Actions

This rather dramatic sub-heading should not be allowed to colour the student's attitude towards the likely outcome of a real forced landing without power. Properly handled, and in all but impossible situations, there need be no damage whatsoever. However, by now it will be clear that flying is based upon a concept of 'braces and belt'; everything possible is done to ensure safety and insure against the unlikely.

Fuel and ignition will have been turned off during the pre-landing vital actions and the mixture will have been pulled back into the idle

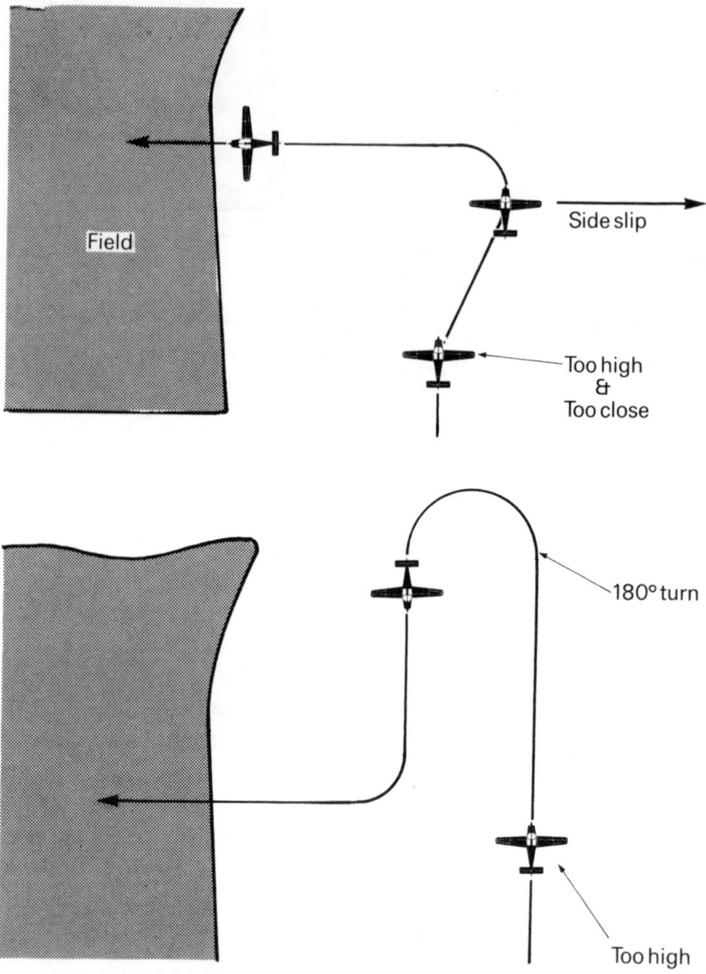

Fig. 53 Two additional methods of losing excessive height before the final turn. Procedure illustrated in the lower drawing should be used only in extreme cases.

cut-off position. The battery master switch must be left on when the aircraft has electrically operated flaps. When the flaps have been fully lowered on the approach the master switch will be turned off.

The harness of all occupants must be tightened and the door(s) must be unlatched so that, if the landing is heavy and there is fuselage damage, the door structure will not become distorted, preventing exit from the aeroplane. Unlatching of the doors must be left until short

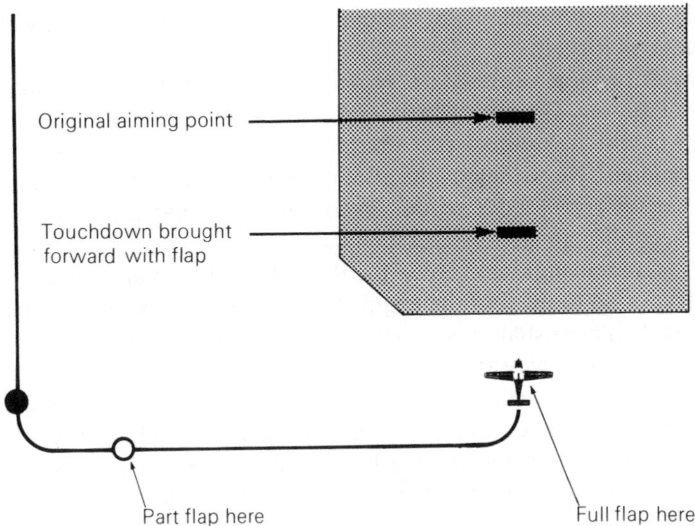

Fig. 54 Bringing the touchdown point forward with flap. The original aiming point, located approximately one third of the way into the field is NOT the touchdown point as is often believed.

finals; some aircraft suffer from disturbed airflow when the doors are open to the point that elevator and perhaps rudder control is seriously affected.

The Landing

During the final stages of the approach certain ground features will be revealed for the first time. There may be a ditch or a low, wire fence which was too small to have been seen before. Gentle avoiding action should be taken.

A normal glide landing should be made, aiming to hold off until a nose-high attitude has been attained and touch-down speed is at its lowest. Brake should be used to bring the aeroplane to a halt. Rabbit holes, branches or other hazards can be avoided with the nosewheel steering during the final stages of the landing roll.

Action after Landing

When the landing roll has been completed it will be necessary to safeguard the aeroplane. If possible place someone in charge of it while the local police, ATC and the landowner are notified. Tie the machine down for the night, remove loose items of value and insert the control lock(s).

The foregoing explanation represents an ideal forced landing without power. There may be times when it is impossible to complete a full circuit because of the aircraft's position in relation to the selected field. Nevertheless, the pilot must aim to be over the marker point at 1000 feet (above ground level) and his method of arrival will vary according to circumstances.

When practising this exercise the throttle should be opened every 500 feet or so to prevent the engine from cooling or the plugs from fouling. It is not usual to land in the practice field and power will be needed to climb away at the end of the exercise. A cold engine could turn practice into reality. In any case the throttle should be opened smoothly to prevent the engine from cutting when climbing away. It is, of course, important that the Forced Landing Without Power is not practised below the point where there is a low-level infringement (Rule 5).

Exercise 17A. Forced Landing Without Power.

AIM — *To rectify an engine failure and, if necessary, land safely in a field.*

AIRMANSHIP — *Height: sufficient to allow 2500 ft over training area*
Knowledge of terrain altitude/Rule 5
Engine: failure simulation, carb. heat, regular warm-up
Location: away from animals, noise sensitive areas or towns
Lookout
Suitability of weather

AIR EXERCISE

Immediate action after failure

Engine has failed:
- *Prevent nose dropping*
- *Attain best gliding speed in level attitude*
- *Allow nose to adopt gliding attitude*
- *Re-trim at best glide speed*
- *Look for suitable field*
- *Assess wind strength and direction*

If no field:
- *Turn downwind*
- *Begin search*
- *MAYDAY (if time)*
- *Warm engine every 500 ft*

Field selection/engine checks

Locate field
Fix 1000 ft point
Glide downwind
Find cause of engine failure:

Ignition	*ON*
Fuel	*ON*
Try another tank	
Fuel pump	*ON*
Carb. heat	*HOT*
Primer	*LOCKED*
Mixture	*RICH*

If engine will not start:
Pre-landing checks – BUMPF

Simulate –		
	Ignition	*OFF*
	Fuel	*OFF*
	Mixture	*ICO*
	Harness	*TIGHT*

Check: *Airspeed*
Height
Downwind leg: wingtip tracing landing path

Circuit and landing

CHECK height and distance from 1000 ft point
Adjust downwind leg to arrive over point at 1000 ft

At 1000 ft point:
- *Turn onto base leg*
- *Flap as required*
- *Assess wind strength and turn in or out accordingly*
- *Fix touch-down point, third way into field*
- *Turn finals*
- *Add flap when sure field is within reach*

On short finals:
- *Full flap*
- *Master switch OFF*
- *Unlatch door(s)*
- *Avoid obstacles*
- *Land tail down*

Safeguard aircraft/Advise ATC and Police

Flight Practice

For the purpose of this exercise imagine that on a cross country flight the engine has failed at a height of 2500 feet.

COCKPIT CHECKS

(*a*) Carb. heat HOT
Fuel pump ON
Engine warm-up at intervals.

(*b*) Trim for best gliding speed.

OUTSIDE CHECKS

(*a*) Altitude: sufficient to allow 2000 feet above ground level.

(*b*) Location: over a suitable forced landing area.

(*c*) Position: to be altered for each practice in relation to the forced landing field.

Forced Landing Without Power procedure

(*a*) The engine has now failed. Convert cruising speed into distance but do not allow the speed to decrease below gliding speed. Place in the gliding attitude and re-trim.

(*b*) Determine the wind direction and select a field.

(*c*) Turn towards the marker point which must be to one side of the downwind boundary and aim to arrive over the point at 1000 feet. Plan as normal a circuit as possible.

(*d*) While gliding towards the marker point try and find the cause of engine failure. Check: switches on, fuel on, try another tank, electric pump on, carburettor heat hot, primer locked and mixture rich. If unable to find the cause of engine failure, carry out the pre-landing vital actions: switch off ignition (simulate), turn off the fuel (simulate), place the mixture control in idle cut-off (simulate) and tighten the harness.

(*e*) Make a distress call giving call sign, position and intentions.

(*f*) During practice, open the throttle every 500 feet or so to keep the engine warm.

(*g*) When over the 1000 ft point turn on to the base leg and aim to land about one third of the way into the field. Reassess the strength of the wind by the amount of drift and keep close to the downwind boundary of the field. If too low turn towards the field and if too high turn away or sideslip away from the boundary. Apply part flap as required.

(*h*) When nearly opposite the approach turn into wind aiming to overshoot slightly. When sure of reaching the field lower full flap or sideslip off superfluous height. Turn off the battery master switch before landing and unlatch the door(s).

(*i*) After touch-down bring the aeroplane to halt with the brakes and if necessary swing the aeroplane across wind to avoid obstacles.

(*j*) Safeguard the aircraft and notify the home airfield and local police.

NOTE. During Forced Landing Without Power practice the aircraft must not be allowed to descend below the legal limit (Rule 5).

Exercise 17B
Forced Landing with Power

Background Information

The advent of lightweight radio equipment of advanced design and enhanced capabilities has, to a considerable extent, made this exercise less relevant in modern aviation than it was perhaps twenty or more years ago.

Circumstances which may make necessary a **Forced Landing With Power** are:

(*a*) Shortage of fuel.
(*b*) Approaching darkness when the pilot is not qualified to fly at night.
(*c*) Uncertainty of position.
(*d*) Bad weather.

A good pilot does not run out of fuel or plan a flight that may end after dark when the licence is limited to daylight flying. Uncertainty of position is, these days, only a temporary situation since radio navaids in even the simplest trainer are capable of guiding a lost pilot to a known point. Also there are a number of air traffic services (e.g. **Lower Airspace Advisory Service** which can assist aircraft flying between an altitude of 3000 feet QNH and, over most of the UK, FL95). The Distress and Diversion (D & D) service, which operates on the **International Aeronautical Emergency Frequency** (121.5 MHz) can provide a lost pilot with comprehensive 'get you home' service. So unless there has been a radio failure no pilot flying over the UK and most of Europe or North America need ever become totally lost.

Bad weather is another matter. In the main, forecasting has reached the stage where the element of surprise is now rare. If there is an unexpected deterioration in the weather the obvious steps to take are:

1. Divert to another airfield.
2. Return to base.

If, for reasons of extraordinary circumstances or poor flight planning neither of these alternatives can be taken, possibly for want of sufficient fuel, a Forced Landing Without Power will have to be made in the best possible landing area.

Selection of Landing Area

Selection of a suitable field was outlined in the previous exercise and while the initial choice will be influenced by the same considerations in both the forced landing with and without power, in this exercise the availability of power will enable the pilot to make a more detailed examination of the approaches and surface before landing.

Surface Recognition of a Forced Landing Field

Depending on the amount of fuel available and the weather conditions at the time it will be possible to look around for a suitable landing area. Requirements are very much the same as those listed on page 248–9 of the previous exercise.

Assessing the surface from the air is not easy but the following colour guide will assist in so far as the initial selection is concerned.

Grassland: Dark green to brownish green according to dampness. Surface may be mottled.

Stubble: Buff colour with regular cultivation lines visible at lower levels.

Ripe Grain: Buff to golden brown (according to crop type). Distinctive wave pattern can be set up by the wind, indicating its direction.

Young Grain: Dark green in various shades.

Root Crops: Regular lines, particularly visible when crops are young.

Ploughed Field: Dark to red-brown according to area. If no other choice, land with the furrows.

Marsh: Dark green with much darker areas, pools of water and nearby streams. A last resort.

Heath: Mottled brown-green denoting gorse or heather.

Usually on high ground. Look out for rocks on the approach.

Beach: Firmest sand will be within a few yards of the water's edge. Can provide a good landing provided it is possible to taxi away from the incoming tide.

Wind Assessment

Smoke near the intended landing area is ideal but cloud shadows or the movement of crops can be a help. If all else fails remember the wind direction at time of take-off.

During the inspection runs over the field (shortly to be described) there will be an opportunity to check on wind strength and direction.

R/T Call

If the forced landing with power has been caused by:

(*a*) Shortage of fuel,
(*b*) Weather conditions beyond the capabilities of the pilot,
(*c*) An aircraft malfunction other than complete loss of power,

a PAN call should be made on the frequency being used at the time (e.g. FIR. Radar Advisory Service etc.) giving the intentions and location of the proposed landing field. Radio procedures of this nature are described in Chapter 11 of Volume 2.

Initial Inspection of Approach and Landing Area

Assuming an emergency landing has been decided upon, speed should be reduced to low safe cruising. This will vary from type to type but 60–70 kt would be suitable for most light single-engine aircraft. 15° of flap should be lowered, since this will make the aeroplane fly in a more nose-down attitude for any particular speed, thus improving forward visibility. By reducing speed in this way it will be possible to view the surrounding area calmly and allow more time to plan the circuit.

Having found a likely field a run into wind should be made over it at several hundred feet although it may be preferable to choose the longest run the field has to offer, even if this is slightly out of wind. Furthermore if there is an incline, a landing uphill even if downwind is often the best plan provided the wind is light.

During the initial approach, which should be made to the right of the intended landing path so that it can be viewed by looking along the left of the nose, the position of high-tension cables, telephone wires, tall trees or any potential hazards should be noted. Now will be the time to assess the strength of the wind, check its direction by any drift that may be present, and determine if the field is large enough. Remember the aeroplane will eventually have to be flown out when the weather improves.

Over the field, look out for obvious obstructions; wire fences, ditches, tree trunks and animals.

Identification of Circuit Turning Points

Should the weather be bad, turning points, e.g. farm buildings, ponds, etc., should be selected on which to establish a circuit. Because the aeroplane is flown at low safe cruising speed the throttle must be opened slightly while making turns, power being reduced after rolling out. The direction indicator may be used to help maintain direction between turning points which must be selected to keep the circuit small and, where possible, the field in sight. Some pilots find it of assistance to set the direction indicator to zero while in the landing direction. At each turning point four 90° turns can then be made on the instrument with comparative ease.

Detailed Surface Inspection

After the first circuit has been completed another approach is made, this time with a view to inspecting the surface. Like the previous run this is best made to the right of the intended landing path so that the landing area can clearly be seen by looking along the left of the aircraft. At this stage the state of the approaches to the field will be known. The aeroplane is lowered to near the 'hold-off' position while the ground is inspected at low speed. There should be no holes, ruts, large stones, or steep inclines. When the inspection run has been completed a final circuit is made again using the turning points.

Because a good straight approach is an essential prelude to a short landing it is advisable to fly a little further downwind on this final circuit before turning on to the base leg. Speed should be reduced, flaps lowered and prior to turning on to the approach, the throttle should be opened slightly or the nose depressed. The throttle friction should be slackened slightly (when applicable) so that power adjustments can be made with ease.

The Short Landing

The Short Field Landing technique has already been explained in Exercise 13 (page 202). In that exercise it was assumed that the pilot was landing at a small airfield or private airstrip. In the case of Forced Landing With Power the surface is unknown except for an inspection from the air, albeit at low level.

During the final approach into the field, which should be on a descending path to ensure a touch-down near the downwind boundary, obstacles not previously noted may present themselves and the pilot must be prepared to avoid these with gentle turns.

On short finals the door(s) should be unlatched and, during the hold-off, the fuel should be turned off, mixture placed in idle cut-off, ignition off and the battery master switch should be turned off. Even if the aircraft turns over there will then be little risk of fire and the door(s) will not jam.

The landing should be made in a tail-down attitude to ensure touch-down at the lowest possible speed. The wheel/stick must be held fully back to safeguard the nosewheel and brake may then be applied to bring the aeroplane to a halt. Avoiding action should be taken if holes, rocks or other debris reveal themselves.

Action After Landing

No attempt should be made to re-start the engine (which will have stopped following fire precaution action during the landing) with a view to taxying. Having made a successful Forced Landing Without Power it would be unforgivable to damage the aeroplane by taxying into a ditch or other obstacle.

The parking brake should be set, control lock(s) fitted, loose articles removed and, if possible, the aeroplane tied down. Try and place someone in charge of the aircraft before contacting the local police and ATC.

Before attempting to fly out of the field, when weather and other conditions permit, the surface must be carefully examined on foot for potholes, soft patches and the like. At the same time its length should be paced and equated with the Owner's/Flight/Operating Manual. Rather than risk a take-off in marginal conditions newly qualified pilots should enlist the services of an experienced pilot.

Practising Forced Landing With Power

When the exercise is being conducted in open country care will have to be taken not to annoy local residents or frighten animals. The sight of a light aircraft circling the same field repeatedly is enough for some people to contact the police and complain.

Sometimes flying schools are able to make an arrangement with a local farmer which allows aircraft to make the two inspection runs, including the one at hold-off height. There is nothing to stop such flying training in open country when the owner of the land has given permission. However, landings in an unlicensed field could raise legal difficulties since flying training must be conducted from proper airfields. The problem is dealt with by teaching the inspection runs at a suitable field, leaving the short landing element until the return to the aerodrome.

When no arrangements have been made with a suitable landowner aircraft may not descend to within 500 feet of persons or property (Rule 5), and care must be taken not to infringe this requirement in law.

Pre-flight Briefing

Exercise 17B. Forced Landing With Power.

AIM — *To select a suitable field, inspect its approaches and surface and land safely when circumstances demand the termination of a flight.*

AIRMANSHIP — *Height: not below legal limit*
Engine: fuel pump on, temperatures and pressures 'in the green'
Location: open country, away from buildings or animals
Lookout
Suitability of weather

AIR EXERCISE

Selecting the field	*Inspecting the field*	*Circuit and landing*
Look for suitable field: *Size* *Approaches* *Proximity to road/habitation* *Colour of surface* *Free of animals* *Climb-out area – free from tall structures, power lines, etc.* *Relationship to wind direction:* *LOOK FOR:* *Smoke* *Cloud shadows* *Crop movement* *If no indications use take-off direction* *Set up low safe cruising speed*	*Inspect approaches:* *Descend to 300 ft* *Fly into wind to right of intended landing path* *Check for:* *Power/telephone lines* *Tall trees/chimneys, etc.* *Drift* *Field length* *Obvious obstructions* *Climb away and select turning points* *Inspection of Surface:* *Fly around small circuit at low safe cruising, adding power for turns* *Establish turning points* *Make low approach to right of landing path* *Check surface for ditches, fences, branches of trees, etc.* *Climb away for another circuit on selected turning points* *Use DI if necessary*	*Fly further downwind* *Pre-landing checks* *Turn base leg and plan a Short Field Approach* *Add flap in stages until FULL FLAP is set* *Unlatch the door(s)* *Cross low over boundary* *Touch-down point near field boundary* *Round out and check:* *Fuel OFF* *Mixture ICO* *Ignition OFF* *Master Switch OFF* *Avoid obstacles* *Land tail down/low speed* *Steer away from obstacles* *Safeguard aircraft* *Report to police* *PRE TAKE-OFF FIELD CALCULATION/ INSPECTION*

Flight Practice

For the purpose of this exercise imagine the cloudbase to be 600 feet with deteriorating visibility.

COCKPIT CHECKS

(*a*) Reduce to low safe cruising speed, lower 15° of flap and re-trim.
(*b*) Check sufficient fuel to carry out several circuits around the selected field.
(*c*) Fuel pump ON/carb. heat as required.

OUTSIDE CHECKS

(*a*) Altitude: below cloud base (in this case fly at 500 feet above ground level).
(*b*) Location: select suitable field with regard to wind direction and colour of surface (in Training Area).
(*c*) Position: downwind of selected field.

AIR EXERCISE

Field Selection, Field Inspection and Circuits on Turning Points

(Full procedure to be demonstrated at the airfield or authorized landing ground.)

(*a*) Fly over the field at about 300 feet to the right of the intended landing path. Notice any high trees, overhead power cables or other obstacles which may affect the approach or climb away. At the same time judge if the field will be long enough for the landing and, when conditions permit, the take-off. Look out for fences and animals. In conditions of poor visibility use the direction indicator which may be set to zero when in the landing direction.
(*b*) Check for drift.
(*c*) When the full length of the field has been flown, climb to circuit height – in this case 500 feet – and when level flight has been resumed find a feature on the ground which can be used as a turning point. Open the throttle slightly throughout all turns and reduce to low safe cruising power on rolling out.
(*d*) At the end of the crosswind leg select another ground feature and turn so a downwind leg can be flown while keeping the field in view.
(*e*) After the field has passed the trailing edge of the wing, select a further turning point and turn onto the base leg.
(*f*) Reduce speed slightly and turn onto the approach, lowering the

aircraft so that it can be flown near the ground to the right of the intended landing path.

(*g*) Study the ground and look out for holes, ruts, large stones or steep inclines which must be avoided.

(*h*) When the length of the field has been flown climb away and complete another circuit using the previously selected turning points, this time flying further downwind before turning onto the base leg. Complete the vital actions.

(*i*) On the base leg reduce speed and lower the flaps. Ease the throttle friction and add a little power to control the rate of descent.

(*j*) When nearly opposite the landing path open the throttle slightly or depress the nose for the turn on to the approach, according to the amount of height in hand.

The Short Landing

(*a*) Carry out a standard short field landing approach.

(*b*) On short finals unlatch the door(s).

(*c*) Aim for a touch-down just over the field boundary.

(*d*) After the round out: fuel OFF, mixture ICO, ignition OFF, master switch OFF.

(*e*) Land in a tail-down/low speed attitude and avoid obstacles on the ground.

(*f*) Brake to a halt, safeguard the aircraft and report incident to the police/ATC.

Exercise 18
Pilot Navigation

Background Information

It is assumed that the student has read Chapter 2, *Navigation* in Volume 2 of this series before embarking on this exercise. As this is a pilot handling manual only essential background information is given here although, in some cases, information may be repeated to provide a degree of revision.

The motorist is confined to roads and this very limitation provides him with a ready means of finding his way from A to B. By using signposts it is possible and often practical for the motorist to cover long distances without reference to a map. Conversely the aeroplane enjoys a high degree of freedom of movement both in relation to the ground and height above it, and it is this very freedom which presents a problem. No longer guided by roads and signposts and sometimes without visual reference to the ground, the pilot must turn to navigation in one form or another if the aircraft is to arrive at its destination on the planned route.

The main navigational systems are:

1. Dead Reckoning (DR).
2. Astronomical (Astro).
3. Radio and Guidance Systems.
4. Pilot Navigation.

Briefly, **Dead Reckoning Navigation** (known as DR) was the system used in civil and military aviation for many years until the development of radio navigation aids (**Navaids**). It required the services of a specialist navigator and, in so far as civil aviation is concerned, DR navigation is now extinct. On the other hand pilots sometimes use a form of **Mental DR**.

Astronomical navigation (Astro) is based upon star observations by sextant which are referred to astronomical navigation tables. These in turn provide the necessary information to enable a navigator to transfer the star readings on to the plotting chart – an impractical task for a pilot committed to the controls of an aeroplane. Like DR navigation **Astro** is now extinct.

Radio navigation, together with **Inertial Guidance Systems** which depend upon super-accurate gyroscopes, is the mainstay of modern navigation. The airlines of the world are largely dependent upon radio navigation and while miniaturized equipment is available for light aircraft, its function and use (explained in Volume 3) is a development which must come later in the student pilot's training, after mastering the method which has been evolved for the pilot-navigator. This is called **Pilot Navigation** and it represents a simple, practical way of flying from one point to another without the use of special equipment, although obviously its scope is greatly enlarged when even the simplest radio facilities are to hand.

Pilot navigation is dependent upon three fundamental requirements:

1. The ability to read a map.
2. An understanding of the effects of the wind on the aircraft's progress in relation to the ground and how to make allowance for the resultant drift, etc.
3. The accurate use of the compass, direction indicator and watch.

Although the aeroplane has been used for all manner of purposes, utility and pleasure, its main task has always been one of transport and that must entail the art of navigation.

Pilot navigation of an aeroplane from one airfield to another or perhaps several destinations is divided into two phases:

1. Flight planning.
2. In flight.

Flight Planning

This phase of a flight embraces all preparation on the ground. The more thorough the preparation, the easier will be the pilot's task in the air. The obvious way to fly from 'A' to 'B' is to draw a straight line on the map and fly along the route represented by it. Unfortunately, there could be a number of factors which prevent such an idealistic method from being used. These will become apparent in the following text.

Route Selection

When selecting the route it will be necessary to take account of the following factors:

1. Route and terminal forecasts: Bad weather may have to be avoided.

2. Terrain clearance: Some parts of the world have mountains higher than a comfortable cruising level for light aircraft.
3. Controlled airspace: Some may not be entered under any circumstances.

While planning the route these factors will have to be taken into account and, if necessary, the ideal shortest distance between two points will have to be modified. Often such detours around, for example, prohibited areas, add surprisingly little to the number of miles that have to be flown.

Route selection is, like most skills, one that develops with experience.

Route Forecast

One of the first steps to be taken before planning the flight is to obtain a weather forecast.

At airfields with a teleprinter terminal route forecasts are readily available and Chapter 3 of Volume 2 explains how to read the printouts which are frequently updated. Information available to pilots is comprehensive. Pilots may obtain this by using the Automatic Telephone Answering System. Details obtained may be entered on suitable forms. For example, in the UK there is the CAA MET 0–7 form which has been devised for the General Aviation Visual Flight Forecast Service (described in Chapter 3, Volume 2).

In addition to forecasts pilots may obtain **Actuals**, reports of existing weather conditions at major airfields and the weather expected in the near future. These can be obtained on the **VOLMET** frequencies in the UK and most of Europe. When visiting minor airfields it is first necessary to check their weather by telephone.

The area forecast may indicate an active cold front lying across part of the intended route and it will then be necessary to plan around it, taking into account the direction of the front's movement.

Terrain Clearance

Most flying accidents that occur en route are caused by flying into high ground that is obscured by cloud. Pilots without an instrument qualification should not be flying in cloud; when high ground is in the vicinity such behaviour is foolhardy in the extreme.

Pilots with an instrument qualification often do spend a high proportion of their flying time in cloud and to guard against risk of flying into unseen high ground their flight plans will include a **Terrain Clearance Altitude** which has a safety margin between it and the highest ground within 30 nm on either side of the route. The safety margin varies from 1000 feet for flights up to 5000 feet and 3000 feet while flying above 20,000 feet. Although these procedures are primarily intended for instrument flight the Terrain Clearance Altitude concept is of value to all pilots. The map should be studied carefully, all high ground and man made structures (television masts, etc.) noted for a distance of at least 20 nm on each side of track, and 1000 feet added to arrive at a **Safety Altitude** (the VMC equivalent to the instrument flying Terrain Clearance Altitude previously mentioned).

Safety altitude should be regarded as the minimum altitude to which the pilot may descend in the event of a lowering cloudbase.

Selection of Cruising Level

Ideally light aircraft without turbochargers are most fuel efficient when flown at 8000 to 12,000 feet, maximum range occurring at the higher figure, a faster cruising speed being attained at the lower altitude. Such cruising levels are not always available to the pilot because of weather and controlled airspace considerations.

For early training flights over relatively short distances altitudes of 2000 to 4000 feet are usually selected but to remain VMC it will be necessary to keep clear of cloud (1000 feet below if the flight is above 3000 feet) so a cruising level between safety altitude at the lower limit and adequate vertical clearance from cloudbase at the upper limit will have to be chosen.

Controlled Airspace

The reasons for, and various categories of **Controlled Airspace** are described in Chapter 1 of Volume 2. For the purpose of this exercise students should be reminded that some controlled airspace extends from ground level to a stated upper limit:

1. **Control Zones:** These may protect one or more aerodromes and most are under permanent IFR (Rule 21) although Special VFR Clearance can often be obtained under suitable weather conditions.
2. **MATZ: Military Aerodrome Traffic Zones** surround service

airfields. When a number of adjacent MATZs overlap they are said to form a **Clutch** and one airfield will be designated the controller for aircraft requesting transit permission.

3. **SRZ: Special Rules Zones** surround some of the more important civil and military aerodromes. Some of them are outside the airways system.

Although it is often possible to obtain permission to fly through items 1, 2 and 3 there are occasions when heavy traffic may preclude this and the training flight will have to be planned to avoid such areas.

Some controlled airspace starts at a designated altitude and extends to an upper limit:

1. **TMA: Terminal Manoeuvring Area,** or more usually known as **Terminal Control Area.** These cover areas of intense air traffic activity, e.g. London TMA, Manchester TMA, Paris TMA, New York TMA etc. Their lower and upper limits are shown on maps and charts.
2. **Airways:** Joining the TMAs and Control Zones are the **Airways,** aerial corridors 10 nm wide which extend from a specified altitude to an upper limit. The lower limit steps down as the airway nears a TMA or Control Zone.

While it is usually possible to route below items 1 and 2, permission may be obtained to cross through an airway by prior arrangement in VMC. A radio call 10 minutes before entering the airway is required and aircraft granted permission are usually expected to cross at a radio facility.

Danger, Prohibited and Restricted Areas

Certain areas, many of them used for military purposes but some designated for the purpose of preserving wild life, are marked on maps and charts. **Danger** and **Restricted Areas** may be either **Permanent** or **Notified** (hours of operation for these are published in AERAD and Jeppesen Charts as well as the various state publications such as the UK *Aeronautical Information Publication* (formerly *Air Pilot*).

When planning the route these areas will have to be avoided unless the flight is at a time when the Danger/Restricted Area is inactive. Aircraft may not fly through **Prohibited Areas.**

Selection of Alternate Airfield(s)

Contingency plans must always be made to cater for an unexpected weather deterioration at the destination airfield. Although weather forecasting is greatly improved compared with the standards of fifteen or even ten years ago in some parts of the world the onset of fog can be balanced on a knife edge; a change in dew point or temperature can rapidly turn a clear sky into very poor visibility.

One or more **Alternates** (alternative airfields) should be selected so that a diversion can be made in response to an actual weather report from the destination which is below limits for the pilot and/or the aircraft. The alternate could be to left or right of track or a suitable aerodrome may be overflown on the way to the original destination. This would have the advantage of being a known quantity so far as weather is concerned.

Preparation of Maps

Of the three scales in common aeronautical use (quarter million, half million and million) the 1:1,000,000, being of small scale, presents insufficient detail for relatively slow flying light aeroplanes whereas the 1:250,000 (quarter million) probably offers too much, particularly in built up areas where many roads and railways converge.

The 1:500,000 (half million) series of aeronautical maps is ideal, offering a compromise between clarity and useful detail. Such maps may be purchased in plain paper or with a plastic coating which will take markings from a soft pencil or various felt pens. The advantages of plastic coated maps are their durability and the fact that tracks etc. may easily be erased. The other advantage of half-million maps *vis-à-vis* quarter-million is that each sheet covers a larger area and fewer maps need be carried in the aircraft.

Measuring Track

For the purpose of training imagine a simple flight from Bloggsberg to Newtown. There are no controlled airspace or Restricted/Danger/Prohibited Area problems and the highest ground is a range of hills 550 feet amsl (above mean sea-level). The **Required Track (Tr Req)** is drawn between departure and destination airfields and its bearing measured, using a meridian near the centre of track on which to place the protractor to allow for the fact that they converge towards the poles **(Convergence)**. In that way an average reading will be obtained (Fig. 55).

Distance to be flown (i.e. length of track) may be measured by using one of the calibrated plastic rules that are available with suitable scales for quarter-million, half-million and million series

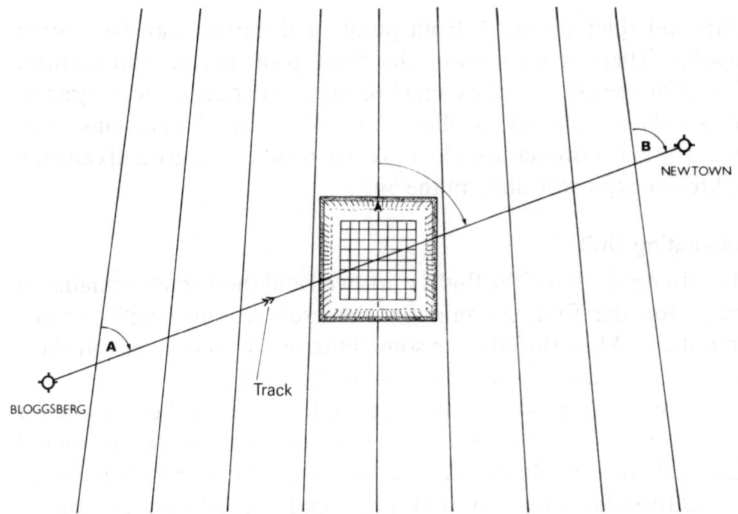

Fig. 55 Measuring an easterly/westerly track. Compare angle of meridian A with meridian B. By measuring track near its centre an average angle will be found.

maps. Alternatively, meridians at degree intervals are marked in minutes of latitude, each minute representing 1 nautical mile.

Track bearing and distance will be noted on the **Flight Plan** form which will be described in a later section.

Time/Distance Marks

When the track has been drawn it may be marked at intervals of 10 nm distances or, having calculated the ground speed, five or ten minute marks. There is no right or wrong method; much depends on the one found most convenient by the pilot concerned. Another alternative, one which has the advantage of simplicity and which does not depend on pre-calculated groundspeeds, which may or may not prove to have been correct, is to place a short mark across the track at the quarter, half and three-quarter position. By noting the time between each quarter an accurate up-dated **Estimated Time of Arrival (ETA)** may be calculated.

Highlighting Main Ground Features

Study the track and pay particular attention to any ground features of obvious map reading assistance. Typical would be a railway crossing the track, a lake or an airfield. Such features may be ringed on the map and their distances from point of departure can be written nearby. There is no reason why these principal ground features should not be listed on the **Flight Log** prior to departure, leaving a few lines between entries to allow for additional observations. Any preparation of this nature which is carried out on the ground can only lighten cockpit workload in the air.

Estimating Drift

It is often said that if the flight starts well and the aircraft remains on track for the first ten minutes the cross country will proceed smoothly. While this may be something of an over-statement there are obvious advantages to starting on the right foot.

To assist in checking that correct allowance has been made for wind conditions **Fan Lines** should be drawn at the departure end of the track at 5 and 10 degree intervals (Fig. 56). Some pilots like to repeat these fan lines at the destination end where they can be used to help the pilot estimate the number of degrees heading alteration that may be needed to reach the airfield.

Corrections are made using the '1 in 60' procedure described in

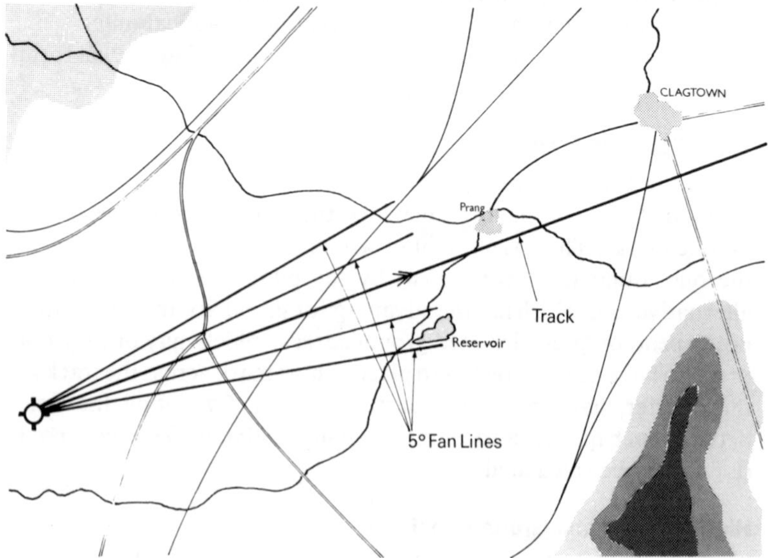

Fig. 56 Use of 5° fan lines for determining track error during the early stages of a cross-country flight.

Chapter 2, Volume 2, but the value of these fan lines is that they enable the pilot accurately to estimate the number of degrees **Track Error** when there is a departure from track in the early stages of the flight. If, for example, the aircraft is over the reservoir shown in the illustration the track error would be 9° to the right of track. A simple method of regaining track would be to alter heading 18° left and steer the new heading for the same length of time as was flown on the original heading. After regaining track there would be a further heading alteration, this time 9° to the right.

Folding Maps

To allow convenient handling of maps in the cabin they should be folded to show 20–25 nm on each side of track. When the track runs close to the edge of the sheet the adjoining map must be carried and folded so that the relevant area is shown. Some pilots like to join the two maps with paper clips so that they present the complete picture. With practice and a little patience good results can be achieved.

Warning

Features on the ground change as new motorways are built, woods are cut down and railways fall into disuse. Also controlled airspace alters from time to time and so do Danger, Prohibited and Restricted Areas.

Before devoting time to drawing in tracks and fan lines the date of the map should be checked. The instructor will also show students how to obtain up-to-the-minute information which may affect the route through the NOTAM and Air Information Circular services.

Radio Frequencies

Apart from the local communications frequencies, before the flight VHF frequencies for the destination aerodrome, the alternates and those of any sections of the air traffic control service relevant to the flight should also be included. Typical examples of these are:

1. Control Zones near the intended track, even when they will not be entered.
2. Any radar service likely to be of assistance.
3. Any airfields near track that should be advised of the proximity of the flight while passing nearby.

For the purpose of gaining a PPL a cross country flight based on map reading and pilot navigation *without* the assistance of radio navaids will form part of the qualifying requirements. Aids such as

VOR and ADF should therefore not be used. They should be regarded as 'there if needed' and only to be operated in the event of becoming lost.

Navigation and Fuel Calculations

Having prepared the map and measured the track it now only remains to calculate the missing information required for the flight. This may easily and quickly be accomplished with the aid of one of the **Navigation Computers** that are available with a wind scale and rotating screen on one side and a circular slide rule on the other. Use of the computer is explained in detail in Chapter 2 of Volume 2.

One of the first actions to be taken prior to a cross country flight is to obtain a weather report (explained on page 271) and this will include the wind velocity at various levels. When the one relating to the intended altitude for the flight has been selected the following calculations will be made:

1. *True Heading* (Hdg.T): This is obtained by applying the wind velocity to the track direction previously measured.
2. *Magnetic Heading* (Hdg.M): Arrived at by applying local magnetic variation to Hdg.T.
3. *Ground Speed* (G.S.): This is True Airspeed after it has been affected by the wind velocity. It will be shown on the navigation computer while True Heading is being obtained.
4. *Flight Time* When the ground speed has been found it is a simple matter to apply this to the circular slide rule and compute time of flight. If 'quarter marks' have been added to the track line, time to these can also be noted on the map.
5. *Estimated Time of Arrival* (ETA): This is found by adding the flight time to the time of departure. A word of warning here. The first part of the flight will include a climb to the cruising level at a speed lower than cruising. For this reason early training flights may start from overhead the departure airfield.

Calculating Fuel Requirements

Fuel consumption expressed in terms of gallons per hour will be listed against percentage power settings and pressure altitudes. The information can be found in Section 5 of the aircraft manual. If the aircraft is of American manufacture figures quoted will be US gallons (1 imperial gallon equals 1.2 US gallons). The conversion from Imperial to US gallons or litres is quickly handled on the circular slide rule (explained in Chapter 2, Volume 2) but for simplicity there is no reason why all calculations should not be carried out in US gallons unless it is flying-school policy to convert to Imperial units.

Normally training aircraft embarking on a cross country flight would start with full tanks. However, at a later date the pilot may fly a more advanced aircraft which cannot carry full fuel and a full cabin. It is therefore important to understand how fuel requirements should be calculated.

Apart from the obvious requirements of fuel for the taxi out, engine checks, take-off, climb, cruise and descent, certain contingencies must be allowed for and a typical fuel calculation would be made as follows:

Requirements

Bloggsberg to Newtown	122 nm
Ground speed	105 kt
Altitude	4000 ft

According to the aircraft manual fuel consumptions for the aircraft are:

Engine start, taxy, power check and take-off	0·9 gall
Climb to 4000 ft	1·0 gall
Still air distance to 4000 ft	6 nm
Time of climb	5 min
Fuel consumption at 75 per cent power at 4000 ft	7 gall/hr
Cruise time:	
122 nm less 6 nm climb =	116 nm
Ground speed 105 kt =	66 min
66 min flying at 7 gph =	7·7 gall

A summary of fuel required is therefore:

Engine start to take-off		0·9
Climb to 4000 ft		1·0
Cruise at 4000 ft		7·7
45 min reserve (diversion etc)		5·3
NOTE: For flights with no convenient alternative greater reserves may be required.	Total fuel req.	14·9 gall
	say	15·0 gall

Fuel for the landing phase is not separately accounted for because during the descent into the circuit power will be reduced for the last part of the cruise. However, high temperature has an adverse effect on fuel economy and most aircraft manuals recommend adding 10 per cent to the fuel requirements for each 10°C increase above standard temperature (15°C at sea level).

Weight, Balance and Performance Check

In most light trainers it is practically impossible to overload the aeroplane, even with full tanks and a full cabin. Likewise it is difficult to place such aircraft outside their centre of gravity limits. However, on the basis that one day the student may fly a more advanced aircraft that can be overloaded and also incorrectly loaded so that it is out of balance this aspect of pre-flight preparation must be understood.

The next consideration is airfield performance. Is there sufficient take-off run/distance for the aircraft? Here again, most flying training is conducted from adequate-size airfields and there is never any doubt about whether or not the airfield can provide sufficient:

(*a*) **Take-off Run**: This is the distance required to lift off.
(*b*) **Take-off Distance**: Distance required to reach a height of 50 ft above ground level.

The subject of weight, balance and airfield performance is fully explained in Chapter 8 of Volume 2.

ATC Clearance Flight Plan (CA48)

A detailed flight plan, using form CA48, must be filed when it is intended to fly in controlled airspace in IMC or at night, or at any time when the controlled airspace is under permanent **Instrument Flight Rules**. It must also be filed if the pilot wishes to make use of the **Advisory Service** which is provided in certain **Advisory Areas**. Furthermore, a flight plan is required whenever the route entails crossing from a UK **Flight Information Region** to another (see Chapter 1, Volume 2).

For most training flights there is no need to **File a Flight Plan** using form CA48 but the pilot must advise ATC of the intentions so that if he does manage to run out of fuel or become lost they will at least have some idea of where to look for a stranded aircraft in a field. ATC will require to know:

Aircraft registration
Destination
Number of people on board
Flight time and aircraft endurance.

This is known as **Booking Out**.

Flight Log Preparation and Check

Pads of flight log forms may be purchased and so can suitable knee boards which carry the form, ball points and a stopwatch.

There are a variety of these forms but in the main they are divided into two portions:

1. **The Flight Plan,** with spaces for pilot's name, aircraft registration, intended route and alternates and details of fuel, range and endurance. Wind velocities for different altitudes, expected temperatures, altimeter settings and radio frequencies are all noted in the spaces provided. Immediately below are spaces for a number of legs (i.e. from/to), the minimum altitude (safety altitude) tracks, TAS, Hdg.M, ground speed, distance, flight time and amount of fuel which will be used. This section of the flight plan will be filled in while computing the details mentioned earlier.
2. **The Flight Log** which is for use in the air. It has spaces for time, position, observations, Hdg.C (i.e. Hdg.M corrected with the aircraft's deviation card). There are also spaces for groundspeeds found during the flight by timing the aircraft over a known distance and converting the figures to knots with the aid of the circular slide rule. Nearby are boxes for ETA and fuel remaining.

A typical flight plan/flight log form is shown in Fig. 57 but others are available which offer more writing space when this is required.

Checking the Flight Plan

It is very easy, particularly in the early stages of training, to make errors while computing the flight plan. The wind may be applied in reverse direction, track could have been misread from the protractor and so forth.

Before starting the detailed work of measuring track, applying the wind and computing heading, groundspeed and the other details it is a good plan to look at the track, estimate its direction, relate the wind and decide whether drift will be to the left or right. An estimated heading should be written down along with an approximation of how long the flight should take. Such estimates can, with practice, be remarkably accurate but, more important, they provide a comparison for use with the computed flight plan as an insurance against having made a fundamental error.

In the early stages of cross-country flying the student's flight plan will be checked by the flying instructor.

In Flight

Having carried out all the preliminary checks on ATC restrictions, prepared the map(s), obtained a weather report and completed the flight plan as described previously the pilot may now walk out to the aircraft, carry out the pre-flight checks, start up, do the engine run-up and make ready for take-off.

ATC Clearance and Altimeter Settings

If the flight is of a local nature, or one not requiring the mandatory filing of a form CA48, having booked out the flight may start immediately the aircraft is cleared onto the runway.

Although it does not concern a pilot under training it is, perhaps, relevant to mention here that instrument qualified pilots leaving an airport for an airways flight usually receive an **Airways Clearance** over the radio while waiting to take off.

Altimeter Setting Procedure

Altimeter settings are described in Chapter 3, Volume 2 but in essence the pilot will take off using:

QFE: With this setting the altimeter will read aircraft **Height** above aerodrome level.

After take-off the setting will be changed to:

QNH: Each of the **Altimeter Setting Regions** over the UK will advise a QNH. When set on the altimeter it will indicate **Altitude** above mean sea-level.

If the flight is conducted above **Transition Altitude** the altimeter must be adjusted to **Standard Setting.** Transition altitude varies in different states. In the UK it is at 3000 feet over most of the country but within the London TMA, for example, it is 6000 feet. Above transition altitude the altimeter is set to 1013.2 mbs. and aircraft fly, not at heights or altitudes, but at **Flight Levels**. These relate to hundreds of feet, thus FL50 means that an aircraft should fly with its altimeter indicating 5000 feet. The procedure, which is akin to synchronizing watches, ensures that all aircraft flying above transition altitude are using the same altimeter setting.

The logic of this seemingly complicated arrangement is that when taking off and landing the pilot can set QFE and the altimeter will then show height above the airfield. For en route flying up to transition altitude or while climbing to the cruising level and descending towards the destination, QNH, when set, will make the altimeter read altitudes above msl. The value of this is that all high ground and tall structures such as television or radio masts are shown on maps with their height above sea level which, in other words, relates to QNH. Finally, when flying above transition altitude all aircraft will have the same altimeter setting, an important safety factor when ATC allocates flight levels to a number of aircraft flying in the same area.

Log Keeping

It is important that all observations written in the flight log include the time. There may be moments when two similar ground features are shown on the map, one perhaps 20 miles ahead of the other. By looking back at the time of the last feature noted with certainty and writing down the time when over the feature in doubt it is relatively easy to decide which is correct; one of the features might mean the aircraft is achieving a groundspeed of 110 kt whereas the other would have needed a groundspeed some 50 kt or more faster than the capabilities of the aircraft.

The flight log should be confined to brief notes. On no account devote too much time to the task since this can only result in a neglected lookout. Many pilots under training find that while writing in the flight log there is a tendency for the aircraft to wander off the required heading. More comprehensively equipped aircraft have an autopilot; light trainers do not. For pilots in such aircraft the answer to the problem is:

1. Accurate trim.
2. Avoid gripping the wheel/stick – that way trim is masked and there is a tendency to roll on bank while the head is down writing up the flight log.

FLIGHT PLAN

Pilot A.PENZER A/C G-ABCD Date ETD 11/10

	Route Clearance	Checked by
	RHOOSE – STAVERTON – ELMDON (LAND) ELMDON – RHOOSE	
VAR:	Alternate COVENTRY – BRISTOL LULSGATE	[illegible]

Fuel Carried	Consumption	Range	Endurance
17 Galls	5/hr	255 NM	3 Hrs

Alt.	W/V	Temp. °C.	IAS	QNH	Facility	Freq.	Facility	Freq.
2000′	120/20	+11°	83K	1008	STAVERTON TWR	122·9	BRISTOL HOMER	127·25
4000′	130/25	+7°	81K	1010	STAVERTON RADAR	125·65	BRECON VOR	116·3
					B'HAM APP.	120·5	RHOOSE NDB	363·5
					COVENTRY TWR	119·25		

From	To	Min. Alt.	Flt. Level	Tr. (T)	TAS	Hdg. (M)	G/S	Dist.	Time	Fuel
CARDIFF/RHOOSE	STAVERTON	1500		055°	85K	076°	75K	NM 53	42½	3½
STAVERTON	B'HAM/ELMDON	3000		025°	85K	048°	85K	37	26	2¼
B'HAM/ELMDON	CARDIFF/RHOOSE	2000		223°	85K	214°	81K	87	64½	5½
						TOTALS:-				

FLIGHT LOG

Time	Position / Observations / Messages	Hdg. (C)	New G/S	ETA	Fuel Left
11.15	AIRBORNE Climb to 3500′	076°	75K	12·02½	
11.20	S/H STAVERTON				
11.45	RIVER WYE TMG 057° A/H 4°P	072°	76K	12·02	
12.02	STAVERTON S/H ELMDON	050°	85K	12·28	13½
12·12½	RAIL/RIVER Cross on Track				
12·29	ELMDON Descending. Landed 12.38				
13·55	AIRBORNE Climb to 3500′				
14.05	S/H RHOOSE	215°	81K	15.09½	
14.38	ROSS 2NM STBD. T.E. 3°P A/H 6°S	221°	"	"	
14.51½	USK 1NM STBD.				6½
15.10	RHOOSE Descending. Landed 15.15				

SIGNATURE: A. Penzer

Fig. 57 A typical Flight Plan/Flight Log.

Use of VHF

The old saying 'see and be seen' is, to some extent, complied with by maintaining a constant lookout and by using the anti-collision beacon and/or strobe lights. However, when the route passes near the approaches or climb-out path of an airfield it is both prudent and a matter of airmanship and courtesy to announce that 'Golf-Bravo Charlie Delta Echo will be abeam your extended centreline at two-four (or whatever the time). Any known traffic?'

Likewise, the various radar services, civil and military, are able to offer valuable advice on other traffic, even when this is not in radio contact at the time. A typical message would be 'Golf Charlie Delta there is traffic in your ten o'clock position moving from left to right, height unknown'.

The extent to which the student will be expected to use VHF while on cross country will vary from school to school. During early flights all attention will be devoted to map reading, flying accurate headings and maintaining the flight log. Later in training use of VHF will be gradually introduced.

Position Confirmation

Map reading, like any other reading, whether it concerns notes on a sheet of music or just words in a book, requires practice before it can become fluent.

The degree of skill required to map read from the air varies considerably with the type of terrain. In open, featureless country with few railways the task is relatively easy. There is the railway and it is the only one for fifty miles. When the flight is over a part of the country crossed with railways and motorways, a profusion of towns, some of them over-spilling into one another, both ground and map can be very confusing.

Map Reading Techniques

The aim should be to read from map to ground anticipating major features that were ringed before take-off. Occasionally a seemingly important feature may appear near the aircraft but it is not shown on the map. It is at times such as this that an inexperienced pilot may feel unsure of position, yet the reason could simply be that the motorway, bridge or reservoir has only recently been built and it will not appear until the next series of maps is printed.

Ideally the aircraft should be flown slightly to the right of track so that ground features may be seen by looking along the left. No attempt should be made to map read every yard of the countryside, rather the aim should be to fly an accurate heading.

Except for industrial areas where there can be a maze of railway lines these can be of prime value to map reading along with coasts. Rivers, canals and lakes come a close second. Roads other than motorways can be confusing from the air since major and secondary roads look very much the same unless they are dual carriageways. Woods can be useful although subject to changes in shape and extent due to forestry activity. In some parts of the world airfields can be confusing because of their close proximity to one another but, in the main, these can provide useful **Pinpoints** while map reading.

The aim should be to use combinations of features. For example, Fig. 58 shows an aircraft approaching a road followed almost immediately by a railway line. A little to the left the two cross. In this case the required track passes over this road–rail crossing but the pilot is to the right of track. Had the railway line been crossed before the road the aeroplane would have been to the left of track.

Figure 59 is an example of two similar villages, both with a nearby railway line, each with the same road running through and another branching in an easterly direction. Which village? Fortunately, one has a nearby airfield so, in this case the aircraft is about to fly over Barmby-on-the-Moor.

Features may be seen on either side of the aircraft (hills, masts, towns, airfields, etc.) and these too are an aid to map reading.

Throughout the flight the map should be held so that the track

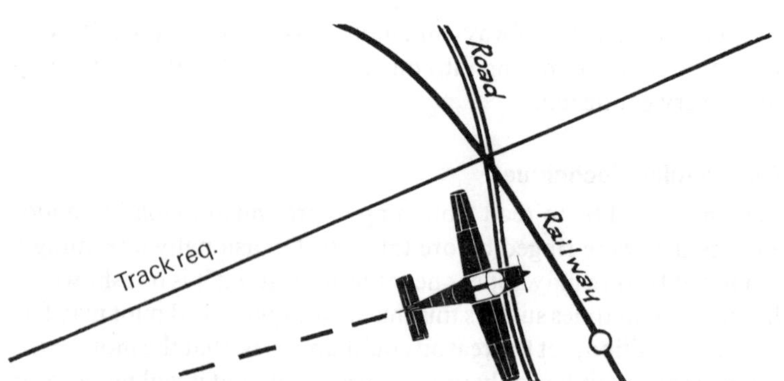

Fig. 58 Using a road/rail crossing to confirm position, left or (in this case) right of track.

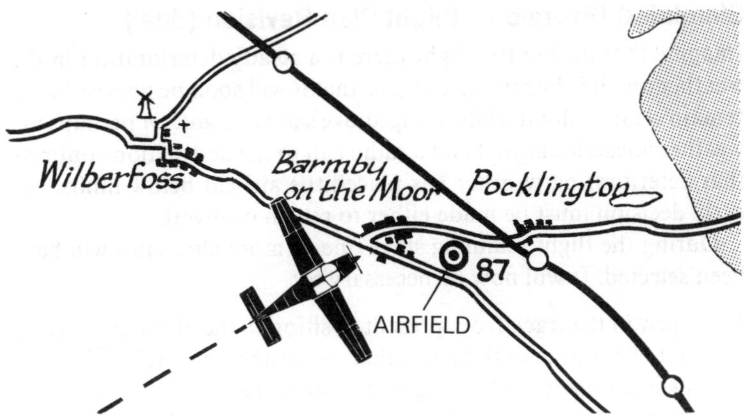

Fig. 59 Which village? Barmby on the Moor is confirmed by its adjacent airfield.

coincides with the direction of flight so ensuring that features on the map will appear in the same position relative to the pilot as those on the ground. This is called **Orientation**.

Use of Hills and Mountains as Ground Features

A prominent hill or mountain to one side of track can provide a useful means of confirming progress. However, what may appear to be a large hill or even a small mountain at ground level can look insignificant at an altitude of 3000 feet or more.

Checking Groundspeed and Revising the ETA

From time to time heading will have to be adjusted to maintain track as the aeroplane flies through regions of changing wind conditions. As the flight progresses and major pinpoints are noted in the flight log along with time overhead it will be possible to set distance flown against minutes flown on the circular slide rule and obtain a revised groundspeed. A particularly convenient time to do this would be if the quarter or half-way point was coincident with a suitable ground feature.

When a revised groundspeed has been obtained this can be used to up-date the ETA.

Simulated Diversion – Flight Plan Revision (dual)

Imagine that during the flight there is a steady deterioration in the weather and it is becoming obvious that it will soon be impossible to remain clear of cloud while flying above safety height. Or perhaps the weather ahead looks bad and a radio call to the destination confirms that deteriorating weather has placed the airfield below limits. An early decision must be made either to return or divert.

During the flight planning stage one or more alternates will have been selected. It will now be necessary to:

1. Draw in the track from present position to the alternate (having first checked over the radio that its weather is suitable).
2. Measure the track bearing and its distance in nautical miles.
3. Compute the new heading using the known wind velocity.
4. With the new groundspeed, compute time to the alternate.
5. Advise ATC of changed flight plan.

All of these functions are simple enough in themselves but they represent something of a task for a single pilot sitting behind the controls. However, this part of training will be conducted with an instructor and the student will be able to devote attention to revising the flight plan.

With practice it is possible to carry out items 1 to 4 while flying solo and there are available small calibrated rules which have a built-in protractor. These are ideal for in-flight navigation.

Experienced pilots will often make a diversion by estimating the new heading, taking into account the expected drift. Students will find rewarding a talent for estimating angles.

Engine Handling, Pitot/Carburettor Heat and Airframe Icing

Apart from routine checks on the engine readouts a close watch on fuel consumption is essential. Obviously the mixture will have been leaned in accordance with manufacturer's recommendations.

At intervals during the flight the carburettor heat should be fully applied for a few seconds then returned to cold. This must be done even on a bright, warm day. After returning the control to COLD if the engine RPM are slightly higher than they were before application of carburettor heat that is a sign of induction ice.

Pitot Heat

In the presence of moist air and low temperatures there will be a risk of ice forming on the pressure head or pitot head. This will cause the following instrument malfunctions:

1. Loss of the ASI information.
2. Fixed or incorrect altimeter reading.
3. Fixed or incorrect VSI reading.

Symptoms will vary according to whether or not there is a pressure head and static vent or a combined pressure/static (pitot) head.

Most light aircraft are fitted with an electric pitot heater and this should be switched on *before* entering an area of rain which may be cold enough to freeze. The outside air temperature (OAT) gauge will give advance warning of freezing risk. If ice has formed on the pitot head 30 seconds or more may be required to remove it. Here again, this is a case of prevention being better than cure.

Airframe Ice

The process of airframe ice accretion is explained in Chapter 3, Volume 2.

The weather forecast will, when appropriate, include a **Freezing Level**. Student pilots will not be exposed to icing risk but after gaining a PPL yesterday's student may fly without supervision and, to any pilot the message has got to be *on no account fly an aircraft in severe icing conditions unless it has full ice protection.*

Some types of ice (clear or glazed) will form very rapidly and the aircraft will soon become unmanageable, mainly because of the breakdown of airflow over the wings and tail surfaces.

Less severe but potentially dangerous is frost/ice on the windscreen. This can occur while descending from very cold air through moist, warmer regions. When the windscreen becomes obscured by ice the cabin heater must be turned fully on and the selector moved to SCREEN. Some aircraft have a direct vision panel let into the side window or left of the windscreen and this can, if necessary be opened for the landing.

Action when Uncertain of Position, with and without Radio

Pilots in urgent need of assistance (e.g. totally lost, low on fuel or caught in bad weather) may use the Distress and Diversion Service which covers the United Kingdom and is operated from both the London and the Scottish Air Traffic Control Centres. The service relies for its information on airfields equipped with Radar/VDF

(**Very High Frequency Direction Finding** is a ground based radio aid which is described in Volume 3 of the current series of *Flight Briefing for Pilots*).

For training purposes, pilots may practise **Emergency** calls using the prefix Pan (*not* **Distress** calls which entail the use of the prefix **Mayday**). Such exercises, which, radio traffic permitting, are welcomed by D & D, should start with a call on 121.5 MHz using the words:

'Drayton/Scottish Centre [according to whether the aircraft is north or south of latitude 55N] Golf Bravo Charlie Delta Echo [i.e. the aircraft callsign] request practise Pan.'

In a real emergency the message would start with the prefix 'Pan Pan Pan' transmitted on 121.5 MHz. Radio Telephony procedures are explained in Chalter 11 of Volume 2.

Lost Procedure without the availability of Radio

If, for any reason, radio is not available to the pilot well proven techniques for dealing with a lost situation may be used. There are a number of reasons why a pilot may become lost. An accurate heading may not have been steered, the direction indicator could be out of alignment with the magnetic compass (it must be re-set every 10–15 minutes to cater for **Wander**) or poor map reading could be the cause.

Unless a major blunder has occurred the aeroplane is unlikely to be farther from its correct position than a distance equivalent to 10 per cent of the mileage flown. This means that after flying 100 nm the aircraft will be within a radius of 10 nm from the intended position when lost, and *pro rata*. This is called a **Circle of Uncertainty** and its use confines the search area when lost to a proportion which is related to the length of the flight. The longer the time in the air since the last position of certainty, the greater the possible error and therefore the larger the circle of uncertainty.

When lost the following procedure should be adopted:

1. Log the time when 'Lost Procedure' is commenced. Check safety altitude.

2. Check the fuel contents and if there is little remaining prepare for a forced landing with power (Exercise 17b).

3. Check the compass reading and when applicable ensure the DI is correctly synchronized and functioning properly.

4. Do not wander about aimlessly but hold the pre-arranged heading, height and airspeed.

5. Establish a circle of uncertainty by drawing a free-hand circle on

the map around the present DR position which can be estimated from the time marks. The radius of the circle will be approximately one-tenth of the distance flown since the last pinpoint which was recognized with certainty.

6. Reverse the normal procedure by reading from ground to map, paying particular attention to important features which are bound to be shown on the map while looking out for prominent landmarks within the circle of uncertainty. The features may not be recognized immediately but they should be logged together with the time so that a chain of good pinpoints is built up, e.g. 11.05 large town, 11.12 river, 11.15 main railway, etc. In this way a picture can be built up which will fit within the framework of the circle of uncertainty.

When the aircraft is known to be more or less on Track but the destination fails to appear on time, the flight plan should be adhered to and the heading continued after ETA for a further 10 per cent of the calculated flight time. This is in accordance with the circle of uncertainty procedure.

It is possible that the circle of uncertainty may cover featureless country, when an alternative procedure is preferable. This makes use of a prominent feature which is known to be on one side of the aircraft or possibly ahead of it. A **Line Feature** (railway, river or canal, coastline, motorway, etc.) should be chosen so that a heading alteration can be made in the certain knowledge that it is bound to be crossed at one point or another. A localized pinpoint such as a town or an airfield is too easily missed when lost. To quote an example, imagine a flight from Exeter airfield to Southampton Airport. Mid-way the pilot becomes lost and decides to use the line feature method of establishing his position after going through items 1–4. In this case he knows that the south coast is on his starboard side although it is out of sight. By turning south he is bound to reach it although the point of interception will probably be unknown. Over the coast he should turn left and log features as they are passed together with the time and the heading of the coastline, until a series of landmarks has presented a picture which will enable him to relate his position with the map.

On some occasions it may be known that the aeroplane is somewhere in an area bounded by two converging railway lines which eventually meet in a large town. These line features will 'funnel' the pilot towards the town and it will merely be necessary to fly on while looking for a railway which is converging with the path of the aircraft. By following this railway and looking out for the other line as a confirmation, the pilot will be guided towards the town and a new heading to destination can then be estimated.

Finally the pilot who is lost should realize that, provided there is plenty of fuel and daylight, an airfield is bound to appear sooner or later in most parts of the country, even if it is disused, and a skilful forced landing with power makes any good-sized field an airfield for a light aeroplane.

While early pilot navigation exercises should be confined to map reading without the use of radio navaids, a pilot who becomes lost should not hesitate to enlist the help of the Air Traffic Control Service using the method previously described.

At the Destination

While approaching the destination the aircraft must be made ready for circuit joining by carrying out the following check:

	F R E D A
Fuel:	Sufficient and on correct tank.
Radio:	On correct frequency or frequencies.
Engine:	Temperatures and Pressures 'in the green'.
D.I.:	Synchronised with the compass.
Altimeter:	On correct setting.

Some ten minutes before entering the destination circuit the pilot will contact the airfield ATC and obtain the following information:

1. Airfield weather
2. Wind velocity
3. Runway in use
4. Airfield QNH and QFE.

In the absence of specific joining instructions the following circuit procedure will be adopted:

1. The aircraft will fly over the destination airfield at a height well above circuit traffic. Usually this will be 2000 feet on the QFE.
2. Having established which runway is in use and the direction of circuit traffic the pilot will fly to the **Dead Side** of the airfield, i.e. on the side opposite to circuit traffic. When there is a left-hand circuit this will entail positioning to the right of the runway when heading in the landing direction.
3. A wide, sweeping turn will be made in the circuit direction while the aircraft descends to circuit height. It is essential that the descent takes the form of a descending turn because this will ensure that there is no risk of letting down onto another aircraft;

in a turn one can see the let-down area – straight ahead it is possible to hit another aircraft.

4. At circuit height fly crosswind so that the aircraft passes the upwind end of the runway. This will ensure that there is no risk of conflicting with aircraft taking off since they would not have reached circuit height at that point.
5. Continue crosswind and integrate with the circuit maintaining a careful lookout, particularly while turning downwind in case other aircraft are already on that leg.

Booking In

At the end of the flight, after the aircraft has been parked, the pilot must **Book In**. Usually all that will be required is:

1. Aircraft registration
2. Point of departure
3. Next intended landing
4. Name of pilot.

There is, of course, usually a landing fee to pay.

Pre-flight Briefing

Exercise 18. Pilot Navigation

AIM — *To fly safely from one aerodrome to another.*

AIRMANSHIP — *Flight planning/route selection*
Safety altitudes
Radio facilities
Weight and balance/performance calculations
Suitability of weather

AIR EXERCISE

Flight planning/departure	*En route/diversions*	*Circuit joining procedure*
Weather information *Route selection/NOTAMS etc.* *Terrain clearance/safety height* *Controlled airspace, etc* *Selection of alternates* *Preparation of maps* *Listing radio frequencies* *Navigation/fuel calculations* *Weight and balance/performance check* *ATC liaison/clearance* *Prepare flight log* *Pre-flight checks* *Take off and climb to overhead* *Altimeter setting* *Set heading* *Check progress with fan lines* *Adjust heading if necessary*	*Lookout/routine radio calls* *Maintaining accurate heading* *Re-set DI at intervals* *Read from map to ground* *Note pinpoints and time of observation in flight log* *Check for carb. ice at intervals* *Check mixture/fuel state* *Check engine Ts & Ps* *Monitor track maintenance* *Use '1 in 60' method to adjust if required* *Check groundspeed and update ETA* *DIVERSION:* *Draw track to alternate and measure* *Check safety height* *Compute new heading, GS and ETA* *Advise ATC of changed flight plan*	*Radio call to obtain:* *Joining instructions* *Airfield weather* *Wind velocity* *Runway in use* *QNH & QFE* *If no special joining instructions:* *Fly over airfield at 2000 ft* *Position over dead side* *Let down in descending turn in direction of circuit* *Fly across upwind end of runway* *Careful LOOKOUT, then join circuit on the downwind leg*

Flight Practice

Flight Planning

Note: It is assumed that the aircraft documents (C of A, C of R, insurance, maintenance release, etc.) have been checked and found to be in order and that the aircraft is fuelled and within its weight and balance limits.

(*a*) Study controlled airspace, danger and prohibited areas and plan to fly around them. Determine the minimum safe height in relation to high ground. Radio facilities and frequencies en route are usually obtained at this stage.

(*b*) Obtain a route and terminal forecast and a wind velocity applicable to the height(s) intended during the cross-country.

(*c*) Draw the required track on the map together with 5° fan lines at each end. Estimate the track, distance, drift and time of flight and note on a separate sheet of paper for comparison with the calculated results later. Measure the track and enter its bearing and distance on the flight plan together with the following information: W/V, height, TAS, variation and any other details called for on the printed form.

(*d*) Compute Hdg.T and G/S and add this information to the flight plan. Calculate the time of flight and using the relevant variation convert Hdg.T to Hdg.M. Repeat this procedure for each leg of the flight when the cross-country involves more than one. Fold the map(s) to a convenient size. Compare computed results with your estimates.

(*e*) Study the track and either indicate time marks at 5- or 10-minute intervals or divide it into four equal parts. Ring important pinpoints and on the basis of the calculated G/S note the elapsed time to each of these. At this stage, select alternative airfields for use in the event of enforced changes of plan (weather, fuel shortage etc.).

(*f*) Calculate fuel requirements. Check weight and balance. Check performance (runway length needed).

(*g*) Complete the flight plan and book out.

(*h*) Carry out a thorough preflight inspection of the aircraft checking fuel contents, Rotating Beacon, strobes, pitot heat and presence of first aid kit and serviceable fire extinguisher.

(*i*) Convert Hdg.M to Hdg.C by referring to deviation card. (The difference between TAS and IAS can usually be ignored on light aircraft at heights in the region of 2000 feet, but at greater altitudes and high temperatures it must be taken into account.)

In Flight

(*a*) Climb to the pre-determined cruising altitude and so manoeuvre the aircraft that it passes slightly to the right of the airfield when on heading. Log the time S/H and by adding the calculated time of flight note the ETA.

(*b*) Steer the heading accurately and after 5 to 10 minutes flying, look for a good pinpoint. If this is on Track proceed with the pre-arranged heading; if not determine the track error by using the 5° Fan lines. Log the time over the pinpoint and note the elapsed time for use during the correction. Double the track error and A/H (alter Heading) by that amount towards track. When track has been regained, halve the alteration of heading so that the aeroplane maintains the required track. In the absence of a good pinpoint track will be regained after flying the first correction for the number of minutes taken to reach the position which was off track.

(*c*) Check the DI with the compass every 10–15 minutes and reset whenever necessary.

(*d*) Make radio contact when passing near airfields or as required.

(*e*) Do not map read all the time but concentrate on steering an accurate heading between the pre-selected ground features.

(*f*) Maintain the log by noting the time at each pinpoint. Log any alterations of heading.

(*g*) At the half-way mark check the time and if necessary revise the ETA.

(*h*) In the absence of joining instructions from control, when nearing destination, provided there is no high ground, reduce power and descend to 2000 feet while maintaining the airspeed constant. Complete the field approach checks F R E D A.

(*i*) Obtain landing instructions or when there is no radio fly over the signals area and check:

1. Landing direction.
2. Direction of circuit.
3. Obstruction or special signals.

On the 'dead side' of the airfield descend on a curved path to circuit height keep a good lookout for other traffic. Ascertain the airfield height from the map and level out at 800 or 1000 feet above it according to local rules.

(*j*) Join the circuit on the cross-wind leg close to the upwind end of the runway and complete the landing in the usual way.

(*k*) Report arrival to airfield control.

Simulated Diversion – Flight Plan Revision

(Note: *this exercise will be practised with a flying instructor in the aircraft).*

(*a*) Imagine the weather is deteriorating and it will soon be impossible to remain clear of cloud while staying above safety height.
(*b*) Draw a track between the present position and the alternate.
(*c*) Measure the track bearing and length. Note in the flight plan.
(*d*) Using the wind velocity compute True Heading and groundspeed, steer the new heading, log the time, then calculate flying time to the alternate. Note this information in the flight log along with Magnetic Heading and ETA.
(*e*) Check fuel state and safety height.
(*f*) Advise ATC of the revised flight plan and ETA.
(*g*) Join the circuit as before and book in.

Action when Uncertain of Position, with R/T *(to be practised by arrangement with D & D).*

(*a*) While on a cross country it becomes necessary to seek assistance to determine position and reach an airfield.
(*b*) If a ground station is already in contact advise the nature of the problem on that frequency.
(*c*) When R/T contact has not been made with ATC select 121.5 MHz and put out a PAN call (e.g. lost, low fuel, etc.). Give the following information:
   1. Aircraft callsign and pilot's qualifications.
   2. Type of emergency and assistance required.
   3. Approximate position based on the last one known with certainty.
   4. Flight level/altitude.
   5. Heading.

(*d*) Follow instructions given by the Emergency Controller.

Action when Uncertain of Position, without Radio
(2 methods)

(*a*) Log the time 'Lost Procedure' is commenced.
(*b*) Check the fuel and decide if there is sufficient to continue flying: if not prepare for a forced landing with power.
(*c*) Check the DI and if it differs from the correct heading ascertain the direction and the amount of error, thus indicating the locality of the Tr. Req. in relation to the aeroplane. Compare the DI with the compass and make sure it is functioning correctly.

(*d*) Throughout these checks maintain the pre-determined heading.
(*e*) If Compass Heading and DI are correct and the position of the aircraft in relation to track is unknown adopt one of the following methods.

1. Circle of Uncertainty

(*a*) Estimate the present DR position had the flight continued as planned, either with reference to the time marks or by calculating the distance flown since the last certain pinpoint.
(*b*) Using the DR position as a centre draw a circle with a radius which is roughly one-tenth of the distance flown since the last known position. The aircraft is likely to be within that area. Check minimum safe altitude.
(*c*) Look for any prominent features within the circle of uncertainty and attempt to find them on the ground. Log any major pinpoints seen. Continue reading from ground to map and endeavour to build up a sequence of features which will fit within the circle of uncertainty.
(*d*) Proceed until a feature is recognized without doubt, when it will be possible to estimate a heading to destination, or failing this until an airfield appears.

2. Line Feature Method

(*a*) Study the map on either side of Track and ahead of the farthest possible position of the aircraft. Endeavour to find a good line feature – a river, canal or main railway line will do provided it extends for some distance.
(*b*) Alter heading so that the line feature will be intercepted at approximately 90° and continue flying until it appears.
(*c*) Dependent upon the location of the feature in relation to the destination, turn and follow the line, logging all important pinpoints together with the heading of the aircraft at the time. Build up a chain of features which will eventually fit the map.
(*d*) Continue until a position is recognized without doubt and estimate a heading to destination, or failing this until an alternative airfield appears.

Exercise 19

Instrument Flying (Appreciation)

Background Information

Throughout this book reference has been made to instrument indications. However, all the exercises so far described will have been demonstrated by the instructor and practised by the student in VMC. The importance to a pilot of outside visual references will by now be clear yet, in many countries, experience shows that when a Private Pilot's Licence is gained and confidence grows this very fundamental truth of flying is sometimes forgotten. Many are the incidents, recorded and unrecorded, where inexperienced pilots have continued flying into weather conditions that demanded the use of instruments, sometimes with very serious consequences. Often these situations are the result of 'familiarity breeds contempt'; over-confidence born of the mistaken belief that because the aircraft is fitted with a full flight panel and various radio aids 'I could cope in an emergency'. The purpose of this exercise is to impress upon the student pilot that, without proper tuition and adequate practice, solo instrument flying is not to be attempted. On no account should it be regarded as a do-it-yourself activity. Equally an acceptable level of instrument flying proficiency is within the reach of all students capable of attaining a Private Pilot's Licence.

So important is instrument flying to the operation of aircraft under modern air traffic conditions that a separate rating must be gained by pilots wishing to fly within certain categories of controlled airspace and in IMC. These are:

The Instrument Rating

The IMC Rating (recognized in the UK only)

The instruments are explained in Chapter 9, Volume 2 of the new series *Flight Briefing for Pilots* (and Volume 2 of the old series). The IMC Rating is covered in the new series Volume 3.

This chapter is confined to a brief explanation of the technique of instrument flying. In the UK 4 hours of air instruction are set aside for the purpose of giving the student a working knowledge of the subject, how to interpret instrument readings and methods of flying without outside visual references. At the same time this part of the course is intended to impress upon the student the dangers of untutored instrument flight in low cloud or poor visibility.

The Instrument Panel

By international agreement the main instruments used for determining an aircraft's performance and attitude are grouped together in a standard presentation known as the **Basic 'T' Flight Panel**. Until more recent times it was the practice to outline the Basic 'T' instruments in the manner shown in Fig. 60. Although the various instruments can in themselves differ very considerably in general design and the amount of information presented, the basic 'T' layout is, in essence, common to all modern civil aircraft irrespective of size. However, the introduction of electronic flight decks has started a move towards totally new presentations.

Function of the Instruments

By now the information provided by the various instruments will be understood by the student pilot. A demonstration was included in the **Straight and Level Flight** exercise (page 65) and in fact reference has been made to the instruments in most of the exercises that comprise the PPL syllabus. However these references were made for the purpose of:

(*a*) familiarizing the student with the instruments and

(*b*) as a cross check on outside visual references for the attainment of accuracy.

So far no attempt will have been made to fly with sole reference to the instruments and this is quite another matter, bringing into play skills that are the result of correct instruction and practice under supervision.

With the rapid development of radio navaids, each with their own information presentation, a situation was reached where it became difficult to accommodate all the instruments on the panel. Thus was born the **Integrated Display**, in which basic attitude and heading instruments became combined with radio navigation readouts. At first these instruments were confined to large transport and military aircraft but now they may be seen on light, single-engine tourers. However, most flying school trainers are confined to simple instruments which provide **Raw** information (i.e. attitude, heading

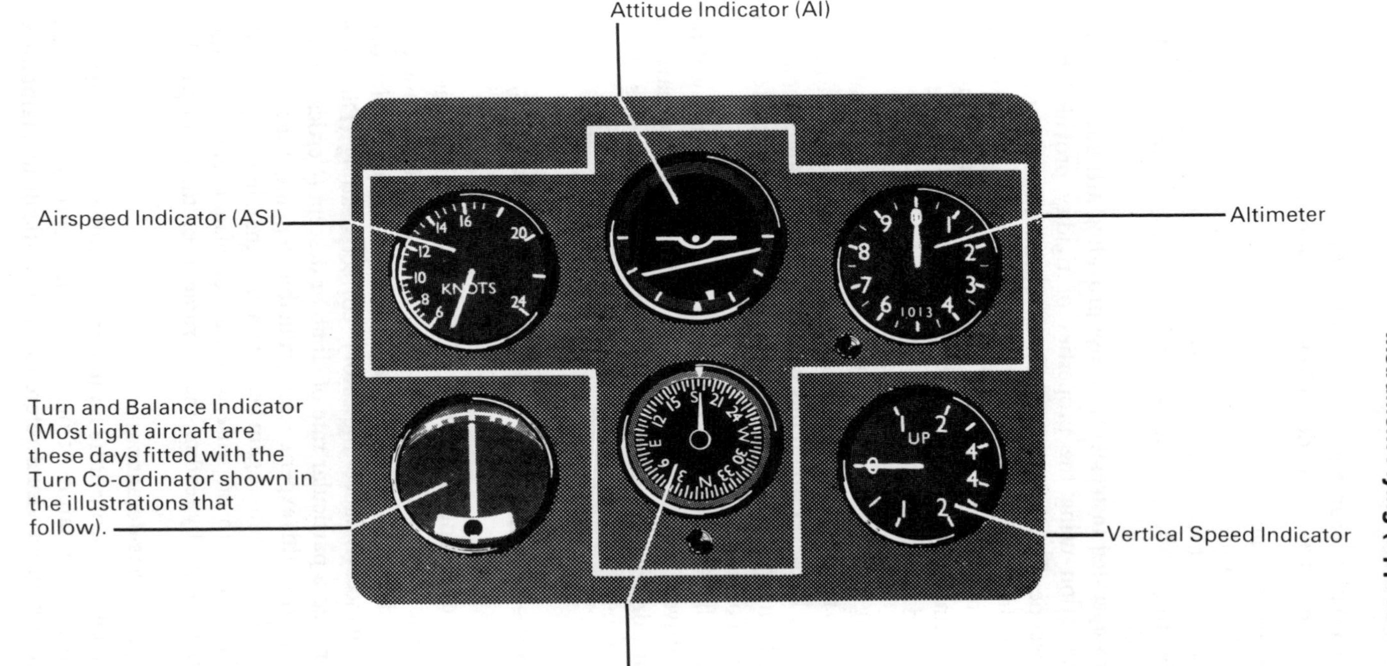

Fig. 60 'Basic T' Flight panel. More modern instrument panels rarely have the 'T' outline and Figs. 61 to 65 show the type of instruments currently fitted in light aircraft.

etc.) and a typical example is shown in Fig. 61. The instruments on the flight panel may be divided into two categories:

1. **Pressure Instruments**. Those which provide their information by sampling static or pressure air (or both) from the pitot head or in some aircraft the pressure head/static vent. The Airspeed Indicator, Altimeter and Vertical Speed Indicator are in this category.
2. **Gyro-operated Instruments**. Those providing attitude and turn information using the properties of rigidity provided by a gyroscope. This may be driven by vacuum usually from an engine-driven pump, or by electric power. The Attitude Indicator, Direction Indicator and the turn portion of the Turn Co-ordinator or Turn and Balance Indicator are the gyro-operated instruments.

By reading the instruments, singly or in combinations which vary according to the task in hand, the aircraft may be flown to fine limits of accuracy. However, to do this while in cloud demands complete trust in the instruments and the ability to relax so that there is no tendency to over-control.

Usually, the Attitude Indicator and Direction Indicator will be vacuum driven in light aircraft and it is the practice to have an electrically driven Turn Indicator as an insurance against loss of the vacuum system.

Importance of Instrument Indications versus Reactive Sensations

At this stage of the course a student will have used the instruments as an aid to achieving precision flight, e.g. turns at the correct angle of bank (Attitude Indicator), turns at a particular rate (Turn Indicator), correct balance (Balance Indicator), correct heading (Direction Indicator) and a particular rate of climb or descent (Vertical Speed Indicator). While this experience is a valuable stepping stone towards instrument flight it must again be emphasized that, in itself, this will not enable a pilot to dispense with outside visual references and control the aircraft by reading the instruments on the flight panel.

Physiological Considerations

Human limitations constitute an important problem of instrument flying. Such is the nature of our balance mechanism and nervous system that without visual references it is difficult to maintain a required attitude unless full use is made of muscle response. For example with both feet on the ground an upright position may easily be maintained with the eyes closed. Raise one foot off the ground and the task becomes more difficult. Spin around several times with the eyes closed, then try standing on one foot. Most people find this impossible because the balance mechanism has been disturbed.

Likewise in an aircraft a pilot denied outside visual references due to cloud or poor visibility would be unable to hold a required heading or maintain level flight unless capable of interpreting the instruments. And this in itself is only part of the problem. Carry out a series of steep turns in one direction then level out and the nervous system will insist that the aircraft is turning in the opposite direction although it is in straight and level flight. Experienced pilots learn to put their trust in the instruments before them. On the other hand a pilot with few flying hours is a natural prey to his own nervous system and the very instincts that may protect him on the ground could well put him at risk in the air.

This chapter and the air exercise that follow are not intended to frighten, rather they are aimed at instilling a little realism into the minds of those otherwise admirable students of limited experience who, on occasions, may perhaps suffer from a little overconfidence.

Provided the exercise has been demonstrated correctly by the flying instructor and accepted in the correct spirit by the student pilot the end product should be a determination to obtain an instrument qualification at the first opportunity after gaining a Private Pilot's Licence.

The Scan

It is helpful to regard the Attitude Indicator as the **Master Instrument** since it provides information on aircraft attitude in the pitching and rolling plane. It also has graduations giving the number of degrees bank and sometimes pitch angle.

Information presented by the Attitude Indicator is supplemented by the other instruments in the following manner.

Pitch Attitude. When a required performance (cruise, climb or descent) is set up by adopting the appropriate attitude with reference to the Attitude Indicator, fine adjustment will be made with the aid of readings from the ASI, VSI and Altimeter.

Bank Attitude. Control of heading for the various modes of flight (left or right turn; straight flight), will be achieved by adopting the appropriate angle of bank, or wings – level attitude on the Attitude Indicator, supplemented by the Turn Indicator (for rate of turn), Ball (for balance) and Direction Indicator (for heading).

It will now be apparent that the aircraft cannot be flown by watching a single instrument. Furthermore, when the additional factor of height is introduced, involving both the VSI and the Altimeter, it becomes obvious that a systematic method of devoting shared attention to the instruments is essential to good instrument flying.

The technique is to move the eyes back and forth, to and from the Attitude Indicator, involving those instruments that are relevant to the mode of flight at the time. For example, to climb on instruments the pilot would add power, keep the wings level (AI), prevent yaw (ball), raise the nose to the approximate climbing attitude (AI), adjust to achieve the correct airspeed (ASI) and so forth. These techniques will be described shortly but the method of reading the instruments just outlined is known as **Selective Radial Scan**, selective because only those instruments relevant to the mode of flight are included in the scan at that time.

Basic Instrument Flying Technique

Before describing straight and level flight, climbing, descending and turning on instruments it is first necessary to talk in general terms about techniques which can, if properly implemented, reduce pilot workload while flying on instruments.

A. Accurate use of the trimmer(s) will materially ease the task and save having constantly to correct a too high or too low situation.
B. Only if the wheel/stick is lightly held will best use be made of accurate trim. Grip the controls and the natural resistance to any departure from the required elevator position will not be felt.
C. All changes from one flight mode to the next must be introduced *gradually*. When making a turn, for example, gently roll on bank and be prepared to stop it increasing. Never adopt steep bank angles or extreme nose-up/nose-down attitudes.

Straight and Level Flight (Fig. 61)

Having set the correct power and accurately trimmed the aircraft it will be necessary to maintain:

1. The required altitude.
2. The required heading.

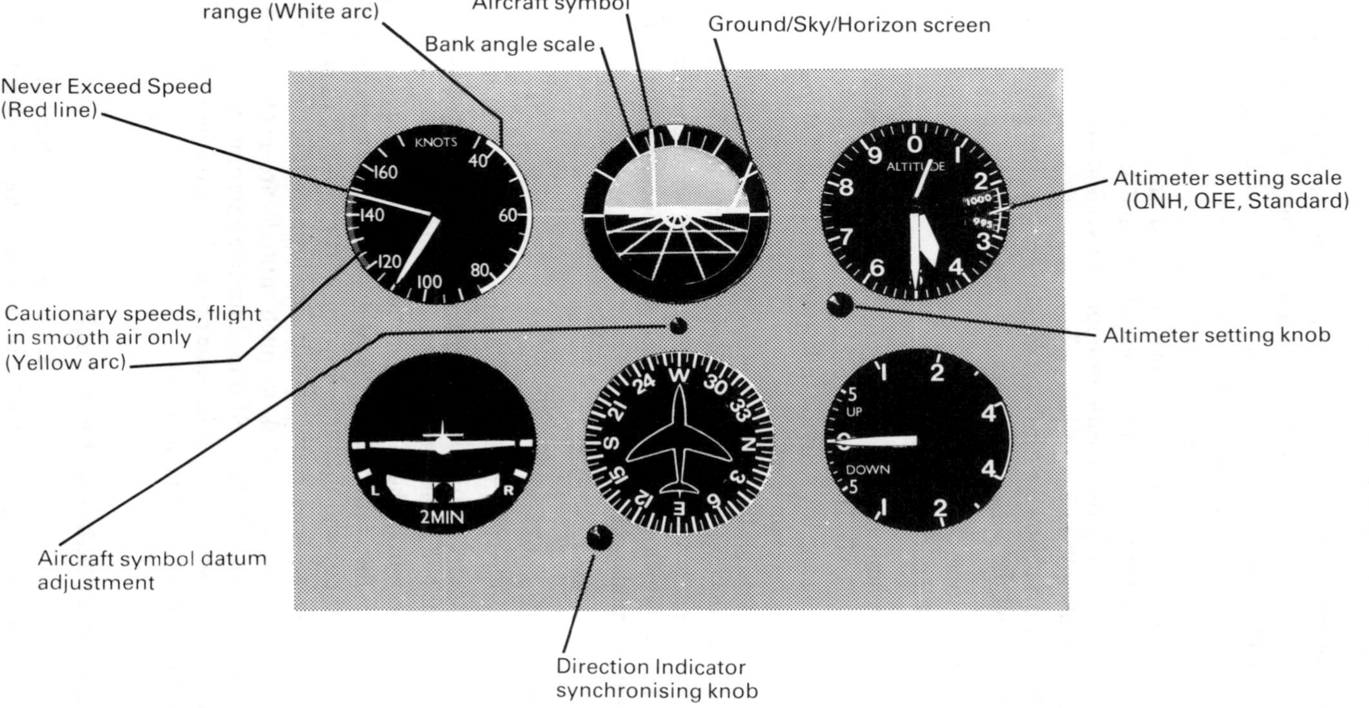

Fig. 61 Straight and Level Flight with the aircraft heading 270° at 110 kt and 4500 ft.

Provided the aircraft is correctly trimmed and the power setting is suitable for the speed required short term warnings of a departure from height/altitude/flight level will be indicated on the VSI. This is a very sensitive instrument but in the presence of several up/down departures from level flight, induced in quick succession, it can become out of step with reality due to a small degree of instrument lag. It is a valuable instrument nevertheless and more advanced aircraft are usually fitted with an improved version known as an **Instant Reading VSI**.

Over a short period the Altimeter will show a persistent gain or loss of height and the pilot must check:

(*a*) Power setting.
(*b*) Trim at the correct airspeed.

Heading maintenance is almost entirely dependent upon keeping the wings level (AI). The slightest bank angle will start a turn towards the lower wing and the Attitude Indicator must be utilized to ensure that the wings remain level.

It is also possible that, due to tension, the pilot may be holding on rudder making it necessary to apply opposite aileron in an attempt to keep the wings level. Such an incorrect technique would be indicated by the ball which would be displaced in the opposite direction to unwanted rudder pressure.

Climbing (Fig. 62)

As power is added for the climb the Balance Indicator (ball) must be kept in the centre with rudder applied to counteract the effects of engine torque and additional slipstream. When the propeller rotates clockwise (seen from the cabin) right rudder will be required.

The nose is then raised, using the AI to establish the approximate climbing attitude and the same instrument is monitored to ensure that the wings are level. A few seconds are allowed for the airspeed to settle, then fine adjustments to pitch attitude are made before trimming the aircraft. The ASI is used for fine adjustments.

The VSI will settle and then indicate rate of climb and the Altimeter will show a constant gain in height.

Return to Straight and Level Flight

At the required altitude the nose is lowered with reference to the AI. As cruising speed is approached the throttle is brought back until the

Fig. 62 Climbing at 70 kt. The aircraft is heading 045° and gaining height at 750 ft/min. Note position of the Attitude Indicator. Because the pilot is holding on insufficient right rudder the ball is displaced to the right.

required RPM are indicated. Throughout the various stages it will be necessary to monitor heading (DI), wings level (Turn Indicator and AI), and height/altitude/flight level (Altimeter).

Finally the aircraft must be carefully trimmed.

Descending (Fig. 63)

From straight and level flight a powered descent is started by reducing power to the RPM estimated to provide the required rate of descent. Yaw is prevented (ball) with rudder and the wings are held level (AI).

As the airspeed nears that required for a powered descent the nose will drop slightly below the horizon (AI). At this stage the trim should be adjusted to hold the required speed. The rate of descent can now be checked (VSI). If it is too high a little more power should be added and the nose raised slightly. Speed and altitude must be constantly checked (ASI and AI). When the descent rate is not high enough power will have to be reduced slightly and the nose lowered a few degrees, again holding a steady airspeed.

Fig. 63 Descending at 70 kt. Heading is 030° and rate of descent is 500 ft/min. Note position of the Attitude Indicator.

The return to straight and level flight is made in much the same way as from the climb.

Turning (Fig. 64)

All instrument turns should be limited to no more than a Rate 1 and there are times when a Rate ½ turn is useful.

Starting from straight and level flight bank is slowly applied, using the Attitude Indicator to establish the correct angle which, for a Rate 1 turn may roughly be calculated as follows:

$$\text{Indicated Airspeed} \div 10 + 7$$

Thus at 110 kt IAS the approximate angle of bank would be:

$$110 \div 10 = 11 + 7 = 18°.$$

Bank angle is adjusted to bring the Turn Indicator needle onto the Rate 1 mark (too high a turn rate – take off bank: too low a turn rate – add bank) and the VSI/Altimeter must be checked to ensure a steady height. The ASI will show a slight decrease in airspeed and the Balance Indicator should be centred. The Direction Indicator will be

Fig. 64 Turning to the left. Note position of the Attitude Indicator.

showing a constant change in heading, readings increasing while turning right, readings decreasing when turning left.

Return to Straight and Level Flight

Some 10° before the new heading is reached the wings should gradually be levelled with reference to the AI. There is often a tendency for height to be gained at this stage because, during the turn the usual back pressure on the wheel/stick will have been applied to maintain height. This must be relaxed as the wings roll level.

It may be necessary to make small heading corrections after rolling the wings level.

Climbing and Descending Turns (Fig. 65)

These are achieved in much the same way as other turns, using the same techniques to ensure a smooth transition from one mode of flight to the next.

Fig. 65 Climbing and Descending Turns on Instruments.

Engine Instrument Scan

Although the foregoing text has been devoted to the Flight Instruments those for the engine must not be neglected.

While instrument flying periodic checks to confirm that all temperatures and pressures are normal should be made. And if instrument flying has been made necessary because of weather conditions most likely the aircraft will be in cloud. Then, regular checks for carburettor ice take on even more importance.

Finally, it should be remembered that RPM settings are affected by changes in airspeed, consequently, when a particular power has been set on the throttle RPM must be checked and, if necessary, adjusted after the airspeed has settled to the required figure.

Instrument Limitations

Although there is a detailed explanation of the instruments and their limitations in Chapter 9, Volume 2, here they are given in brief terms.

Airspeed Indicator. Other than a need to convert IAS to TAS the instrument is free of temperament. However, when changing airspeeds time must be allowed for the aircraft to accelerate or decelerate; like any vehicle it has inertia.

Altimeter. At light aircraft rates of climb and descent the altimeter will give more or less instant readings. At high rates of climb or descent the altimeter can lag by several hundred feet.

VSI. Can give erroneous readings when the aircraft is subjected to rapid departures from level, up and down, over a short period. However the instrument settles quickly.

Direction Indicator. Since a number of factors or errors cause the instrument to wander over a period of time it must be re-synchronized with the magnetic compass at intervals of 10–15 minutes.

Turn and Balance Indicator. This simple and reliable instrument is of particular value because its readings can be used when other gyro-operated instruments may have become inoperative following an extreme aircraft attitude.

There was a time when this was the only gyro instrument and, with practice, it is possible to fly the aircraft accurately with the aid of no more than a Turn and Balance Indicator, Altimeter, ASI and Magnetic Compass.

This is known as **Flight on the Limited Panel**. It is not a requirement for the PPL course. In most light aircraft the Turn and Balance Indicator has been replaced by the Turn Co-ordinator presentation illustrated in Figs. 61 to 65.

Attitude Indicator. Apart from slight bank and pitch inaccuracies which occur during a turn through 180°, and which correct themselves during the next 180° if the turn is continued, the instrument has no limitations other than its Toppling Limits which are now described.

Toppling Limits of Instruments

Because of the mechanical problems inherent in providing a gyro with complete freedom of movement while at the same time ensuring a supply of air to drive it, all early gyro instruments and indeed some modern examples have limiting stops to keep the rotor within its air jets. When some extreme aircraft attitude causes the gyro to come up against its stops some instruments will behave in a manner similar to a child's spinning top in the last stages of slowing down. The resultant movement of the gyro will cause the Attitude Indicator to fluctuate, random fashion, from maximum left bank to maximum right, at the same time indicating violent changes in pitch attitude, while the Direction Indicator will spin intermittently. These indications will continue after the aircraft has regained straight and level flight and while the DI may be re-set with its heading knob an Attitude Indicator will require 10 minutes or more for the automatic corrective device to stabilize the instrument. When, following some extreme flight attitude, gyro instruments behave in the manner described they are said to have **Toppled** and the angular limits where a gyro reaches its stops are known as **Toppling Limits** (expressed in degrees). Although some modern instruments have complete freedom of movement throughout 360° in both the lateral and longitudinal planes many do not and since toppling is the result of allowing an aircraft to attain some unusual attitude an out-of-hand situation could very well result in loss of indications from the Attitude Indicator and the Direction Indicator, a further danger of untutored instrument flying.

While it is not intended at this stage that the student should have a detailed knowledge of instruments it should be noted that the turn-needle gyro has freedom of movement in one plane only. Like the other gyro instruments, it has limiting stops, but when these are reached the gyro does not topple. In other words the Turn-and-Balance Indicator will continue to provide valuable attitude information after the Attitude Indicator has toppled. While this is one of the values of the instrument, interpreting its indications and translating these into control movements calls for a higher level of skill than does instrument flight with the aid of an Attitude Indicator. The task is not one that should be attempted by pilots without previous experience of flying on the **Limited Panel**.

Instrument Flying Training

To a considerable extent the cost of gaining skill in both basic instrument flying and the various radio procedures can be minimized by making use of a simulator. Those used for airline training are very realistic to the smallest detail but there are now some excellent simulators which are intended for light and general aviation pilot training.

In the air it will be necessary for the student's external vision to be curtailed so that all flying must be with reference to the instruments. The two methods in widespread use are:

(*a*) **Hoods**: These take the form of a cap which has a large hood or peak. When worn by the student this obscures the view outside, concentrating the vision onto the instruments.
(*b*) **Screens**: Slatted screens, so arranged that the instructor or safety pilot can see out of the aircraft although the student is unable to, and must therefore rely entirely on the instruments.

When under training there will, of course be a flying instructor in the aeroplane so the student will be absolved from the usual need to maintain a keen lookout. During practice a qualified pilot who is not a flying instructor may act as **Safety Pilot** for the purpose of avoiding collision and controlled airspace.

In the UK training for an Instrument Rating may only be conducted at certain approved flying training establishments and screens used during the test must likewise be approved.

Conclusions

The exercise described in this chapter is no more than an introduction to instrument flying. Pilots who intend making full use of their newly attained qualification are strongly advised to obtain an IMC Rating (at present confined to the UK). The training entailed represents an excellent stepping stone on the path to an Instrument Rating which is an essential qualification for any pilots intending to fly on a regular basis in the course of their business or profession.

Exercise 19. Instrument Flying (Appreciation)

AIM — *To develop the ability of reading the relevant instruments and controlling the aircraft in the absence of outside visual references.*

AIRMANSHIP
- *Height: sufficient for the exercise*
- *Airframe: configuration as required; screens/hood for student*
- *Engine: temperatures and pressures 'in the green'*
- *Security: no loose articles in cabin*
- *Location: not in controlled airspace, etc.*
- *Lookout: Instructor to act as safety pilot*
- *Suitability of weather*

AIR EXERCISE

Basic I/F techniques

Relax
Selective radial scan
Use of trimmer
During straight and level flight:
- *Wings LEVEL to hold heading*
- *Power/Attitude/Trim to hold altitude*
- *Power/Attitude/Trim to hold speed*
- *Remember Power & Attitude for different speeds*

From S & L to climb or descent:
- *Adjust power and adjust wheel/stick as follows:—*
- *Change–Check–Hold–Adjust–Trim*
- *Remember Power/Attitude for Climb/Descent*

Turning:
- *Limit Bank to Rate 1*
- *Rolling in anticipate height loss*
- *Rolling out anticipate height gain*
- *Anticipate new heading by 10° when rolling out*

Manoeuvres and their related instruments

Manoeuvre	*Instruments to scan*	*Information*
Straight and Level	*Attitude indicator* *ASI* *Direction indicator* *Altimeter/VSI* *Turn indicator* *Ball*	*Pitch and roll* *Airspeed* *Heading* *Height/Altitude* *Hdg. deviations* *Balance*
Climbing and descending	*Attitude indicator* *RPM indicator* *ASI* *Direction indicator* *VSI* *Altimeter* *Ball*	*Pitch and Roll* *Power* *Airspeed* *Heading* *Rate of climb or descent* *Height/Altitude* *Balance*
Turning	*Attitude indicator* *Turn indicator* *Altimeter/VSI* *Direction Indicator* *Ball*	*Angle of bank and pitch* *Rate of turn* *Height/Altitude* *Heading* *Balance*

Flight Practice

COCKPIT CHECKS

Check the instruments for serviceability before take-off.
Arrange screens or visors.

OUTSIDE CHECKS

(*a*) Altitude: Sufficient for practice. Climb into smooth air when possible.
(*b*) Location: Not over towns, airfields or in controlled airspace.
(*c*) Position: Check in relation to a known landmark.
(*d*) Lookout: Safety pilot to maintain lookout.

AIR EXERCISE

Note: While the following exercises must be taught by a suitably qualified flying instructor, they may be practised with a non-instructor safety pilot.

Straight and Level Flight

(*a*) Apply power for cruising flight. With reference to the attitude indicator adopt the cruising attitude by placing the symbol in the correct position on the horizon. Cross-check fore and aft level with the airspeed indicator which should indicate normal cruising speed.
(*b*) Watch the altimeter and correct any loss or gain of height by adjusting power. The vertical-speed indicator will give an early warning of incorrect power setting provided the airspeed has settled. Re-trim at the new power setting.
(*c*) Check lateral level with the attitude indicator and direction on the direction indicator. Make any necessary corrections with aileron and rudder. Trim out all control loads when power has been correctly adjusted for level flight.
(*d*) Check the direction indicator with the magnetic compass every 10–15 minutes.
(*e*) Continue the selective radial scan to maintain straight and level flight.

Changing airspeeds

(*a*) Decide upon an airspeed below normal cruising. Reduce power to a setting estimated for that speed.

(*b*) Maintaining lateral level on the attitude indicator and direction on the direction indicator, as the airspeed decreases gradually raise the symbol above the horizon. Watch the altimeter and prevent a gain in height by correct use of the elevator control while changing attitude. Hold the aircraft steady at the new attitude. Note the position of the attitude indicator and power setting. Remember these for future use.

(*c*) Allow sufficient time for the airspeed to change at the new attitude and, when it has settled to the required figure, re-trim. Check the balance indicator (ball).

(*d*) Watch the altimeter and if necessary adjust power to maintain height. Note the new position of the symbol in relation to the horizon.

(*e*) Now select a high cruising speed and repeat the procedure by increasing power and placing the symbol slightly below the horizon line. Try and remember the position of the attitude indicator and related power settings for various airspeeds.

Climbing

(*a*) Apply climbing power and raise the nose so that the approximate climbing attitude is indicated on the attitude indicator. Check the ball for balance.

(*b*) Maintain lateral level on the attitude indicator and check the heading on the direction indicator.

(*c*) Adjust the climbing attitude until the required climbing speed is attained. Re-trim the aircraft. The rate of climb is now shown on the vertical speed indicator.

(*d*) Continue the radial scan to ensure a steady climb on the correct heading.

Resuming level flight

(*a*) Shortly before the required height progressively lower the nose to the cruising attitude with reference to the attitude indicator.

(*b*) Maintain lateral level on the attitude indicator and check the heading on the direction indicator. Refer to the altimeter and keep the height constant during the transition from climbing to cruising speed.

(*c*) When cruising speed is reached, reduce power accordingly. Re-check both airspeed and height, making power and attitude corrections as required. Re-trim when the aircraft has settled in cruising flight.

Descending

(*a*) Decide upon a rate of descent at an appropriate airspeed and adjust the engine controls to give the power estimated for the performance. Check the ball for balance.
(*b*) Adjust the attitude of the aircraft with reference to the attitude indicator, allowing the airspeed to settle. Adjust the attitude to give the required airspeed for the descent. Re-trim for the new attitude.
(*c*) Maintain lateral level on the attitude indicator and check the heading on the direction indicator.
(*d*) When the aircraft has settled check the rate of descent on the vertical-speed indicator and should an alteration be necessary adjust power accordingly. Maintain a constant air-speed during power adjustments by slight changes of attitude. Re-trim when necessary.

Resuming level flight

(*a*) As the new flight level is approached apply cruising power. Hold the descending attitude, thus allowing the airspeed to increase towards cruising speed.
(*b*) Check lateral level on the attitude indicator and heading on the direction indicator.
(*c*) At the required height resume the level flight attitude with reference to the attitude indicator.
(*d*) Make the usual adjustments to power, attitude and trim to achieve balanced cruising flight.
(*e*) Now gain height and practise the descent with varying amounts of power and flap.

Level Turns

(*a*) Watch the attitude indicator and adopt a bank suitable for a Rate 1 turn.
(*b*) Check the rate of turn with the turn needle and if necessary adjust the angle of bank accordingly. Maintain the ball in the centre with sufficient rudder to balance the turn.
(*c*) Watch the altimeter and prevent loss of height by gentle backward pressure on the stick.
(*d*) When the aircraft is turning at a steady Rate 1 without gain or loss of height, note the position of the symbol in relation to the horizon and memorize for future turns. Note the airspeed is slightly less than normal cruising.
(*e*) Maintain an accurate height and turn rate by using the radial scan.

(*f*) To stop the turn, roll out in the usual way until the attitude indicator indicates level flight. Check both height and airspeed.

Climbing and Descending Turns

(*a*) From a steady climb apply sufficient bank for a Rate 1 turn using the rudder to prevent slip or skid. Do not allow the airspeed to decrease. Be prepared for a marked tendency to overbank.

(*b*) To stop the turn, take off bank and make the attitude indicator indicate the climbing attitude. Reduce to climbing power.

(*c*) Now practise descending turns at various power settings with and without flap.

Instrument Limitations (Toppling)

In the open and clear of cloud

(*a*) Deliberately place the aircraft in an extreme nose-up attitude when the direction indicator and attitude indicator will topple.

(*b*) Return to level flight. Both direction indicator and the attitude indicator are now inoperative and in practice instrument flight would have to continue on the limited panel. Note the turn needle is still working correctly.

(*c*) The direction indicator can be caged and re-set immediately but the attitude indicator will require up to ten minutes to re-erect.

Note:

This exercise can only be demonstrated when the aircraft is fitted with instruments that have toppling limits. Some attitude and direction indicators have complete freedom of movement in pitch and roll, consequently, it is not possible to topple these instruments by adopting extreme attitudes.

Special techniques

Exercises 1 to 19 in this manual are required to meet the syllabus of training for a UK Private Pilot's Licence. This last chapter deals with special techniques which are not part of these requirements. They are included because some pilots may wish to fly types of aircraft that demand special handling. There is also an emergency exercise in this chapter which extends the *Forced Landing Without Power* exercise (17A) and which is a training requirement in Australia.

Subjects covered in this chapter are:

1. Hand starting
2. Handling tailwheel aircraft
3. Ditching.

The only occasions when the pilot of a modern light aircraft may have to hand start would be the result of battery incapacity. Vintage aircraft, on the other hand, often have no electric starter and since there are many Tiger Moths and other designs of that period still flying Section 1 of this chapter is included.

Tailwheel aircraft, although no longer in the majority, continue to be made by a number of manufacturers. Also, most of the pre-war or vintage aircraft are of tailwheel design. The special handling requirements of such aircraft are described in Section 2.

Although Exercise 17A deals with engine failure while over land, the same emergency, should it occur when flying over water will demand rather different treatment on the part of the pilot. Ditching is explained in Section 3.

After gaining a PPL many students feel they have reached the end of the learning process. In fact it is only after the licence has been issued and the pilot is able to fly alone that experience, that finest of all teachers, can play its part. The three topics dealt with in this chapter should be regarded as extensions to the private pilot's basic knowledge.

Section 1. Hand-starting

Whatever the design of the aircraft, old or modern, before any attempt is made to hand-start it is essential to ensure that the person responsible for **Swinging the Propeller** will be standing on firm ground. Loose gravel or any surface where there is a risk of slipping (wet grass, mud, etc.) is a potential danger and a serious accident could result from loss of balance.

The aircraft should also be sited so that its tail is pointing away from open windows, hangar doors or other aeroplanes which may be damaged by slipstream after the engine has started.

Engine Priming

Most aircraft of modern design are fitted with a plunger-type primer and engine priming will be conducted in exactly the same way as in an aeroplane with electric starting.

Some aircraft of pre-war design are primed by flooding the carburettor. This is achieved by opening the engine cowling on the carburettor side and holding down the float needle until fuel spills through the overflow tube which extends below the aircraft. When the aircraft is of low-wing design flooding entails operating the fuel pump via a small extension lever protruding from the pump body while a ring toggle is pulled to depress the carburettor float. The ring must be released when fuel flows onto the ground.

Pilot-Starter Liaison

During hand-starting the pilot and the starter must work as a team. The pilot will say out loud:

'Brakes ON'
'Switches OFF'
'Fuel ON'
'Throttle CLOSED'

As each item is called it must be checked by the pilot. These important checks must on no account be recited automatically without first ensuring that they really are ON, OFF or CLOSED as the case may be.

The starter will then repeat 'Brakes ON, Switches OFF, Fuel ON, Throttle CLOSED'.

When ready to swing the propeller the starter will call:

'Throttle set'
'Contact'

The pilot will open the throttle to the correct position for starting. This is usually half to one inch forward of fully closed. The appropriate magneto for starting will then be turned on and the pilot will respond:

'Throttle set'
'Contact'

The starter will then pull the propeller through compressions until the engine fires.

Starter Magnetos

To ensure that a spark occurs at the correct instant for ignition of the mixture magnetos are timed to create a spark some degrees before the piston reaches the top of the cylinder. To guard against the propeller kicking back as the starter pulls each blade over compression the magneto used for starting is usually fitted with a device which does not advance the spark until after the engine is running at 800–1000 RPM.

A magneto rotating at the low speed of a propeller being pulled over compression by hand is less able to produce a powerful spark than one being turned by an engine running at 1500–2500 RPM. To provide a spark of sufficient intensity for starting while the propeller is being turned at low speed two methods are used:

1. A high intensity magneto may be fitted.
2. A device incorporated in the magneto drive winds up like a clock spring and then flicks the magneto at relatively high speed, so producing a spark of good intensity just as the piston passes top dead centre. These are known as **Impulse Magnetos.**

Whatever the method adopted only the switch controlling the starter magneto should be made live when the starter calls 'Contact'. The other switch is turned on after the engine is running.

Aircraft without Brakes

When no brakes are fitted chocks must first be placed in front of the wheels before hand starting commences. They should also be used when the brakes are known to be less than fully effective.

Failure to Start caused by Over Rich Mixture

Generally a cold engine is easier to hand start than one that is warm and only recently shut down. A starting problem with a warm engine is often caused by a build-up of fuel vapour in the induction system or the cylinders themselves.

If, after a number of attempts to start the aircraft the engine refuses to fire (assuming it has either been correctly primed or the engine is already hot) the following procedure will have to be adopted. It may be initiated by either the pilot or the starter. In the example to follow the pilot, who perhaps knows this particular engine better than the starter will call:

'Switches OFF'
'Throttle OPEN'
'Blowing OUT'

Each item is carefully checked as it is announced.
The starter will then repeat 'Switches OFF, Throttle OPEN, Blowing OUT'.

The propeller is pulled backwards through 6 to 8 compressions blowing out the over rich mixture. Then the starter will shout:

'Throttle SET'
'Contact'

The pilot will bring back the throttle to the starting position and switch on the starting magneto before repeating 'Throttle SET, Contact'. Usually the engine will fire after the propeller has been pulled through several compressions.

Sucking In

Some pilots and starters go through the ritual of 'Sucking in', turning over the engine with the ignition off after priming. There is little point in this because there is always the possibility of making the engine over rich. In any case, had the ignition been on it is more than likely the engine would have fired while it was being turned over by hand.

The practice of turning off the ignition after each swing of the propeller is potentially dangerous because in the 'Off' – 'Contact' – 'Off' – 'Contact' exchange that follows it is very easy to become confused and inadvertently have a live magneto while both pilot and starter are under the impression that it is safe. The only time the starter should call 'OFF' is when, for any reason, the propeller has stopped in an awkward position for the next swing.

Swinging Technique to be Adopted by the Starter

To guard against the possibility of being struck by the propeller when the engine fires the starter should adopt these precautions:

1. Stand at approximately 45 degrees facing the propeller, not directly in front of it.
2. Hold a blade that is above head level so that it is in a good position to allow the arm freedom to pull down. Never attempt to turn the propeller by pushing a blade that is below propeller shaft level.
3. Pull the blade sharply through compression and allow the arm to follow through so that the hand continues in an arc, down and away from the propeller's plane of rotation.
4. While pulling the arm down and through in a sweeping motion the weight of the body should transfer from the leg nearest to the propeller to the one farther away. This will tend to swing the body away from the propeller each time it is pulled over a compression.
5. Never swing a leg up towards the propeller arc in an effort to gain additional momentum while starting. Additional power for the swing is best obtained as described in 3 and 4.

The hand should rest on the propeller blade with the fingers only partly round the trailing edge. This is a precaution against the possibility of the engine kicking back. Metal propellers tend to have relatively sharp trailing edges and the starter should wear an old glove while handling these.

Provided these precautions are taken there is no risk entailed during hand starting. Indeed, there was a time when all aircraft had to be started by swinging the propeller.

Section 2. Handling Tailwheel Aircraft

Although some of the earliest aircraft had nosewheel undercarriages they were gradually discarded in favour of tailwheels. However, during the mid to late 1930s the nosewheel layout was re-discovered and now there are few modern aircraft which embody tailwheel undercarriages.

Apart from the great number of old tailwheel aeroplanes still flying, a few modern designs are being produced to meet special requirements (agplanes, aerobatic aircraft, etc.) and the techniques required while handling these differ greatly from nosewheel flying.

Advantages and Disadvantages of Nosewheel and Tailwheel Undercarriages

Although the nosewheel undercarriage is fitted to most aircraft of modern design tailwheels are not without their advantages and the following tables set out the pros and cons of both types:

Nosewheel undercarriages

Advantages	*Disadvantages*
Directionally stable	Fixed nosewheel undercarriage creates more drag than tailwheel layouts
Firm braking in safety	More easily damaged by rough ground or poor handling than tailwheels
Good visibility while taxying	Heavier installed weight than tailwheel undercarriages
Better than tailwheel types in crosswind conditions	
Take-off and landing easier than in tailwheel types	

Tailwheel Undercarriages

Advantages	*Disadvantages*
Less drag than nosewheels	Directionally unstable and prone to groundlooping while landing
Can withstand rough ground or poor handling better than nosewheel designs	Tendency to swing during take-off
Lighter installed weight than nosewheel undercarriages	Risk of nose-over during firm braking
	Visibility can be poor ahead
	Badly affected by crosswinds while taxying
	During take-off and landing more skill demanded than in nosewheel aircraft

The advantages and disadvantages listed above are manifestations of directional stability and pitch stability.

Directional Stability on the Ground

All vehicles are affected by the relationship between their centre of gravity and the point of wheel contact with the ground. A prime example of instability through bad design is the average airport baggage trolley which will often resist efforts to steer it in the right direction.

Figure 66 shows a typical nosewheel undercarriage. Note that the aircraft's centre of gravity is ahead of the mainwheels (left-hand picture). In the drawing on the right a gust of wind has caused a swing to the left, however this is opposed by the nosewheel and a change in moment arm between the wheels and the centre of gravity. Comparing distances A and B it will be seen that the rolling friction from the right wheel enjoys more leverage than the one on the left. Rolling friction is denoted by the length of the arrows trailing back from the mainwheels. In essence a nosewheel undercarriage is directionally stable because it is self-correcting.

Figure 67 should now be compared with Fig. 66. It relates to a tailwheel aircraft faced with a situation similar to the nosewheel swing just described. In the left-hand picture it will be seen that the centre of gravity is behind the mainwheels. When a gust of wind from the left provokes a swing to the left (right-hand picture) the tailwheel does, to some extent, provide a corrective force. However, many tailwheel assemblies are arranged to disengage from the rudder steering to facilitate parking and after about 25° or so it will freely caster. There will then be no corrective force from the tailwheel. Furthermore, being a tailwheel undercarriage, the left wheel, having moved away from the centre of gravity because of the swing, will exert more rolling friction (through the advantage of increased leverage)

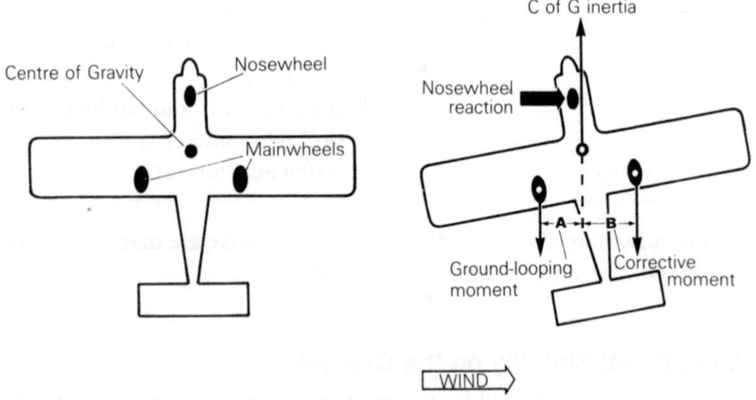

Fig. 66 Stability of a Nosewheel undercarriage in a crosswind.

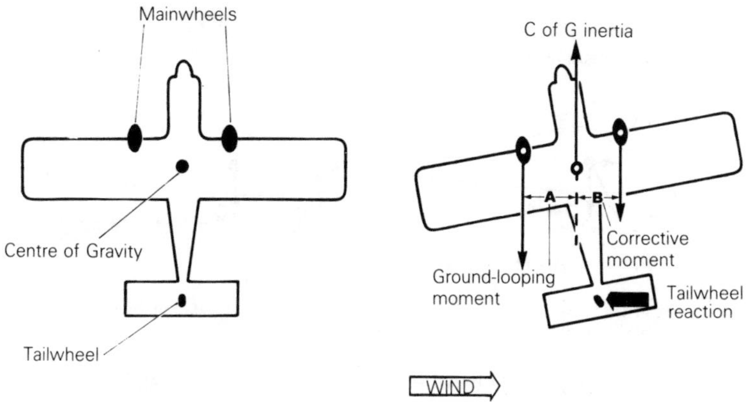

Fig. 67 Instability of a Tailwheel undercarriage in a crosswind.

than the wheel on the right. Tailwheel undercarriages are therefore directionally unstable because, during a swing, there is no self-correction. On the contrary, the geometry and relationship that exists between the wheels and the centre of gravity will actually make the situation worse after a swing has started.

Stability in Pitch

A little thought will reveal that were it possible to arrange a tailwheel undercarriage with the mainwheels near or even behind the centre of gravity directional stability would improve. Unfortunately, such an aircraft would be unmanageable because it would tend to stand on its nose.

Figure 68 shows a nosewheel aircraft. In the right-hand picture it will be seen that the centre of gravity is ahead of the mainwheels. When brake is applied (left-hand drawing) the centre of gravity tends to continue in the direction of movement, i.e. towards the nose of the aeroplane. It still exerts a force directly downwards but the resultant, although it has moved forward, nevertheless remains behind the nosewheel.

The situation just described is rather different in the case of a tailwheel aircraft (Fig. 69). In the right-hand picture it will be seen that while taxying the centre of gravity is behind the mainwheels. When the brakes are applied (left-hand drawing) the resultant force provided through the centre of gravity moves forward of th mainwheels and excessive braking could provoke a nose-over

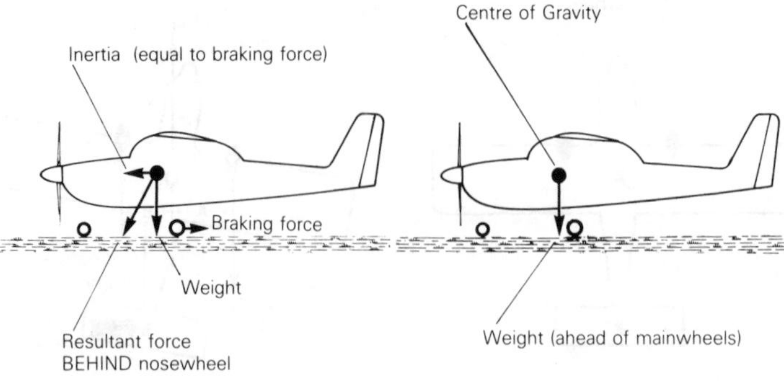

Fig. 68 Stability of a Nosewheel undercarriage while braking.

To provide the best directional stability with least risk of building in a tendency to nose over when brake is applied demands compromise on the part of the designer. It would be fair to say that although some of the earlier tailwheel aircraft were difficult to taxi, a weakness that was compounded by poor, cable-operated brakes that suffered from fade, modern designs offer improved forward visibility in the tail-down attitude, the brakes are usually of the disc type and they are less prone to directional instability. However, to some extent the problems remain and it is because of these problems that tailwheel handling is somewhat different to the management of nosewheel aircraft.

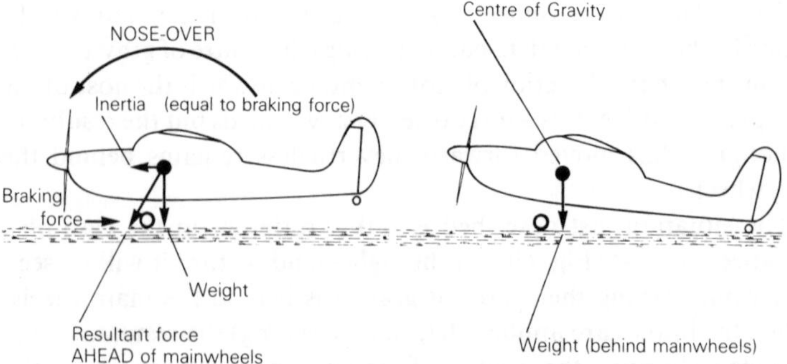

Fig. 69 Risk of nosing over during heavy braking in a Tailwheel aircraft.

Handling Techniques

In general it should be remembered that when flying a tailwheel design:

1. The view ahead is more restricted than in a nosewheel aircraft while manoeuvring on the ground. In the case of some vintage aircraft it is very bad indeed.
2. Because they are sensitive to wind a swing may occur more readily and this can quickly develop into a groundloop.
3. Brake must be applied with caution to avoid lifting the tail and striking the propeller on the ground.

Handling techniques are now described for each phase of flight.

During Engine Starting

Because a sudden burst of slipstream could lift the tail it is essential to fully hold back the wheel/stick during starting. Although the engine would normally run at some 800–1000 RPM immediately after it fires, a misplaced throttle could cause it to start at a higher speed. Should the tail lift there is a risk of striking the propeller on the ground and this is bound to result in extensive damage to propeller and engine alike.

While Taxying

Visibility ahead of a tailwheel aircraft ranges from almost as good as a nosewheel design at its best to total lack of vision at its worst. To guard against taxying into an object ahead it is essential to swing the aeroplane from side to side while looking along the nose in the opposite direction to the turn. This is illustrated in Fig. 70.

Aircraft of vintage design will have cable-operated drum brakes and these must be used sparingly to prevent fade. Some of these aircraft have no brakes. They rely on a tailskid which is quite effective on grass but of little assistance while taxying on hard surfaces.

Although tailwheel aircraft respond well to changes in direction under conditions of low wind, when the surface wind exceeds 15 knots or so difficulty may be experienced in turning crosswind. If the aircraft refuses to turn in one direction a turn the other way through a greater number of degrees should be made, provided there is sufficient room to manoeuvre.

To some extent steering while on the ground can be assisted by using slipstream over the rudder, but care must be taken not to allow

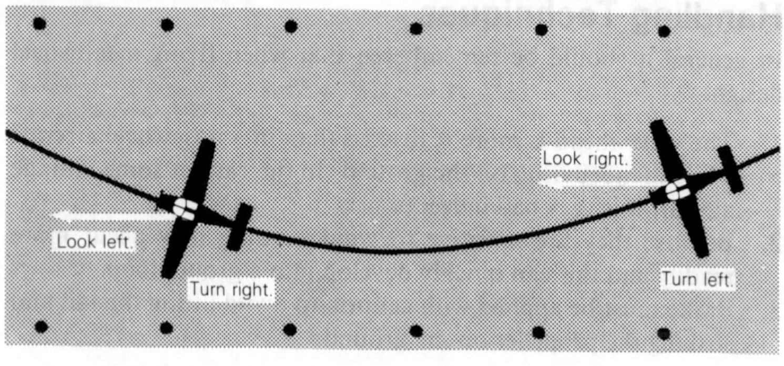

Fig. 70 Clearing the view ahead while taxying a Tailwheel aircraft.

taxying speed to increase unduly while applying additional power to increase slipstream effect.

Because of the need to avoid excessive use of brake and guard against risk of swinging, tailwheel aircraft must be taxied slowly at all times and never for long in a straight line.

During Take-off

Apart from their natural instability while on the ground, tailwheel aircraft are affected by four forces which, during take-off, act in the same direction to cause a swing if this is not prevented. While it is not necessary for pilots to have a deep understanding of these forces a brief description now follows.

1. **Slipstream Effect**: This was explained in Exercise 6, *Straight and Level Flight*. It is caused by the slipstream coiling its way back down the fuselage where it strikes one side of the fin and rudder to cause a swing.
2. **Torque Effect**: As the engine and propeller rotate there is a tendency for the entire aircraft to roll in the opposite direction, rather like a rubber-driven model aircraft being held by the propeller. In flight aerodynamic forces on the wings are more than adequate to counter this torque effect but during take-off there is little airflow over the wings in the early stages of the run. Consequently, one wheel will be pressed harder on the runway than the other, so causing greater rolling friction, almost as though brake has been applied on that wheel.

Slipstream and torque effects are experienced by all aircraft, nosewheel and tailwheel, during take-off. The two that now follow are peculiar to tailwheel designs.

3. **Gyroscopic Effect**: At the start of the take-off run the tail will be on the ground. As it is raised the plane of propeller rotation is tilted forward. A rotating propeller will behave like a gyroscope in so far as the act of tilting it upright (akin to pushing forward the top of the disc) is translated into a reaction at 90° in the direction of rotation.

   Imagine taking off in an aircraft where the propeller rotates clockwise when seen from the cabin. As the tail is raised it is as though a force, moving overhead in the direction of flight, has pushed forward at the top of the propeller disc. The force precesses through 90° to emerge as a reaction forward but acting from the right-hand edge of the propeller disc. This will tend to swing the aircraft to the left.
4. **Asymmetric Blade Effect**: In the early stages of the take-off, while the tail is still on the ground, the propeller shaft is inclined upwards in the direction of flight. Naturally the propeller disc is tilted backwards when seen from the side and this has the effect of presenting the down-going blade at a greater angle of attack than the up-going blade.

   When propeller rotation is clockwise (seen from behind) the down-going blade will be on the right-hand side of the aircraft, consequently, more thrust will be developed on that side and there will be a tendency for the aircraft to swing to the left until the tail is raised. Then both halves of the propeller disc will be generating the same amount of thrust and asymmetric blade effect will cease to exist.

The net result of these four forces is to provide a swing to the left when the engine turns clockwise (e.g. all American piston engines) and a swing to the right in the case of engines that turn anti-clockwise (Gipsy, Cirrus, Renault, etc.).

Technique to adopt during the take-off is as follows:

1. Line up on the centre of the runway and run forward a few yards to straighten the tailwheel. On grass airfields find a point on the far boundary on which to keep straight.
2. Look along the left of the nose, hold the wheel/stick in the neutral position then open the throttle smoothly and fully in one movement.

3. Keep straight with rudder and on no account allow a swing to develop.
4. As speed increases gently ease forward the wheel/stick to raise the tail and adopt the level attitude. Be prepared for gyroscopic effect to cause a swing. (RIGHT rudder for clockwise rotation: LEFT rudder for anti-clockwise rotation).
5. Prevent the tail from rising as speed increases by progressive back pressure on the wheel/stick until the aircraft lifts off the ground.

After lift-off it will be found that tailwheel aircraft behave in much the same manner as nosewheel designs although vintage aircraft have their own special characteristics.

During Landing

Two techniques may be adopted while landing tailwheel aircraft. These are:

1. The wheel landing.
2. The three-point landing.

The Wheel Landing (Fig. 71)

Having flown the usual circuit and approach to arrive over the threshold:

1. Round out over the runway centreline.
2. Close the throttle and allow the mainwheels to touch the surface.
3. Do not allow the tail to lower since this would cause the aircraft to lift off again. A little forward pressure must be applied to the wheel/stick and the level attitude should be maintained until the tail sinks naturally to the ground as speed decreases.
4. As the tail lowers there may be slight reverse gyroscopic effect while the propeller disc tilts backwards. Prevent any tendency to swing with rudder, if necessary assisted by the brakes.

With practice a wheel landing can be very gentle but when the aeroplane is allowed to drop onto its wheels a bounce will follow because, in tailwheel designs, the immediate reaction to such a bounce is nose-up, an increase in angle of attack and a lift-off at low speed. Should this occur power must immediately be added while the aircraft is brought back to a less nose-up attitude in readiness for a second landing or a go around as circumstances demand.

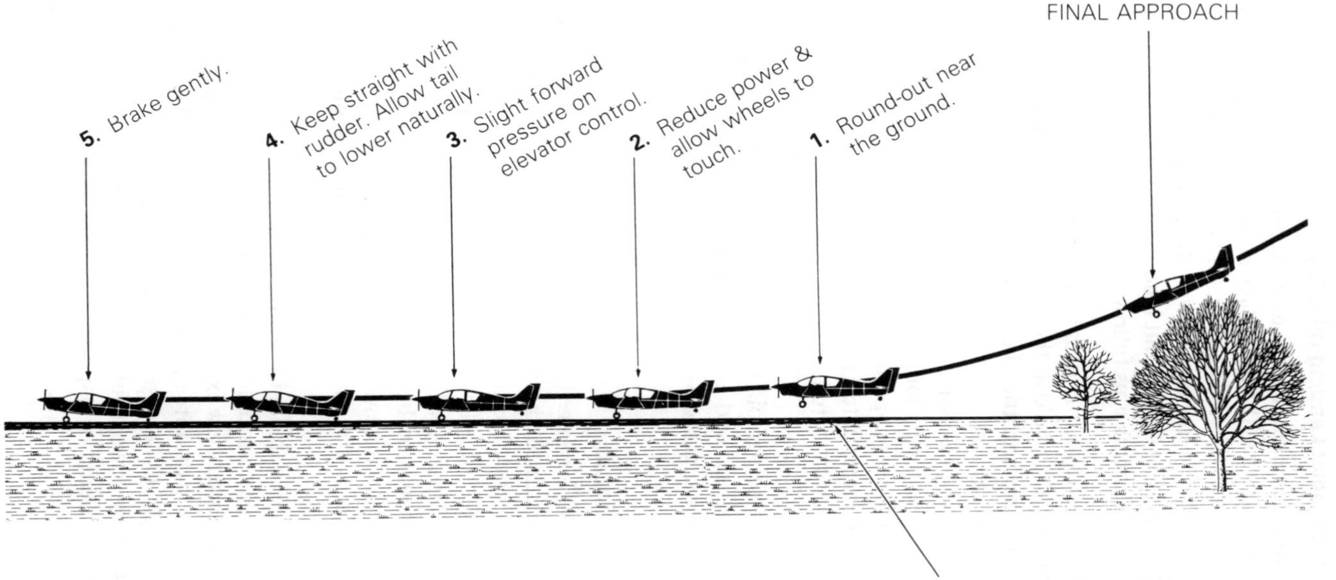
FINAL APPROACH
5. Brake gently.
4. Keep straight with rudder. Allow tail to lower naturally.
3. Slight forward pressure on elevator control.
2. Reduce power & allow wheels to touch.
1. Round-out near the ground.
RUNWAY THRESHOLD

Fig. 71 Wheel landing in a Tailwheel aircraft.

The Three-point Landing (Fig. 72)

In a three-point landing the aircraft touches down in the three-point attitude, namely with the fuselage at the angle adopted while standing on the ground. There is a relatively large angle of attack and touch down speed is somewhat lower than during a wheel landing. To achieve a perfect three-point landing requires some skill of the pilot. Sequence of events is as follows:

1. Having flown the usual circuit and approach arrive over the runway threshold at the correct speed and make the round out.
2. Close the throttle then hold off with progressive backward movement on the wheel/stick. Guard against ballooning (following excessive back pressure) and keep the eyes moving back and forth along the left of the nose in the usual way.
3. When the three-point attitude has been reached the aircraft will almost immediately land on all three wheels. Avoid exceeding the three-point attitude since this will result in a tailwheel first landing.
4. After landing hold the wheel/stick fully back to prevent the tail from lifting and take immediate action to prevent a swing, however slight, developing.

Possible Faults

If the aircraft is allowed to touch down before the three-point attitude has been attained there will be a tendency to hop which can be difficult to control in some aircraft.

When the hold-off has been completed three or more feet from the surface the landing will be heavy but not necessarily dangerously so. The round out and hold-off should both be enacted in exactly the same way as for nosewheel aircraft. The main difference in technique is that the hold-off is continued until the three-point attitude has been attained.

Contrary to popular belief aircraft do not 'stall' onto the ground in the three-point attitude. They 'sink' and that is a totally different matter. With very few exceptions aircraft are designed so that angle of attack while in the three-point attitude is several degrees less than the stalling angle. Otherwise an aircraft that actually stalled immediately before touch-down would drop the nose and possibly one wing. Such an aircraft would be impossible to land in the three-point attitude.

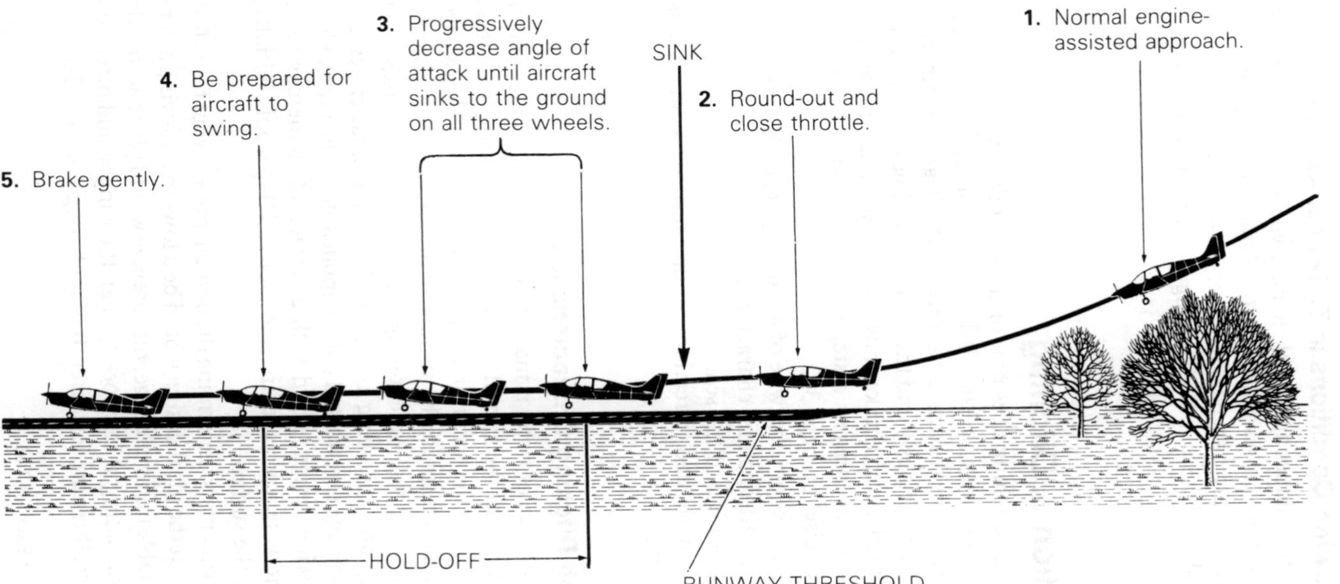

1. Normal engine-assisted approach.
2. Round-out and close throttle.
SINK
3. Progressively decrease angle of attack until aircraft sinks to the ground on all three wheels.
4. Be prepared for aircraft to swing.
5. Brake gently.
RUNWAY THRESHOLD
HOLD-OFF

Fig. 72 Three-point landing in a Tailwheel aircraft.

Crosswind Conditions in Tailwheel Aircraft

All the techniques applicable to nosewheel aircraft while landing and taking off in a crosswind apply to tailwheel undercarriages. However, since touch-down speed is relatively low during a three-point landing these should be avoided during a crosswind in favour of wheel landings.

Section 3. Ditching

Although ditching does not form part of the flying training syllabus in every country, a knowledge of the procedure is required, for example, by the Australian authorities*. Clearly the emergency is one to be taken seriously and therefore an outline of the factors involved is included here. Unlike most other exercises, ditching cannot be practised; nevertheless a step-by-step procedure is given under the 'Air Exercise' on page 342.

The success or otherwise of a ditching depends upon three factors:

1. Aircraft ditching characteristics.
2. Condition of the sea.
3. Strength and direction of the surface wind in relation to condition 2.

Aircraft Ditching Characteristics

The resistance of water is many times that of air at the same speed, so that if a part of the aircraft situated some distance from its centre of gravity enters the water first considerable force will be exerted, which can seriously displace the attitude of the aircraft. For example, a fixed undercarriage will on entering the water cause a nose-down pitch. Full flap on a low-wing aircraft will have a similar effect, even if the undercarriage is retracted (when applicable to the type). Conversely, a high-wing design, fitted with a retractable undercarriage, would enter the water fuselage first and even full flap would initially remain clear of the surface.

Experiments to determine the best airframe layout for safe ditching reveal that the ideal design would be a low- to mid-wing aircraft with a retractable undercarriage; the fuselage would plane across the water before subsiding at low speed, when the wing would offer buoyancy. However, there are few aircraft of mid-wing design and the nearest to

*Australian Exercise 14d.

this ideal has been found to be low-wing monoplanes with retractable undercarriages, followed by low-wing aircraft with fixed undercarriages. A high-wing aircraft with a fixed nosewheel undercarriage is known to offer the worst ditching characteristics; nevertheless there have been a number of recorded cases where such aircraft have been landed in the water without loss of life and with only minor injury.

While it has been established that the larger the aircraft the better will be its ditching qualities it should be remembered that the average light aircraft will sink within 30 seconds to two minutes according to design.

Condition of the Sea

The ever-changing sea is affected by a number of factors. Furthermore, its complex movements may bear little relationship to the prevailing surface wind. Movement of the surface is defined as follows:

Swell. This is the result of a past disturbance, possibly originating some distance away. It may be distorted by nearby land masses or sea currents. Since it is the result of past wind effects, a heavy swell may exist in conditions of zero wind. Swell may best be understood by throwing a stone into the centre of a pond. The long ripples that reach the water's edge some appreciable time later are, in miniature form, the swell, and this movement of the surface of the sea is often referred to as the **Primary Swell**.

Waves. Winds in excess of 5 kt will superimpose on the primary swell a **Secondary System** of waves. This assumes greater importance as the wind speed increases until, at approximately 30 kt, the wave pattern may obscure the primary swell. Whereas the primary swell may better be seen at heights of 2000 feet or above, secondary waves are usually more recognizable below 1000 feet. The importance of being able to determine the direction of swell will be explained under 'Choice of Landing Direction'.

Strength and Direction of the Surface Wind

Obviously smoke or steam from a nearby ship will provide an excellent indication of surface wind speed and direction. Alternatively, cloud shadows will, to a lesser degree, give similar information. In the absence of either observation the only other indication of surface wind is the appearance of the sea itself. Direction may sometimes be ascertained by looking for wind lanes (light and darker strips on the water which are best detected when looking downwind). Wind speed may be gauged as follows:

Wind Speed	*Appearance of Sea*
Light wind	Ripples of a scaly appearance.
5 kt	Very small waves.
8–10 kt	Small waves, some with foam crests.
15 kt	Larger waves with white caps.
20–28 kt	Medium size waves with long foam crests and many white caps.
30–35 kt	Larger waves with white foam blowing across the surface.
Above 35 kt	Wavecrests breaking into large streaks of foam which cover areas of the sea.

Choice of Landing Direction

Before the landing direction may be decided pilots must be able to determine:

1. Direction of primary swell.
2. Surface wind speed and direction.

This has been explained in the previous text and the choice of landing direction should be made as follows:

In a Calm Sea

Obviously these are the easiest conditions and the ditching should be made into wind.

In a Heavy Swell (winds up to 30 kt)

Think in terms of landing in undulating sand dunes. Avoid alighting into a rising swell. When possible, alight parallel to the undulations (Fig. 73 shows the best position).

When the wind is strong enough to cause pronounced drift, the aircraft must alight across the swell and towards the wind (Fig. 74).

In a Heavy Swell (winds above 35 kt)

Such winds will reduce the touch-down speed of most light aircraft to 15 kt or less, so clearly they are of prime importance while ditching. In a high wind, the swell will be shorter and the sea will be broken up into a pronounced secondary wave system, which is important. The

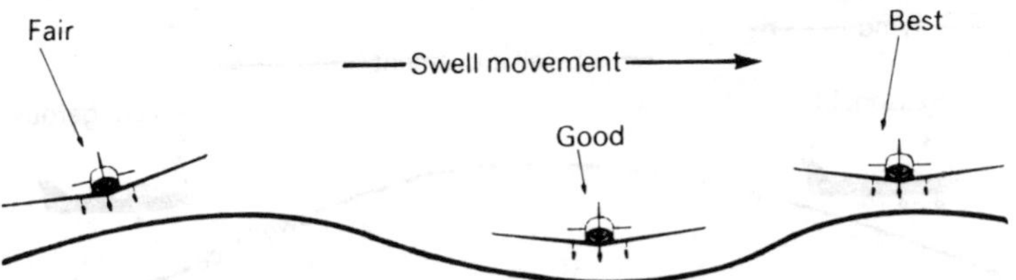

Fig. 73 Alighting parallel with the swell in conditions of light to moderate wind.

ditching must then be made into wind and down the back of a primary swell (Fig. 75).

Positioning for the Ditching

The first consideration after alighting in the water will be speed of rescue. This is bound to be enhanced when a ship is nearby, but it is

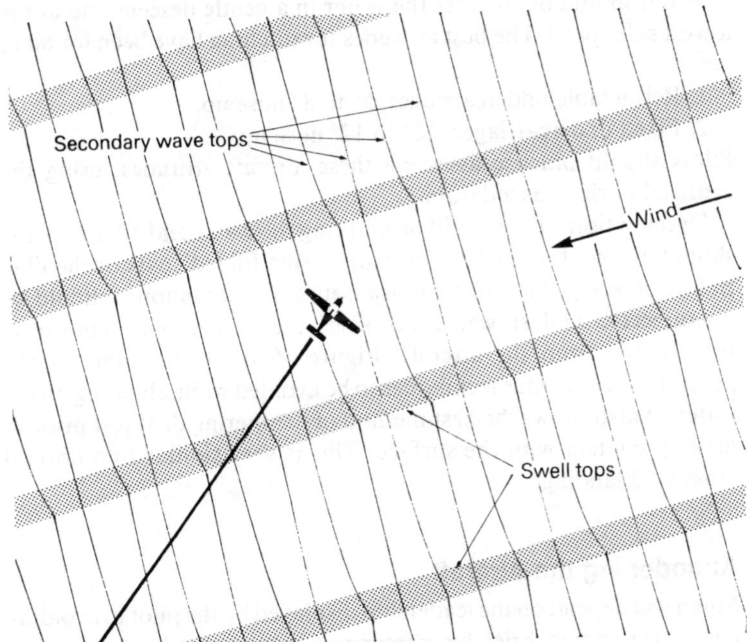

Fig. 74 Alighting across the swell and secondary waves when the wind is between 25–35 kt.

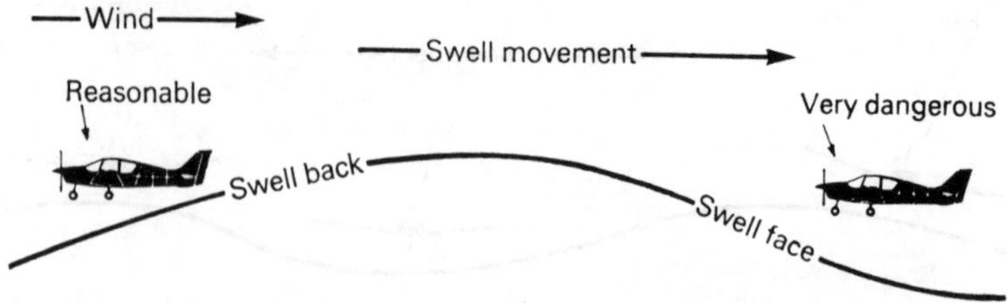

Fig. 75 Winds in excess of 35 kt assume prime importance and it is then necessary to alight across the swell as illustrated.

not generally realized that a ship under way will require a considerable distance to stop, typical figures being half-a-mile for a 5000 ton steamer and up to six miles for a 200,000 ton tanker. So there is little point in ditching alongside a moving ship, while alighting well ahead of a stationary one will prolong the rescue operation.

Alighting on the Water

The aim should be to enter the water in a gentle descent and at the lowest safe speed. The best attitudes for ditching have been found to be:

1. Retractable undercarriages, 5° to 8° nose up.
2. Fixed undercarriages, 10° to 12° nose up.

Pilots should practise assessing these aircraft attitudes, using the artificial horizon as a datum.

Deceleration will be rapid on entering the water and all occupants should protect their heads and faces, using folded coats or the like. Before entering the water, all headsets and microphones should be disconnected and stowed away since these have been known to restrict exit from the aircraft. Figure 76 illustrates some of the potentially dangerous techniques to be avoided while alighting on the water. It also shows the best method of countering drift just prior to making contact with the surface. This is very similar to a normal crosswind landing.

Abandoning the Aircraft

Much will depend on the leadership displayed by the pilot in avoiding panic. He should brief his passengers during the descent so that everyone knows what to do. The door(s) must be wedged open to ensure that it is not jammed shut on impact and, in the case of

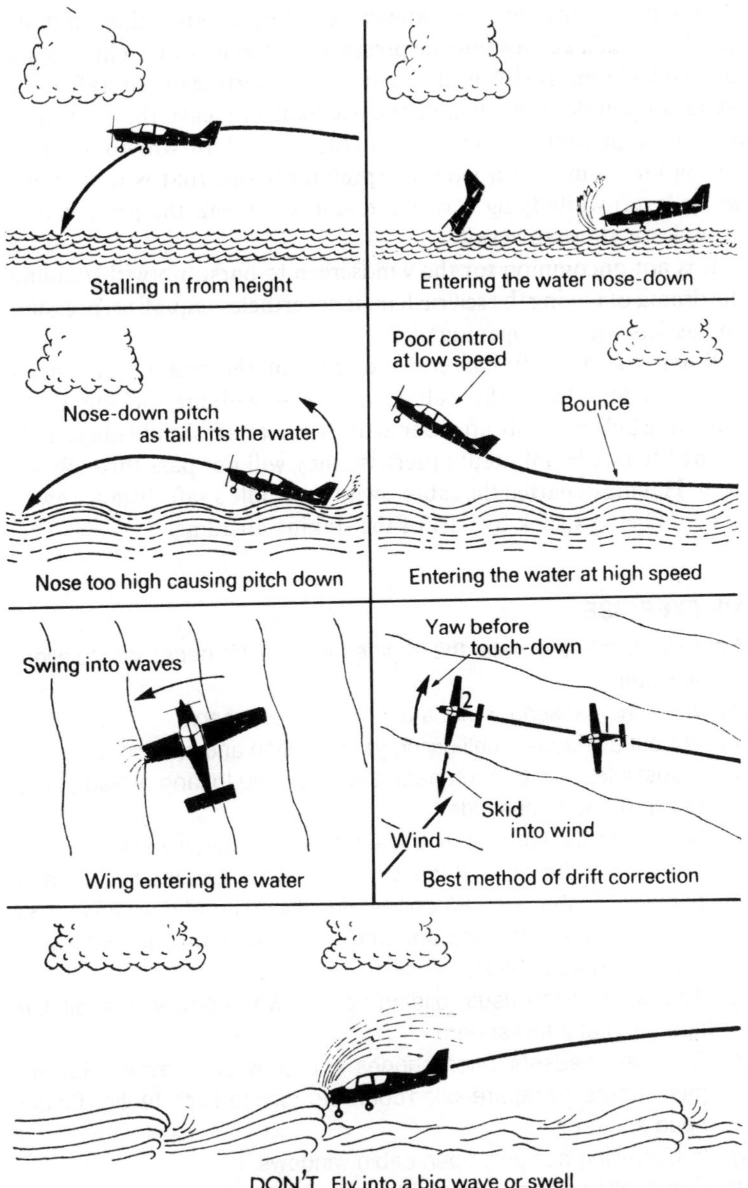

Fig. 76 Considerations while ditching.

high-wing designs, that the water can enter the cabin so that the door may be opened against outside pressure. Remember that a high-wing aircraft will rapidly sink until the wing is supporting the aircraft and a wedged-open door will enable the occupants to leave the aircraft as quickly as possible. When for any reason the door will not open, a strong kick aimed at a window (preferably one that is well above water level) while lying across the seat will break the perspex and provide an alternative exit.

It is not uncommon for the windscreen to burst; notwithstanding the drama of having the sea rush in, it nevertheless equalizes pressure and assists with the opening of doors.

When the aircraft has come to rest in the water, release all harnesses and leave the cabin smartly – without kicking those following behind. Only after departing the aircraft should life jackets or the life raft be inflated: otherwise they will not pass through the door. Having departed the cabin, swim or paddle a safe distance away to avoid being struck by part of the sinking airframe.

AIR EXERCISE

While flying over the water the engine fails and it is necessary to alight on the water:

(*a*) Warn the passenger that a ditching is necessary.
(*b*) Put out a 'Mayday' call giving your position and intentions.
(*c*) If possible, plan to ditch near a ship, aiming to land ahead (if it is moving) and to one side.
(*d*) Determine the surface wind strength and direction and assess the direction of the swell. If the wind strength is less than 25 kt, land parallel with the swell. When the wind speed is between 25 and 35 kt, head partly into wind and across the swell but land into wind when it exceeds 35 kt.
(*e*) Approach at the usual gliding speed (when power is available approach at a low speed).
(*f*) Remove headsets/microphones and stow them away. Get the passengers to prepare rolled coats etc. for use as protection for the head and face.
(*g*) In high-wing designs, open cabin windows.
(*h*) Check all harnesses are tight and that shoulder straps are on.
(*i*) Use flap as recommended (see the Owner's/Flight/Operating Manual).
(*j*) Near the water, open the door(s) and use a briefcase or the like to keep it/them from closing.

(*k*) Aim to land in a tail-down attitude on the crest of a large wave or swell. When this is not possible alight on a downslope. ON NO ACCOUNT FLY INTO THE FACE OF A SWELL.

(*l*) After touch-down, hold back the wheel/stick, wait for the aircraft to come to a halt, then release harnesses, open the door(s) and leave the cabin without delay.

(*m*) In the water, inflate the lifejackets/liferaft, keep together and move away from the aircraft.

(*n*) Make for the nearest ship, using any signalling equipment available.

Appendix 1 – Graphs

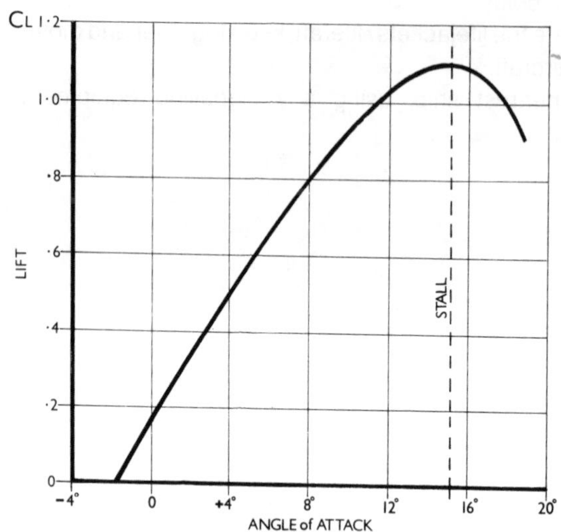

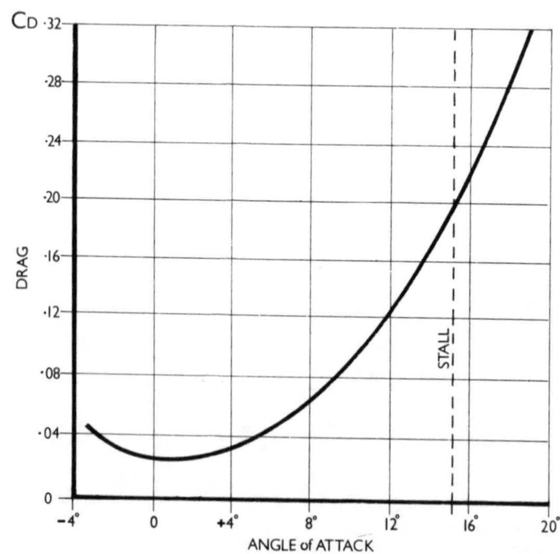

Graph 1. (*Top*) Showing changes in lift at various angles of attack. (*Lower*) Behaviour of drag with changes in angle of attack.

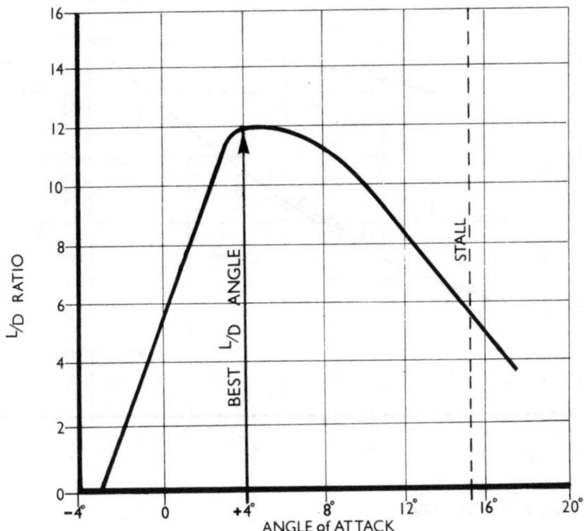

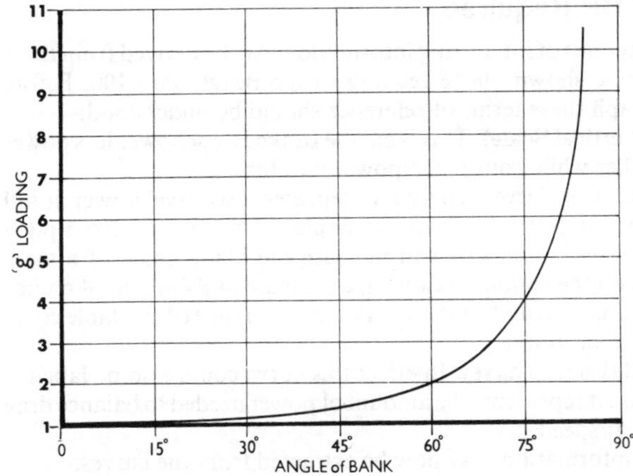

Graph 2. (*Top*) The relationship between angle of attack and lift/drag ratio. (*Lower*) Loading during turns at various angles of bank.
Note that up to 60° Bank loading is moderate, doubling over the next 15° and reaching 10 'g' at 84° Angle of Bank.

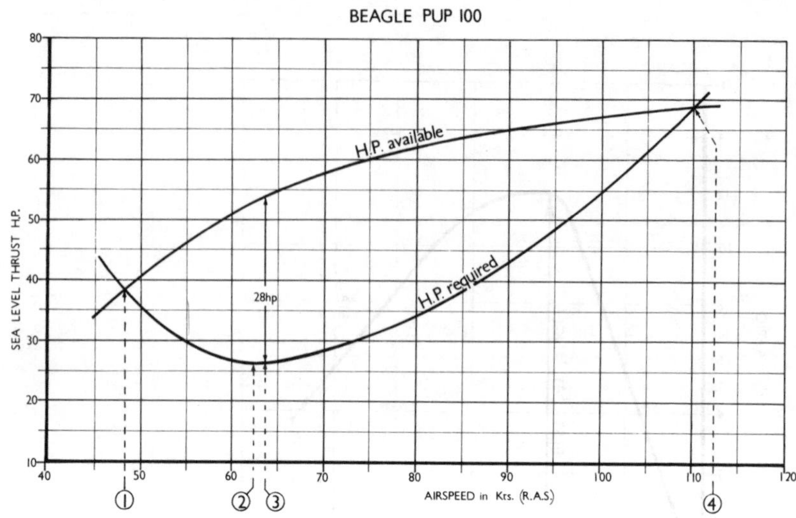

Graph 3. HP available/HP required.

HP Available HP Required

A considerable amount of interesting information may be derived from these curves, the example shown above relating to the Beagle Pup 100. Before examining the graph these terms of reference should be understood:

1. Thrust HP (Vertical Scale). This is engine Brake Horsepower less power lost by the propeller while converting power into thrust.

2. HP Available (Top Curve). This curve represents sea level power at full throttle. Since the aircraft used in this example is fitted with a fixed pitch propeller, reductions in airspeed will increase the blade angle of attack, increase the load on the engine so causing a decrease in RPM and of course BHP. As altitude is gained and air density decreases so the HP available curve will descend down the graph.

3. HP Required (Lower Curve). In effect this curve could also be labelled 'total drag' because it represents the amount of power needed to balance drag at any particular airspeed.

The following information may now be extracted from the curves:

1. Propeller Efficiency. Assuming the engine to produce a net 90 BHP it will be seen that a maximum speed (and power) thrust HP is in the region of 70 THP (i.e. power available for propelling the aircraft) indicating a propeller efficiency of approximately 78 per cent.

2. Minimum Speed (1). Note the sharp rise in horsepower required as speed is reduced below 60 kt. This is because of the considerable increase in induced drag which occurs at high angles of attack (page 60). At the low speed power available is at its lowest and where the two curves meet corresponds to Minimum Speed. A further reduction in speed would cause a loss of height.

3. Maximum Endurance (2). The speed at which total drag is at its lowest value will likewise be the speed which requires the least power for level flight and in consequence the lowest fuel consumption in gallons (pounds or kilos) per hour. This will not produce maximum range (i.e. best air miles per gallon), speed for maximum range being rather higher than speed for maximum endurance.

4. Maximum Speed (4). The HP required curve shows a steady increase as speed is developed above 60 kt thus indicating a similar build up of total drag. When power required reaches the level of power available (position 4) any further increase in speed will cause a descent. In other words maximum level speed is indicated by the point at which the two curves again meet.

5. Maximum Rate of Climb (3). In the chapter on Climbing (page 68) it was explained that power surplus to requirements was used to climb the aircraft. At 64 kt (RAS) there is a surplus of rather more than 28 HP over and above that required for level flight at this speed. Remembering that 1 HP is the equivalent of 33,000 lb raised 1 ft in one minute, the surplus 28 HP may be calculated in terms of rate of climb quite simply:

$$\text{Rate of Climb} = \frac{\text{surplus HP} \times 33{,}000}{\text{weight of aircraft (lb)}}$$

$$= \frac{28 \times 33{,}000}{1{,}600} = 577 \text{ ft/min rate of climb}$$

The Pup is unusual in so far as a similar level of surplus horsepower available for climbing is obtainable over a wide range of speeds. Reference to the curves shows that rate of climb will remain fairly constant between 64 kt and 78 kt RAS, steepest climb gradient occurring at the lowest speed within the range (64 kt RAS corresponding on the Beagle Pup to an Indicated Air Speed of 60 kt).

Appendix 2 – Ground Signals

Runway or taxiway unfit for aircraft landing or manoeuvring.

Area to be used only for take-off and landing of helicopters.

Movement of aircraft and gliders confined to hard surface areas.

Direction of take-off and landing may differ. (Also shown by a black ball suspended from the signals mast.)

A red L on the above sign indicates that light aircraft may land on runway or grass area. (May also be used in conjunction with signal below.)

A yellow diagonal on a red square means special care is necessary in landing, owing to temporary obstruction or other reason.

Landing and take-off on runways only, but movement not confined to hard surfaces.

Landing prohibited — a yellow cross on a red square.

Glider flying in progress. (Used in conjunction with two red balls on the signals mast.)

Right hand circuit in force. The arrow is made up of red and yellow stripes.

Shaft of the T indicates direction of take-off or landing — i.e. towards the crossbar.

Part of the manoeuvring area to be used only for the take-off and landing of light aircraft.

Emergency landing only — two yellow bands on a red square.

Orange/white tent strips alternating with orange/white flags mark bad ground to be avoided while taxying. Airfield or other boundaries may be marked with the orange and white tent strips illustrated.

Yellow and red chequered flag (or board) denoting airfield control is operated by special signals. The letter C (on yellow background) marks pilot's reporting point. Two figures (also on yellow background) give direction of take-off and landing (magnetic) to nearest ten degrees. These figures are usually painted on the duty runway or in the case of grass airfields, on the landing and take-off strip in use at the time.

Appendix 3 – Table of Air Exercises

AUSTRALIAN SYLLABUS

Exercise		*UK Exercise*
1	Aeroplane Familiarization	1
2	Preparation for Flight	2
3	Taxying	5
4	Operation of Controls	4
5	Straight and Level Flight	6
6	Climbing	7
7	Descending	8
8	Turning	9
9	Stalling	10
10	Sideslipping	included in 8
11	Take-off	12
12	Approach and Landing	13
13	Spins and Spirals	11, 15
14	Emergency and Special Procedures	
	(a) Engine failure	17a
	(i) partial	
	(ii) complete	
	(b) Precautionary search and landing	17b
	(c) Action in event of fire	1E
	(i) engine fire	
	(ii) other causes	
	(d) Ditching	17
15	Instrument Flying	*19
	Navigation	18

*Appreciation only. Fully explained in Volume 2. (old series)

CANADIAN SYLLABUS

Exercise		UK Exercise
1	Familiarization	3
2	Aircraft Familiarization and Preparation for Flight	2
3	Ancillary Controls	4
4	Taxying	5
5	Attitudes and Movements	4
6	Straight and Level Flight	6
7	Climbing	7
8	Descending	8
9	Turns	9
10	Flight for Range and Endurance	6
11	Slow Flight	10B
12	Stalls	10
13	Incipient Spins and Full Spins	10B, 11
14	Spiral Dives	15
15	Sideslipping	8
16	Take-off	12
17	The Circuit	12, 13
18	Landing	13
19	The First Solo	14
20	Illusions Created by Drift/Low Flying	16
21	Off Airport Approach Procedures	17a
22	Forced Approaches and Landings	17b
23	Pilot Navigation	18
24	Instrument Flying	*19
25	Night Flying	Ex. 20, Volume 2 (old series)

*Appreciation only. Fully explained in Volume 2. (old series)

SOUTH AFRICAN SYLLABUS

Exercise		*UK Exercise*
1	Cockpit Layout	1
2	Preparation for Flight	2
3	Air Experience	3
4	Effects of Controls	4
5	Taxying	5
6	Straight and Level Flight	6
7	Climbing	7
8	Descending	8
9	Stalling	10
10	Medium Rate Turns	9
11a	Descending Turns	9
11b	Climbing Turns	9
12	Take-off (including engine failure during and after take-off)	12
13	Approach and Landing (including going round again)	13
14	Spinning	11
15	First Solo	14
16	Sideslipping	included in 8
17	Steep Turns	15
18	Instrument Flying	*19
19	Low Flying	16
20	Crosswind Landing and Take-off	12, 13
21	Precautionary Landings	17a
22	Forced Landings	17b
23	Action in the Event of Fire	1E
24	(reserved)	
25	Aerobatics	Volume 2 (old series)
26	Night Flying	Volume 2 (old series)
27	Navigation	18

*Appreciation only. Fully explained in Volume 2 (old series).

Flight Briefing for Pilots. Volume 2 (old series) contains the advanced exercises which are outside the requirements of a UK Private Pilot's Licence in its basic form. These are:

5 Taxying (Twin-engined)
19 Instrument Flying
20 Night Flying
21 Aerobatics
22 Formation Flying
23 Multi-engine Conversion

Index

Index

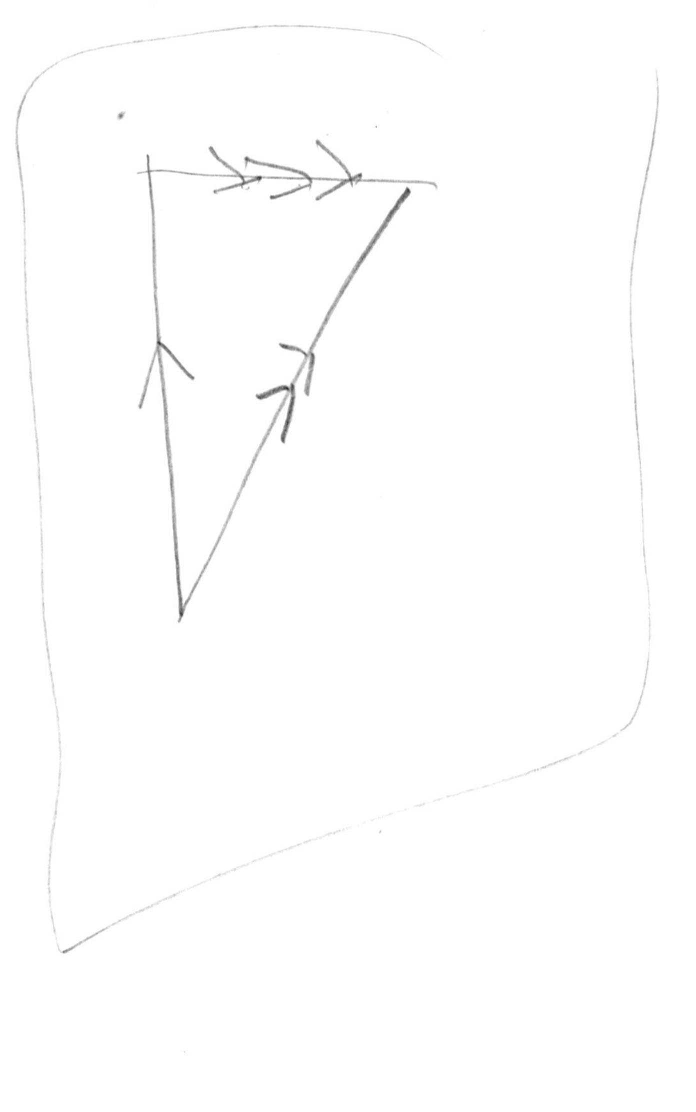

Creative FRIENDSHIP QUILTS

Creative HOUSE

The Template

Creative FRIENDSHIP QUILTS

KAREN FAIL

CONTENTS

INTRODUCTION

One of my very favourite quilts is on the front cover of this book. It is my friendship quilt from The Quilters' Guild of Australia. Each block was made by someone I know well – someone I had worked with while on the committee of The Quilters' Guild from 1990 to 1993.

At the end of my time as president, I was presented with this wonderful quilt. Knowing my busy schedule, the committee had not only sewn the blocks together, but basted the quilt ready for quilting. What an overwhelming gift! It is a constant reminder, not only of the exciting time I had working for the Guild, but of the people who shared my enthusiasm for quilting and became my friends. I will always treasure it.

This was not my first friendship quilt. Years before I had started a friendship group with members of The Eastwood Patchwork Quilters. Intrigued with the stories about friendship quilts in the nineteenth century, we began with great zeal and, by the end of the year, had made twelve stunning quilts. There were still over fifteen quilts to make and mine was one of them. It was hard to be patient. My 'Homespun Ocean Waves' quilt was finally made and, as it was one of the first really scrappy quilts I owned, I love it. Having quilted with this group for some time, I can identify the maker of each block by the fabric used, even though the blocks weren't signed.

There seems to be a certain mystique associated with the making of a friendship quilt. Often this making is done in secret and everyone involved becomes quite conspiratorial. The anticipated pleasure in the giving seems to become an overriding factor, as other projects are put aside and finishing the block on time becomes all important.

My enthusiasm for friendship quilts has not diminished, even though I have been participating in friendship groups for nearly ten years. I continue to revel in the creativity and pleasure that is generated in designing and executing these memory-filled quilts. This book is full of wonderful stories of friends and their quilts, from the very traditional Album block quilt to the outrageous 'Friendly Fungi' quilt. Be inspired!

FRIENDSHIP QUILTS

THE CRAFT REVIVAL

In today's fast-moving society, many of the traditional roles of men and women have been put aside in the Western world. Gone is the stereotypical image of the male as the provider and the woman as the homemaker. Since the Second World War, when women joined the workforce and performed many roles usually reserved for men, change has been inevitable. These days, it is not unusual for women to run their own companies or become engineers, while increasing numbers of men join the nursing profession or become at-home dads.

While welcoming the opportunities that come from being freed from traditional constraints, and recognising that people are more free to make decisions about their lives, the need to interact with other women is still keenly felt by many women. This need has, in part, led to a worldwide revival of interest in the traditional crafts.

For both the career-minded woman and the at-home mother, the chance to be creative as well as cognitive is treasured, and the opportunity to care and share is welcomed in the learning environment. Heirloom sewing, folk art, embroidery and patchwork are all enjoying this unprecedented revival, along with many other crafts. Skills that were once used to make clothes and household items out of necessity are now being learned and used to create heirlooms of the future. Designs developed to camouflage poor-quality furniture today are used to create treasured folk art pieces. Patchwork quilts are made lovingly from carefully selected pieces purchased from specialty patchwork shops, rather than from the fabric in the scrap bag. It seems that women without a traditional family support network and the everyday interaction in a neighbourhood have found ways of establishing similar structures that offer opportunities for creativity and friendship. Craft groups and guilds have sprung up everywhere, providing camaraderie, instruction and the opportunity for 'show and tell' through exhibitions and meetings. The proverbial 'pat on the back' is, of course, an essential part of the creative process.

Quiltmaking has always been recognised as an interactive craft, with a high level of sharing between the participants. Born out of necessity, when quilts provided warmth for many families during the bitter winters in frontier America, quiltmaking encompasses many inviting images. Perhaps it is the very nature of the quilt as a bed cover that conveys feelings of family, closeness, caring and warmth. With this ever-present need for women to provide bed coverings for their families and the drive to create something of beauty in a harsh environment, there were always quilts to be quilted on the frontier. Today, it is difficult for us to imagine the deprivation suffered by women and their children in the

FRIENDLY FANS, *169 cm x 190 cm (65⅞ in x 74⅛ in), 1995. Made by Epping Patchwork Quilters for Elaine Gallen.*

pioneering West, housed in an earth-lined dugout with a draughty roof over their heads, far from family and friends. Their collection of quilts offered them some measure of warmth, both physical and emotional.

'Back when I was a girl, quilts were something that a family had to have. It takes a whole lot of cover to keep warm in one of them old open houses on the plains.' (Patricia Cooper and Norma Bradley Buferd, *The Quilters – Women and Domestic Art, An Oral History*, Anchor Press/Doubleday USA 1978.)

Great excitement greeted the completion of a quilt top. Women in the area would anticipate a quilting bee and wait for their invitation to participate. They would travel considerable distances to be part of a 'bee' and often these events would create an opportunity for new neighbours to be introduced for the first time. While these events were ostensibly to complete the quilting of the quilt, the real benefits for all those who gathered were found in the enjoyment of each other's company, the sharing of hopes and dreams for the future and a time to relax, away from the harsh realities of their daily lives. Such was the popularity of these events that, when the quilting was finished, the men were invited to join the women for a shared meal and to be part of the fun.

Young women in America were all taught to piece blocks from a very early age, while girls in Australia were working on cross stitch samplers. Sometimes, the lessons in piecing began when a daughter was as young as four. Simple triangular and square shapes were assembled to form the block patterns, and the running stitch required for stitching was demonstrated and corrected when necessary. Often at quilting bees, the younger girls were allowed to thread the needles so there was a continual supply of thread for the quilters. Everyone appreciated the necessity of these skills and young women would aim at completing at least twelve quilts in anticipation of their marriage. After marriage, many new wives travelled to isolated areas in the West with their husbands. Their quilts, packed in a trunk, were an essential part of their luggage.

DENISE'S SMOOTHING IRON QUILT, *130 cm x 194 cm (50⅞ in x 75⅝ in), 1995.*
Made by friends and family for Denise East.

DESPITE THE NEED FOR WARMTH IN THESE DESOLATE AREAS, SOME QUILTS REMAINED CAREFULLY FOLDED AT THE BOTTOM OF THE TRUNK AND WERE RARELY, IF EVER, USED. THESE WERE FRIENDSHIP QUILTS. KEPT IN NEAR-PRISTINE CONDITION, THEY WERE A CONSTANT REMINDER OF FAMILY AND FRIENDS WITH THEIR MESSAGES OF LOVE AND FRIENDSHIP WRITTEN CAREFULLY ON EACH BLOCK.

Thousands of friendship quilts were made in America during the nineteenth century, especially from 1840 to 1878. As families moved from the settled villages in the New England area into the newly established frontier towns, the idea of friendship quilts went with them. Simple blocks, pieced from scraps, were soon being made by many women throughout the country. In fact, some of these women would have had their name on several friendship quilts. Friendship blocks were usually of a simple design, not unlike those used for everyday quilts. More often than not, these quilts had a light coloured piece in the centre so that the maker

DETAIL OF 'MY FRIENSHIP QUILT', *150 cm x 180 cm (58½ in x 70¼ in), 1993. Made by friends for Lillian Atkinson.*

DETAIL OF ALBUM QUILT, *170 cm x 225 cm, (66¼ in x 87¾ in), 1987. Made by friends for Annette McTavish.*

could sign her name and write a greeting. A popular choice of pattern was the Album block (above left and right).

There were several ways a friendship quilt could be made. One way, was to send the pattern for the proposed friendship quilt to various family members and friends in the hope that they would make a block from their scraps. Each woman considered it essential to have a scrap bag in which every piece of fabric was saved, because new cloth was expensive and considered a luxury. In the scrap bag would be offcuts from past dressmaking, household linen and furnishings projects, and these would be selected to be included in the friendship block. The blocks would then be returned, either signed or with the maker's name attached. Sometimes, one person with a particularly attractive copperplate handwriting was chosen to write the name of each maker on their block.

Friendship quilts could also be made by only one person from fabric she collected from her family members and friends.

LINDA OTTO LIPSETT IN HER BOOK *REMEMBER ME* (THE QUILT DIGEST PRESS, 1985) DESCRIBES LUCY BLOWER'S FRIENDSHIP QUILT AS A COLLECTION OF BLOCKS MADE BY LUCY OVER MANY MONTHS FROM FABRIC SUPPLIED BY HER FRIENDS AND FAMILY FROM THE MATERIAL OF THEIR FAVOURITE DRESS. LUCY IS REPORTED TO BE PARTICULARLY FOND OF HER QUILT AS SHE WAS ABLE TO PICTURE THE PEOPLE SHE LOVED IN THEIR FAVOURITE OUTFITS.

Sometimes, the maker would only require her family and friends to sign their names to a block, after she had completed all the blocks herself.

During the 1830s and 1840s, the autograph album was extremely popular in America, with friendship being romanticised by such publications as *Godey's Lady's Book*. Poems and thoughts suitable for inclusion in these treasured albums were often included in Godey's, and it was not long before these began to appear on friendship quilts. It was also during the 1830s that indelible inks, made from a mixture of silver nitrate, ammonia and lampblack, had been developed in France, making the process of writing on fabric an attractive alternative to painstakingly embroidering signatures. Previously, the only inks available contained iron which rotted fabric and left a nasty stain. Simple messages like 'Remember Me' or a verse from a favourite poem could now be included, along with the maker's signature. Often, the signatures were hidden among elaborate drawings of scrolls, tiny leaves and flowers. Such was the enthusiasm for these ornate inscriptions on friendship quilts, that stamps and stencils of them were soon developed, with a space in the centre for name signing to make this type of embellishment easier.

RAJAH QUILT, *325 cm x 327 cm (126¾ in x 127½ in), 1841. Made by the women aboard the* Rajah *for Elizabeth Fry.*

Commonly, a friendship quilt was presented to a woman by her family, when she was leaving to make a new life with her husband, or by a community where she had lived for some time. Any event, in fact, could be the stimulus for making a friendship quilt, from births and weddings to leaving for the mission field.

Because these quilts were so treasured, many have survived and provide a lasting record of the inhabitants of a town or the people who made the quilt. Even the names of deceased family members were often included. In some circumstances, these quilts are the only record of the female members of a family as, sadly, the government records only listed the heads of families – usually men.

THE *RAJAH* QUILT

Working together on quilts was not the sole prerogative of the American quilter. As early as 1841, the convict women bound for Australia aboard the *Rajah*, worked on a quilt for Elizabeth Fry. She and her Quaker committee, as part of their prison reform measures, had been providing an opportunity for gainful employment for women from Newgate Prison. These hapless women were given materials for sewing and knitting and were then able to sell the finished items. These measures were very successful and led to Mrs Fry's decision to provide fabric and sewing supplies to the women being sent to the colony in New South Wales. The British Society of Ladies, as Mrs Fry's organisation became known, gave each reluctant traveller a Bible; one hessian apron; one black stuff ditto (bag); one black cotton cap; one large hessian bag; one small bag, containing one piece of tape, one ounce of pins, one hundred needles, four balls of white sewing cotton, one ditto black, one ditto blue, one ditto red, two balls of black worsted, twenty-four hanks of coloured thread, one cloth with eight darning needles, one small bodkin, two stay laces, one thimble, one pair of scissors, one pair of spectacles, two pounds of patchwork pieces, one comb, one knife and fork. With these supplies, it was anticipated that the women would make quilts

REVEREND NADAL'S QUILT, *260 cm square (101½ in square), 1847. Made by parishioners and friends for Reverend Bernard Nadal.*

in the traditional English style, so they would have something to sell at the end of their horrific voyage.

The level of skill among the women who worked on the medallion quilt for Mrs Fry was very varied, and little care was taken to see that exactly the same-sized templates were used or that borders finished in an appropriate spot. It is obvious that many hands worked on the quilt, as the borders change along their length, as does the size of the shapes and their accuracy. Nevertheless, the quilt is quite complex – an accomplishment for those women living in such cramped conditions on board the *Rajah*. Preserving every bit of precious fabric they had, they appliquéd small motifs onto the central square, using the broderie perse technique so popular in England at that time. Surrounding it is a quite complex border of pieced squares and triangles. The final wide border has more broderie perse and appliqué.

Although not signed by each individual who shared in the work, the quilt does bear the following inscription:

'To the ladies of the Convict Ship Committee this quilt worked by the Convicts of the Ship Rajah during their voyage to Van Diemans Land is presented as a testimony of the gratitude with which they remember their exertions for their welfare while in England and during their passage and also as a proof that they have not neglected the Ladies kind admonitions of being industrious. June 1841.'

On receiving the quilt, Elizabeth Fry must have felt that her work with the women prisoners was not in vain, and surely it would have become one of her most treasured possessions.

The women from the *Rajah* do not appear to have maintained their enthusiasm beyond this one medallion quilt, and they failed to embrace Elizabeth Fry's suggestion that quiltmaking could be a viable cottage industry in the developing town of Sydney. Lack of initiative, as well as the extremely poor supply of fabric in the colony, certainly would have contributed to this situation.

WHILE THERE WERE MANY FRIENDSHIP QUILTS MADE IN AMERICA USING SIMPLE GEOMETRIC PATTERNS, SOME QUILTS WERE MORE ELABORATE AND INVOLVED QUITE COMPLEX APPLIQUE BLOCKS. THESE WERE KNOWN AS ALBUM QUILTS, THE MOST FAMOUS OF WHICH WERE THE BALTIMORE ALBUM QUILTS.

It is thought that these superbly executed appliqué designs were the work of a few talented women, like Mary Evans of Baltimore, who completed blocks on commission from wealthy women of the area. These wonderful quilts were often given to those held in very high esteem. Not all appliqué friendship quilts were as elaborate or meticulously worked as the Baltimore Album quilts, but appliqué was often chosen to create very special quilts. A much-loved travelling pastor might be the recipient of a special appliqué quilt from the women of the circuit, who usually made only serviceable quilts for their families.

'We sent out the word by him along the circuit for ladies of other congregations to send a design for the top. He could carry them little appliquéd pieces easy in saddlebags, no weight to 'em. We gathered it all in and put that quilt together. That was a feat in those days. He said he never seen anything so pretty. It was a treasure.' (Patricia Cooper and Norma Bradley Buferd, *The Quilters – Women and Domestic Art, An Oral History*, Anchor Press/Doubleday USA 1978.)

Perhaps this quilt was like the one presented to the Reverend Bernard H. Nadal of Baltimore. His quilt top reflected the trend for complex appliquéd blocks to be included in friendship quilts, if the person receiving the quilt was held in high regard. Most of the blocks have embroidered or inked details, with the red Bible, dated 1847, in the centre square inscribed 'To Rev. Bernard H. Nadal, Baltimore'. Nearly all the inscriptions appear to have been done by the same hand, demonstrating the trend to employ a single person with a good standard of copperplate writing to inscribe all the names and messages. This quilt top is in the collection of the Smithsonian Institution in the United States, a gift of Miss Constance Dawson, great-niece of the Reverend Bernard Nadal.

By 1870, after the Civil War, America was recovering economically and women wanted the freedom to buy, rather than make their bed linen. Although quilts were still being made, the making of friendship quilts became less fashionable. Indeed, the use of cotton fabrics was considered passé. With the introduction of velvets and silks for clothing, these fabrics quickly replaced cotton in quiltmaking, and crazy patchwork became the fad. Australian women usually kept up with the trends overseas and, by 1890, many of them were experimenting with crazy patchwork and the use of silks and satins in traditional quiltmaking. The making of friendship quilts was one trend that was not observably followed in Australia at this time. The quilts made during 1830 to 1870 were in the English style of quiltmaking, using one shape, such as the hexagon, or medallion-style like the *Rajah* quilt. In the early part of the nineteenth century, these quilts were made in cotton, but by 1890, quilts using hexagon and diamond shapes were fashioned in the richer silks and satins now readily available in the colonies.

AUNT CLARA'S QUILT (Above), *150 cm x 200 cm (58½ in x 78 in), 1915; Clara Bate as a young woman* (Top right); **DETAIL OF 'AUNT CLARA'S QUILT'** (Right).

Aunt Clara's Quilt is a remarkable Australian quilt which emulates the style of crazy patchwork with its ornate embroidery stitches, but features a black square surrounded by a regular elongated hexagon as the basic unit, rather than irregularly shaped pieces. It is reminiscent of a pattern recorded in Caulfield's *Dictionary of Needlework* in 1887. The hexagons are cut from a variety of exotic fabrics of many hues, including some plaids and stripes, among the mostly plain fabrics. Each hexagon is embroidered with a symbol of everyday life, including garden tools, vegetables and flowers, garden bugs and spiders, pipes, firecrackers, a telephone, dates and names: the list is endless. Many different techniques and styles of embroidery are used, including satin stitch, French knots and feather stitch. Various trinkets, like clay pipes and tiny handbags, were attached to the quilt and are still preserved, a testimony to the care given to the quilt by the present owners. The quilt was completed in 1915, the last date recorded on the quilt. Family folklore suggests that the quilt is like a diary, recording the daily events in Clara Bate's life, and the comings and goings at her guesthouse, Frankfurt, at Gingkin, New South Wales at the turn of the century. One story, passed down through the family, concerns the recurring pipe motif on the quilt. While travelling on a train, a passenger accused Clara of giving him hayfever and promptly threw her offending flowers out of the window. She retaliated by snatching his pipe out of his mouth and tossing it from the train, complaining of the smell.

The variety of styles of embroidery on the quilt suggest that many hands participated in its making. Perhaps Clara invited her guests to contribute and this is, in fact, her friendship quilt, a reminder of all the interesting people who visited her guesthouse. The quilt was completed by Clara's sister, Emma, during 1915, after Clara's death on 18 December 1914.

CHANGI QUILTS

The elegance and excellence of Aunt Clara's Quilt is in direct contrast to the simple embroidered quilts known as the Changi quilts. After the surrender of Singapore to the Japanese on 15 February 1942, many soldiers and civilians were interned in camps at Changi on Singapore Island. Over three thousand men, women and children were crowded into a facility originally designed for six hundred people, the women arriving by foot with whatever possessions they could carry. It was essential that the women be able to communicate with their husbands in the military camps. Ethel Mulvany, who had worked for the Red Cross, suggested that they make quilts embroidered with their own names and other motifs as a means of letting the men know they were safe.

In the early days of the camp, there were limited supplies for sewing, which some of the women had managed to bring with them – despite their hasty departure from their homes.

ON THE WHITE FABRIC SQUARES THAT WERE DISTRIBUTED, EACH WOMAN EMBROIDERED HER NAME AND SOMETHING ABOUT HERSELF. WONDERFUL IMAGES WERE CREATED WITH LIMITED RESOURCES AND THERE WERE EVEN ATTEMPTS AT HUMOUR, WITH ONE BLOCK READING 'CHANGI HOLIDAY HOME'. THE BLOCKS WERE THEN SEWN TOGETHER BY MACHINE AND THE SEAMS EMBROIDERED.

While these quilts are not typical friendship quilts, they were made with the same spirit of communicating love and care to others. Three quilts are still in existence, although it is thought that several more may have been made. These three were each made from sixty-six squares, and each had an inscription on the back. One was for wounded Australian soldiers, one for the English soldiers and one for the Japanese soldiers. The inscription on the quilt for the Australian soldiers reads:

'Presented by the women of Changi Internment Camp 1942 to the wounded Australian soldiers with our sympathy for their suffering. It is our wish that on cessation of hostilities, this quilt be presented to the Australian Red Cross Society. It is advisable to dryclean this quilt.'

The women hoped that when the quilts were presented to the hospital in the military camp, the men would be encouraged by the patriotic messages. They thought that including a quilt with an inscription for the Japanese soldiers would improve the chances of the other quilts arriving at their chosen destination. On arrival, the names on the quilt were quickly circulated throughout the camp, reassuring many that their wives and children were safe and well. Two of the three surviving quilts are housed at the Australian War Memorial in Canberra, and the third at the Red Cross Training Centre at Barnet Hill, England.

CHANGI QUILT, *130 cm x 203 cm (50⅝ in x 79⅛ in), 1942. Made by the internees of Changi POW camp.*

FRIENDSHIP QUILTS, *160 cm x 220 cm (62½ in x 85¾ in), 1965. Made by Ilma Hinwood.*

AUSTRALIAN FRIENDSHIP QUILTS

Before the revival of quiltmaking in the 1970s, first in America and eventually worldwide, the quilts made in Australia continued in the English tradition. Many of these quilts, made during the first half of the twentieth century, were traditional hexagon quilts. Construction was quite different from the American system where pieces were sewn together using a running stitch. Here, each shape was basted to an exact paper hexagon before the pieces were whipstitched together. When the top was completed, the basting was removed along with the papers.

It is unusual to find an Australian quilter who followed American traditions and patterns during that time; however, one such quilter was Ilma Hinwood, now aged eighty-six, who took up quiltmaking in the 1920s. Greatly influenced by the quilts being made in America, she made twin quilts for her husband and herself. As one set wore, she would make another, completing several pairs of quilts for their twin beds. Favouring appliqué, Ilma chose such interesting themes as wedding customs around the world, and scenes from Louisa May Alcott's book, *Little Women*. She also made two friendship quilts during the early 1960s. Ilma asked family members and friends to embroider or write their name on the small squares of fabric she provided, then she appliquéd these around the central panel of her quilts, which feature flowers and birds arranged slightly differently for each quilt. Her daughter-in-law, Marilyn Hinwood, remembers using the quilts when she visited her mother-in-law. Ilma was not interested in her quilts just being showpieces. She was always keen for her quilts to be used.

FOLLOWING THE RESURGENCE OF INTEREST IN QUILTMAKING IN AMERICA IN THE 1970s, AUSTRALIAN QUILTMAKERS QUICKLY FOLLOWED SUIT, ADJUSTING TO THE NEW TECHNIQUE OF PIECING WITHOUT PAPERS AND WELCOMING THE WIDE RANGE OF TRADITIONAL PATTERNS ON OFFER.

Some Australians, like Heather Madden of Epping, New South Wales, were lucky enough to be living in America during this time. Heather joined a newly formed quilting group, called the Trumbull Piecemakers, while living in Connecticut. When she was about to return to Australia in 1978, the group presented her with a friendship quilt. It features simple pieced and appliquéd blocks joined with cream and brown sashing, and each block is signed. One of the blocks shows a map of the United States marked with the location of the Trumbull Piecemakers.

Tiny Kennedy, from Launceston in Tasmania, is an expatriate American who is the proud owner of a friendship quilt from her friends of the Gabilian Mountain Quilters in San Juan, California. Tiny designed her own house block and each one of the group of twelve picked a month of the year and decorated the house block with that month in mind. Some houses are decorated for Christmas and Thanksgiving, while others welcome spring and summer. Tiny was so thrilled by her blocks that she decided to make a rainbow of triangles as a border. This was all before half-square triangles and quick-piecing had been thought of! 'Friends across the Sea' was completed in time for the first Tasmanian Quilters' Guild's Exhibition in 1984. With many adjustments to be made in a new country, Tiny found her quilt a constant reminder of her friends back in America.

Quilts, like Heather's and Tiny's were made in the tradition of friendship quilts where several friends each made a block and signed it, as a memento for a friend who was leaving. Sometimes, the person receiving the quilt was surprised and delighted with this special gift, as was Heather, but often they were completely involved in the making of the quilt, just like Tiny, who actually designed the block and distributed the instructions for its completion.

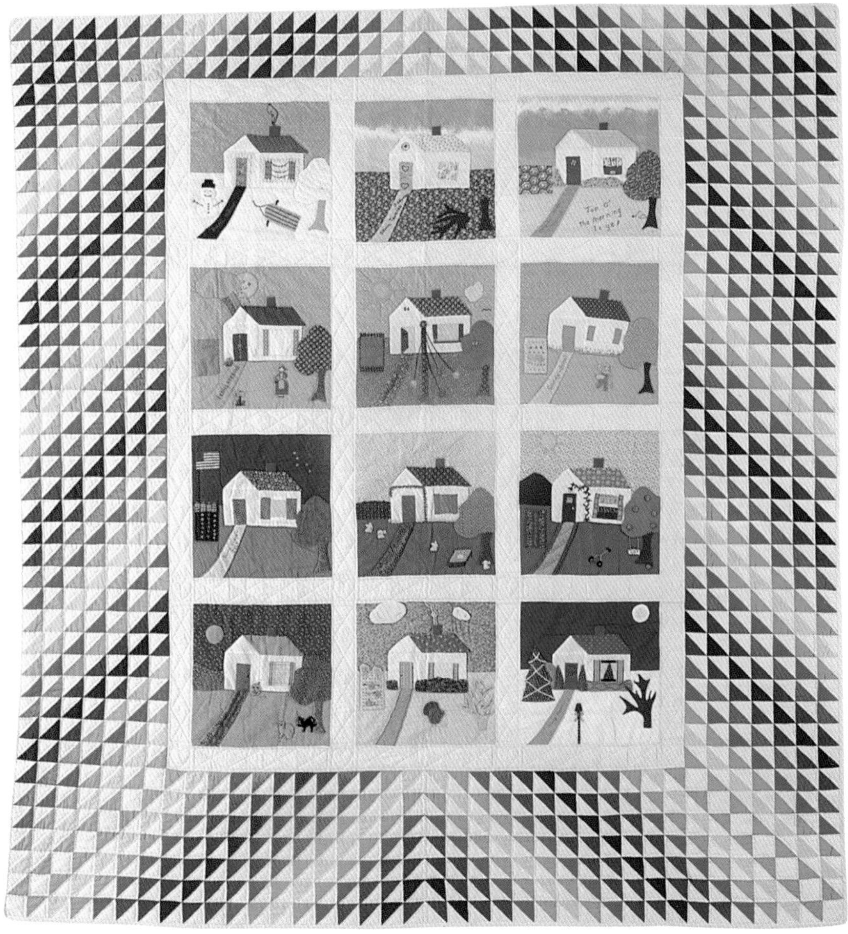

MY AMERICAN EXPERIENCE (Above right), *180 cm x 230 cm (70¼ in x 89⅝ in), 1978. Made by the Trumbull Piecemakers for Heather Madden.*

FRIENDS ACROSS THE SEA (Right), *248 cm x 255 cm (96⅝ in x 99½ in), 1984. Made by the Gabilian Mountain Quilters for Tiny Kennedy.*

As enthusiasm for quiltmaking grew in Australia, so did the making of friendship quilts. Quilters began to form small groups to make each other friendship quilts. Usually, the names of group members were placed in a container and drawn out in turn to determine the order of the making. Then, each member of the friendship group, when it was her turn, would decide on a pattern for her block and perhaps a colour scheme and, within a month, the completed blocks would be returned to her.

In some groups, the package handed out also included the templates and any background fabric that the owner wanted to remain consistent. Other groups preferred to cut out the entire block, then all that was required was for the participants to sew them. Participants were usually asked to sign their blocks and some were even asked to embroider motifs and their names on the completed blocks, such as on Sylvia Fenech's quilt pictured opposite.

These early friendship quilts reflected the basic, simple style of quiltmaking. However, with more experience, the planning of the quilts became quite sophisticated, demonstrating interesting settings, original block designs and clever coordination.

The Western Australian Quilters' Association was formed in 1981 and the idea of friendship quilts rapidly gained acceptance among its members. Faye Cunningham's hat quilt was planned so that each hat would reveal something about the maker. One hat even has a feather from the quilter's budgie attached to it! Faye says the quilt reminds her of the times when she and her friends used to travel to town in hat and gloves. With four of the participants now in their eighties and some having returned to live in America and England, this quilt is one of Fay's treasured possessions.

News about quilts, exhibitions, group activities, workshops and tutors was easily disseminated, once the quilters' guilds were formed in each state. The enthusiasm for friendship groups and the quilts they made was contagious. Groups all over the country set in place the processes for making friendship quilts. These same processes were used to make quilts for charity, following the generous example of quilters of the past.

One Sydney group, the Eastwood Patchwork Quilters, embraced with great gusto the idea of making friendship quilts and, even though they had nearly twenty members, decided to form a friendship group in 1985. As you can imagine, there was a long wait for some, but for those whose name was drawn giving them an early slot, there was great excitement. The first name drawn was Evelyn Finnan's, who chose a scrappy 'Autumn Breeze' quilt from a pattern in the *Quilter's Newsletter*. It was a frantic time for Evelyn, as she only had one month to work out her design and prepare more than twenty packs of instructions. The rules were simple: choose a block with less than twenty-four pieces

FOURTEEN FRIENDS FRIENDSHIP QUILT, *147 cm square (57⅞ in square), 1983. Made by the Western Australian Quilters' Association for Faye Cunningham.*

and provide any templates required and any fabric you want to be consistent in the blocks. The pattern sheet, instructions for making, together with prepared templates and fabric, if required, were packaged and given to each member of the group. Evelyn's design involved two different blocks, so she divided the members into two groups, with each one completing a different block. She provided cream homespun with the templates already marked on the fabric and asked her friends to provide autumn-toned fabrics for the leaves.

THE ADVANTAGE OF WORKING WITH A LARGE GROUP IS THAT YOU RECEIVE LOTS OF BLOCKS; THE DISADVANTAGE IS THAT IT CAN TAKE A LONG TIME FOR YOUR TURN TO COME AROUND, IF YOU ARE UNLUCKY ENOUGH TO BE DRAWN LAST. MOST GROUPS EVENTUALLY CONCEDE THAT TWELVE MEMBERS IS THE OPTIMUM SIZE, AS EVERYONE RECEIVES THEIR BLOCKS WITHIN A YEAR.

After the initial round of friendship quilts was completed, the Eastwood Patchwork Quilters broke into smaller groups, each with twelve members. One group decided to be a challenge group for those who wanted to tackle harder blocks. Jo Petherbridge of Asquith chose Feathered Stars for her challenge friendship block and requested that the blocks be machine-pieced. To obtain the accuracy required, using the machine, was indeed a challenge for everyone involved, but the resulting quilt was spectacular. The challenge enabled many of the group to refine their machine-piecing skills, and Jo was always available with helpful advice.

FEATHERED STARS (Top right),
165 cm x 225 cm (65½ in x 87¾ in), 1991.
Made by the Eastwood Patchwork Quilters
for Jo Petherbridge.

AUTUMN BREEZE (Centre right),
205 cm x 256 cm (80 in x 100 in), 1985.
Made by the Eastwood Patchwork Quilters
for Evelyn Finnan.

FRIENDSHIP RINGS (Right),
238 cm x 262 cm (92¾ in x 102⅛ in), 1995.
Made by the Castle Hill Quilters
for Sylvia Fenech.

USE OF SIMPLE BLOCKS

Traditionally, very simple blocks were chosen for friendship quilts and the Album block, a favourite in the nineteenth century, is still a popular choice. Annette McTavish of Beecroft, New South Wales, has a beautiful Album block quilt, inscribed with names and messages from the members of her group, Just Friends. This group was initially formed to make friendship quilts and has made over one hundred quilts since the group's inception in 1983. Annette's friends chose not to write on the blocks, preferring to embroider their messages instead.

Moon over the Mountain is such a simple block that it is rarely chosen by quilters, even in a beginner's sampler quilt. But Faye Cunningham's quilt, made by members of the Quilters' Network, is a delight. Quilters' Network is organised by Marti Johnson, who lives in Sacramento, California, and communicates with quilters worldwide through a quarterly newsletter. Member quilters from Brazil, Japan, Hungary, Ireland, Norway, Belgium, America, Canada and Australia make 'cuddle' quilts for quilters in trouble, and have fabric and block exchanges. When Faye participated in a Moon over the Mountain block exchange, she never dreamed that she would win the blocks. When she did win, Faye wondered how she could make an interesting quilt from what she thought were uninspiring blocks. Her idea of adding stars and details like

GRANDMOTHER'S FLOWER GARDEN (Above), *204 cm x 360 cm (79½ in x 144 in), 1933. Made by the Northbridge Quilters for Anne Docker.*

ALBUM QUILT (Left), *170 cm x 225 cm (66¼ in x 87¾ in), 1987. Made by American friends for Annette McTavish.*

the witch on a broomstick, has certainly worked and now she treasures her international friendship quilt, made by quilters from America, Canada, England, Austria, New Zealand and Australia.

Anne Docker from Northbridge, New South Wales, chose Grandmother's Flower Garden for her friendship block, using a simple hexagon template.

HER LOCAL GROUP, THE NORTHBRIDGE QUILTERS, HAD DECIDED THAT EVERYTHING HAD TO BE CUT OUT READY FOR SEWING FOR THEIR FRIENDSHIP QUILTS. WHILE IT SEEMED AN ONEROUS TASK TO CUT OUT OVER ONE THOUSAND HEXAGONS, IT WAS EVEN MORE ONEROUS TO CONTEMPLATE SEWING THEM, SO ANNE WELCOMED THE OPPORTUNITY TO HAVE A HEXAGON FRIENDSHIP QUILT.

When she received the completed blocks, Anne realised she had been too liberal with the hot pink fabric. In an effort to tone this down, she added the green diamonds when setting the blocks together. Starting with a simple block, Anne has created a majestic quilt.

Another very simple block design is Mayflower. Mayflower is a very welcome choice for those making friendship blocks, because it is so easy. Helen Sears of Eastwood, New South Wales, received a delightful selection of fabrics when she asked for a scrappy Mayflower block quilt. To maintain uniformity, Helen provided the background fabric which she then quilted heavily to create her beautiful quilt.

INTERNATIONAL MOONS (Top right),
152 cm x 205 cm (59¼ in x 80 in), 1992.
Made by the Quilters' Network for Faye Cunningham.

MAYFLOWER (Right),
180 cm x 210 cm (70⅞ in x 82 in), 1990.
Made by Eastwood Patchwork Quilters for Helen Sears.

UNUSUAL SETTINGS

Even the simplest block can look stunning, when the blocks are set in an interesting way and the finished quilt is heavily quilted. Margaret Scott of Epping, New South Wales, set her simple Crosses and Losses blocks, made in various shades of blue, in a medallion style which allowed generous spaces between the blocks. With such a nautical feel to the quilt, Margaret decided to use sailing ships as the quilting motif, which she quilted beautifully in the large triangles surrounding her central medallion of blocks. A member of her group has a husband in the Navy – his books on sailing ships provided the quilting patterns. Margaret says she really appreciated the input from several members of her group on both the setting of the blocks and her quilting design.

Small groups need not be a bar to generously sized friendship quilts. With only eight members in her group, Bearly There Quilters of Hornsby, New South Wales, Judy Ellis had to make an extra twenty-four blocks herself to complete her beautiful 'Path to Granny Sullivan's House' quilt. The use of two tones of green to set the blocks creates an interesting effect, with fourteen of the blocks set into a dark green border. Judy has lived most of her life in the country, in the little town of Merriwa in the Hunter Valley and has only recently moved to Sydney. As a result, being part of the Bearly There Quilters has special significance for her, creating an instant circle of friends with common interests. When Judy received her friendship blocks, she embroidered each person's name on their block, giving her a constant reminder of their friendship to a 'girl from the bush'. The quilt has pride of place on her bed.

Most friendship blocks are of uniform size, but when there is no size restriction, interesting dilemmas are created for setting the blocks.

The idea for Lauree Brown's friendship quilt came to her as she was recovering from a fractured elbow, spending much of her time sitting at her living room window, gazing at the sky and the view over Launceston, Tasmania. In an effort to capture that image of the night sky, she asked her friends to make any star, comet or planet pattern, using cottons or shiny chintz fabric in red, pink, orange, gold, yellow, white or

SAIL ON (Above), *65 cm square (25⅝ in square), 1994. Made by the Eastwood Patchwork Quilters for Margaret Scott.*

silver, patterned or plain fabric from their scrap bags. Lauree provided the blue background fabric. Each person was invited to embroider her name on the star, comet or planet in a contrasting colour.

Lauree assembled the blocks so that they looked like the night sky, using the blue background fabric she had supplied to join the blocks and fill in the gaps. She then applied a black silhouette of gum trees and the outline of the surrounding hills, festooned with sequins, beading and embroidery to represent the lights in the valley. The hand-quilting was completed in 1991. Lauree called her unusual friendship quilt 'Friends Remind Me of the Stars in Heaven' and feels it is a fitting reminder of so many wonderful friends.

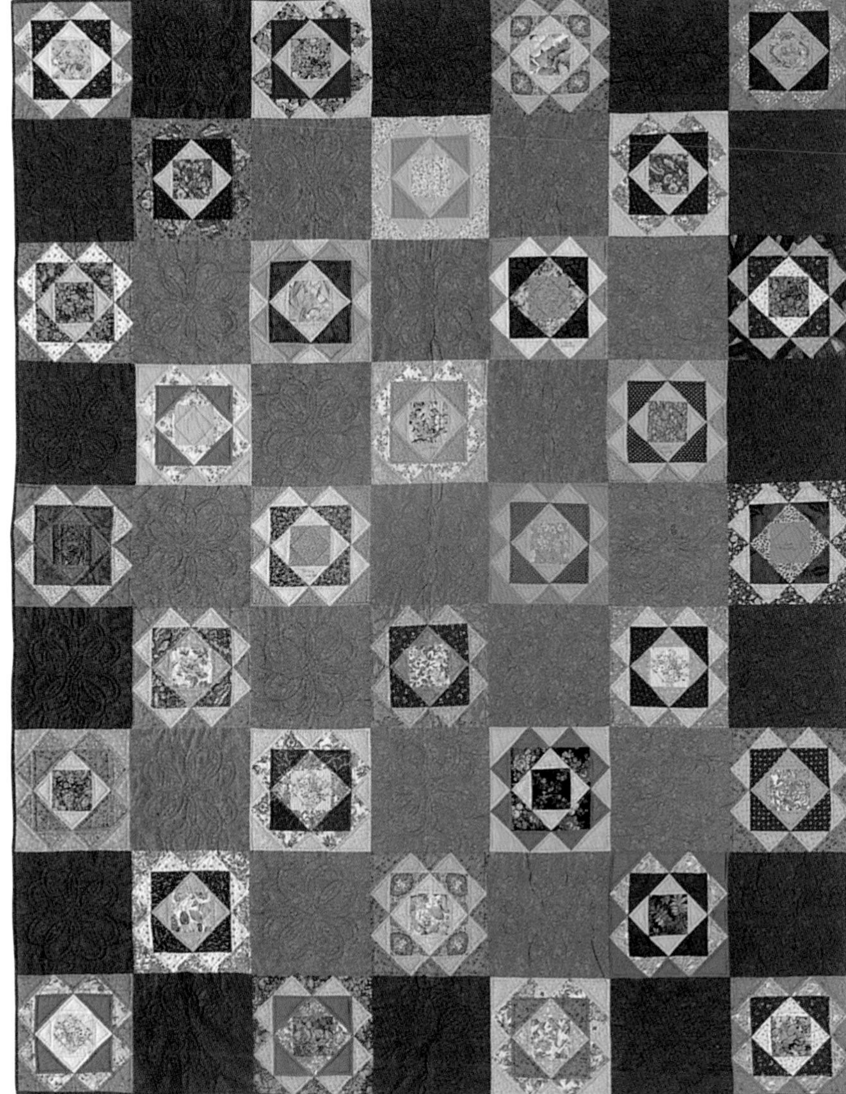

PATH TO GRANNY SULLIVAN'S HOUSE (Above right), *185 cm x 238 cm (72⅛ in x 92⅞ in), 1995. Made by the Bearly There Quilters for Judy Ellis.*

FRIENDS REMIND ME OF THE STARS IN HEAVEN (Right), *217 cm x 226 cm (84⅝ in x 88⅛ in), 1991. Made by the Launceston Quilters for Lauree Brown.*

AN AUSTRALIAN IDENTITY

In the 1980s, Margaret Rolfe and Deborah Brearley began to develop an Australian identity for quiltmaking by publishing original designs featuring Australian wildflowers and animals. These designs proved very popular and many Australian quilters incorporated them into quilts and clothing, especially for children. Occasionally, these designs are chosen for friendship quilts, creating charming quilts which are recognisably Australian.

Margaret Parry from Epping, New South Wales, has made a small quilt, featuring Deborah Brearley's wattle pattern. The pattern is very simple, using only a square in yellow and white to give the impression of Australia's national floral emblem. Margaret provided all the fabrics for the blocks and asked each of the members of Epping Quilters to sign their names in white or yellow thread. This ensured that her friends' names were included on her quilt, without detracting from the design. She arranged the twelve blocks to create a delightful medallion-like centre, with half the wattle pattern being used to create the border. Margaret used a thick wadding and simple quilting, in keeping with the puff of yellow wattle.

WARATAHS (Above), *232 cm x 265 cm (90½ in x 103⅜ in), 1994. Made by the Marion Quilters for Dot Foster.*

WARATAH CUSHION (Below left)

THE BOTTLEBRUSH, ANOTHER ORIGINAL BLOCK DESIGN BY DEBORAH BREARLEY, IS FEATURED IN THE FRIENDSHIP QUILT MADE FOR SANDRA JAMES BY THE EASTWOOD PATCHWORK QUILTERS. THE BOTTLEBRUSH BLOCK USES ONLY A SIMPLE RECTANGULAR SHAPE, AND SANDRA'S FRIENDS HAVE INTERPRETED THE FLOWER EFFECTIVELY BY USING INTERESTING PRINTS AND COLOURS THAT REFLECT THE FEATURES OF THE FLOWER.

With twenty-eight blocks to include in her quilt, Sandra chose a very successful symmetrical placement, adding detailed quilting of the bottlebrush in the blank squares. The whole design is unified by the green fabric Sandra has chosen for the blank blocks, reminiscent of the foliage of the bottlebrush bush.

The Marion Quilters made Waratah blocks for Dot Forster from Clovelly Park, South Australia. Dot comes from England, and she chose this design

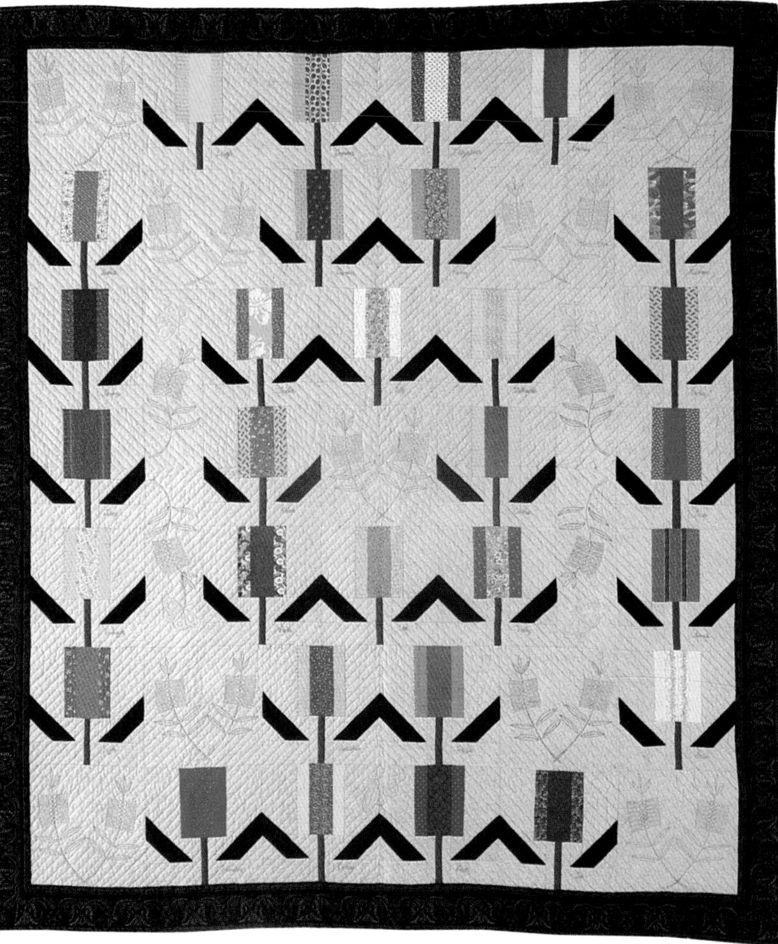

as she wanted a truly Australian quilt. She was also keen to provide all the fabrics for each block, as she had very definite ideas about how she wanted her quilt to look. She arranged fourteen of her fifteen blocks around a rectangular central panel to create a very graphic quilt, which she then quilted beautifully with a grid pattern. The fifteenth block was made into a cushion, on which are recorded the names of all the participants, in the order of the placement of the Waratah blocks in the quilt. Rather than putting the names on the back of the quilt, as other members of her group had done, Dot wanted them on display, and her treasured cushion is always on the brass bed which was bought especially to display Dot's wonderful waratah quilt.

BOTTLEBRUSH BOUQUET (Above left), *200 cm x 230 cm (78 in x 89⅝ in), 1995. Made by the Eastwood Patchwork Quilters for Sandra James.*

DETAIL FROM 'BOTTLEBRUSH BOUQUET' (Top)

WATTLE QUILT (Left), *140 cm x 147 cm (54⅝ in x 57⅜ in), 1995. Made by the Epping Quilters for Margaret Parry.*

CUT-OUTS

Many quilters like the variety and the surprises that come when friends are asked to use their own fabrics for friendship blocks. However, some quilters like to engineer their quilts so that they receive blocks for a specific quilt. Heather Wootton asked her friends at the Eastwood Patchwork Quilters to piece all the baskets she needed to make her basket quilt. Each of the sixteen participants received identical fabrics, already cut out with the seam lines marked, ready for hand- or machine-piecing. When the completed blocks were returned, Heather decided to add appliquéd flowers and leaves on the basket handles.

WITH THE HELP OF A FELLOW EASTWOOD QUILTER, STEPHANIE INTRONA, SHE DESIGNED SOME DELIGHTFUL FLOWERS ENTWINED AROUND THE HANDLE AND REPEATED THE MOTIFS ON THE LARGE CALICO BORDERS. WHEN HEATHER HAD FINISHED CROSS-HATCHING THE ENTIRE QUILT USING A TWO-AND-A-HALF CENTIMETRE GRID, SHE HAD HER DREAM QUILT.

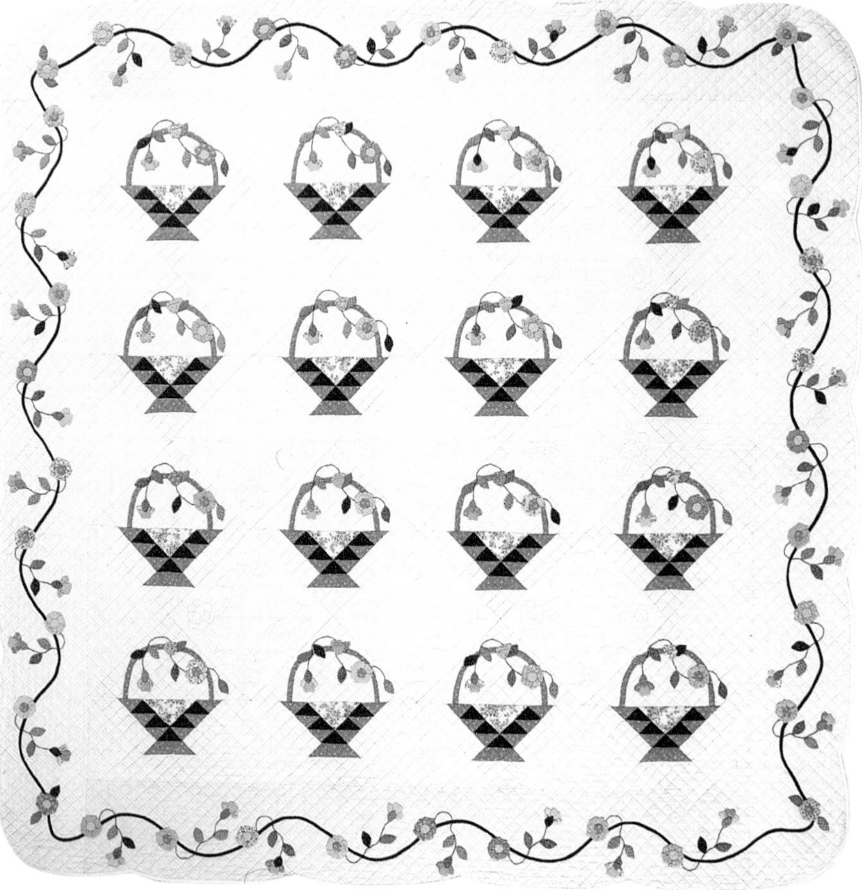

FOX AND GEESE (Top), *142 cm x 243 cm (55⅞ in x 94⅞ in), 1994. Made by the Northbridge Quilters for Robyn King.*

BASKETS OF FLOWERS (Above), *225 cm square (87¾ in square). Made by the Northbridge Quilters for Heather Wootton.*

The Northbridge Quilters of New South Wales always provide cut-outs for their members – each block cut out and ready for sewing. For Robyn King's beautiful 'Fox and Geese' quilt, all twenty-eight blocks were pieced by her quilting friends, from fabric supplied and cut out by Robyn. When the blocks were pieced, she set them on point with plain blocks in between and bordered them with Flying Geese. Providing cut-outs for friends to sew together is proving very popular in today's busy society. With the invention of self-healing cutting boards and rotary cutters, together with specially marked rulers, the process of cutting pieces for an entire quilt is very quick. The Northbridge Quilters find that it takes no time at all to machine the pieces together for their friendship blocks, ensuring that they all get at least one quilt top each year from their friendship group.

Another member of Northbridge Quilters created a very scrappy quilt, still using the cut-out method. While her quilt does not have recognisable pieces from other people's fabric stashes, Sheelagh Thompson loves her scrappy variation on a nine-patch, using her collection of blue and red fabrics with touches of yellow. Using the blue check as the unifying fabric, Sheelagh has created a charming small quilt, machine-quilted in a simple design.

As part of the 1988 Bicentennial celebrations, the Castle Hill Friendship Quilters decided to make every member of the group a 1988 friendship quilt. This was quite an ambitious goal, as there were twenty-four members in the group at the time. One of them, Viive Howe, drew up a roster for the handing out and returning of blocks and, to speed things along, the group decided to provide cut-outs for everyone. Each person added some embellishment or embroidery to their block to give it a personal touch. Angela Langdon, now of Carseldine, Queensland, was involved in this project and decided to choose North Carolina Lily for her block. She handed out kits, containing the cut-out pieces for the block, the pattern and the complete instructions. Angela also included a triangle of Aida cloth on which each member was invited to embroider their name. When the block was assembled, the Aida cloth triangle formed the basket for the lily. The quilt, which Angela now has on her bed, holds many memories of the friendships she made while she was a member of the Friendship Quilters of Castle Hill, New South Wales.

NORTH CAROLINA LILY (Top), *228 cm x 276 cm (89 in x 1075/8 in), 1992. Made by the Castle Hill Friendship Quilters for Angela Langdon.*

SCRAPPY NINE-PATCH (Above), *150 cm square (58½ in square), 1994. Made by the Northbridge Quilters for Sheelagh Thompson.*

HOME SWEET HOME, *127 cm x 190 cm (49½ in x 74 in), 1992. Made by the McLaren Vale Quilters for Pam Waite.*

RECURRING THEMES

The strong link of quilts with hearth and home might go some way to explaining the popularity of the house as a theme for friendship quilts. Marlene Boatwright of Launceston, Tasmania, has a spectacular house quilt made for her by the Launceston Patchworkers and Quilters. Her love of house blocks led her to make this her choice for a friendship quilt. Friends were free to do anything they liked – as long as the blocks were 30 cm (12 in) square and used the blue fabric Marlene supplied for the sky. New skills, such as broderie perse and embroidery, were learned as her friends exercised their creativity. Inspiration for the houses came from many sources, including a National Trust publication, *Tasmanian Midlands*. It took three years for all the blocks to be returned!

EACH BLOCK TELLS A STORY ABOUT ITS MAKER – SOME HAVE WONDERFUL FLOWER GARDENS, OTHERS HAVE QUILTS ON CLOTHESLINES. THE FINISHED QUILT IS HUGE, BUT MARLENE SAID SHE COULDN'T LEAVE ANYONE OUT AND IT WAS WORTH THE WAIT! MARLENE HAS WRITTEN THE NAMES OF ALL HER FRIENDS ON THE BACK OF THE QUILT IN A DELIGHTFUL HOUSE LABEL.

Pam Waite from McLaren Vale, South Australia, also wanted a house theme for her friendship quilt, but she asked for the houses to be set against anything but a blue sky. The variety and uniqueness of her collection of house blocks is typical of the McLaren Vale Quilters. Originally a group of spinners, these ladies decided to make quilts for each other, and now do very little spinning and lots of patchwork! For their friendship quilts, they nominate a theme, which is often just one word – for example, houses, chooks or clowns. The size of the block is not necessarily specified, creating interesting dilemmas when it comes to setting the blocks. Appliqué is often used to interpret the given theme and this is true for Pam's houses with only a few traditionally pieced house blocks evident. To finish her quilt, Pam has added a wonderfully zany border to her collection of zany houses.

Marion Russell of Angaston in South Australia decided to set her house friendship blocks into a house of the same design. Marion is a 'member by post' of her friendship group, which is based in an area approximately two hours away from her home. Marion sent her fifteen prepared blocks to the other members of the group by mail, but she planned to be present at the meeting when the blocks were returned. As she had provided and cut out all the fabrics, she hoped that the group members would add their own personalities to the blocks with their decorations and embellishments. Marion travelled to Victor Harbor to collect her blocks and to meet all the people who had made them. She was delighted with the results – each house is very individual, with the addition of all sorts of embellishments, such as flowers, cats, chooks, spiders and webs, curtains and beautiful front doors. Marion calls her quilt 'House Proud', because she is very proud of her quilt.

HOUSE PROUD (Top right),
165 cm x 210 cm (64⅜ in x 82 in), 1994.
Made by the Victor Harbor Quilters for Marion Russell.

HOUSES OF FRIENDSHIP (Right),
235 cm x 274 cm (91⅝ in x 106¾), 1993.
Made by the Launceston Patchworkers and Quilters for Marlene Boatwright.

DETAIL OF THE LABEL FOR 'HOUSES OF FRIENDSHIP' (Above).

Cats are another recurring theme for friendship quilts. For her 'paper bag' quilt, Yvonne Wooden, another of the McLaren Vale Quilters, chose the theme of 'Cool Cats'. Paper bag friendship quilts are a new idea for friendship groups. Yvonne decided on the theme, created the background for her small quilt, and then put it into a paper bag. Each member of her group did the same with their own quilt design. These bags were then passed around the group, according to a predetermined order, without the originator being aware of what each person had added to her quilt.

WHEN ALL THE 'COOL CATS' WERE ADDED TO YVONNE'S QUILT, AND ALL THE OTHER QUILTS WERE FINISHED, A SPECIAL DAY WAS HELD FOR THE UNVEILING. EVERYONE SHRIEKED WITH DELIGHT AS THE FINISHED QUILTS WERE DISPLAYED TO REVEAL MANY WONDERFUL AND ORIGINAL DESIGNS, LIKE YVONNE'S 'COOL CATS'.

Doreen Carter, although not an inveterate cat lover, also chose cats as her theme for her friendship quilt, to be made by the Marion Quilters. Having just come through a rather difficult period, Doreen decided she wanted to have some fun with her friendship blocks and create a humorous quilt. She borrowed the charming cat illustrations from a quilting friend who had found them in 'some old magazine'. Doreen passed out several different outlines of cats with the instruction: 'Do what you like'. Every cat was different, even those with the same outline. Some even reflected the maker's personality, giving everyone a lot of fun – just as Doreen had hoped it would. She made the corner blocks, and added a butterfly and other embellishments, here or there, to add to the whimsy. The quilt now resides on her bed and, while it is not the usual type of quilt chosen for a master bedroom, it is well used and continues to delight its owner with its variety and sense of fun.

Glenda Olesen of Roleystone in Western Australia was the coordinator of the Western Australian Quilters' Association Bicentennial Quilt Exhibition during 1988. The organising committee had so much fun working together that they decided to form a friendship group, affectionately known as

The '88s. After the first round, when fairly traditional blocks were chosen, the group decided to continue making friendship quilts, but this time to make fun quilts. Glenda chose cats as her theme, providing each member with the Attic Window block with a latticed windowpane. She asked each member to appliqué a cat on the window ledge. When the blocks were completed, a delightful array of pussycats greeted Glenda, including Guy Wackie, the grey cat with its paw hanging down, a combined effort by mother Ngaire (who painted it) and daughter Pippa (who sewed it).

FANTASTIC CATS (Opposite), *153 cm x 223 cm (59⅝ in x 87 in), 1994. Made by The '88s for Glenda Oleson.*

COOL CATS CHALLENGE (Above right), *58 cm x 62 cm (22⅝ in x 24⅛ in), 1993. Made by the McLaren Vale Quilters for Yvonne Wooden.*

CAT-O-LOGUE (Right), *184 cm x 210 cm (71¼ in x 82 in), 1992. Made by the Marion Quilters for Doreen Carter.*

UNUSUAL THEMES

Quilters must spend a lot of time talking over cups of tea or coffee, judging by the number of friendship quilts using this theme. Barb West of McLaren Vale collects teapots, so she had an additional reason to choose teapots as the theme for her friendship wallhanging. In true McLaren Vale Quilters' style, each member produced a unique teapot, using their own innovative design. Three of the designs are based on work by Clarice Cliff, one on an antique bronze pot, and one was designed from a black metal Japanese teapot given to Joan Harnett (one of the group) by a Japanese friend. The teapots were then appliquéd to different-sized rectangular blocks, leaving Barb with the difficult task of assembling them. Black-eyed Susan from Barb's garden completes the jigsaw of blocks. She has chosen an unusual fabric to border the colourful blocks, which have been outlined in black, giving a stained-glass effect. The typed label on the back of the quilt records all these details, and also includes a photocopy of the 'perfect' teapot, more than one hundred and thirty-five years old which holds sixty-one litres of tea and stands seventy-six centimetres tall. It is presently owned by Twinings. Barb has also included on the label a list of what to look for when buying a teapot and how to clean it.

TEAPOTS (Above), *77 cm x 112 cm (34 in x 43⅝ in), 1991. Made by the McLaren Vale Quilters for Barb West.*

FREE RANGE CHOOKS (Below left), *153 cm x 196 cm (59⅝ in x 76½ in), 1995. Made by the Bennethaus Patchers for Bev Bennett.*

The Bennethaus Patchers of Lane Cove, New South Wales, not only make friendship blocks for the members, but they then go on to finish the quilt. Bev Bennett is the lucky owner of 'Free Range Choox'. She delights in the collection of hens and roosters – and they all have names!

DURING ONE HILARIOUS MEETING OF THE GROUP, THE 'CHOOX' BECAME DRIBBLE AND DRABBLE (THE TWINS), DAFFY, HENNY PENNY, ESTER, CELESTE, STICKY BEAK, LABELIA

(THE VEGETARIAN IN THE COOP), DOIDLE, POLLY, SUZANNAH, DOTTY AND POSH BETTY. RANDY ROOSTER, NAPOLEON AND FLAUBERT ARE THE COCKS. CHICKEN WIRE HAS BEEN QUILTED BEHIND THE STYLISED CHICKENS TO CREATE A CHARMING QUILT, AN ADAPTATION OF A PATTERN FROM THE *RED WAGON* SERIES.

POSH FOOD (Above),
166 cm x 214 cm (64¾ in x 83½ in), 1991.
Made by the McLaren Vale Quilters for Inez Ewers.

CLOWNS (Below), *172 cm square (67 in square), 1992.*
Made by the McLaren Vale Quilters for Emma Wood.

The McLaren Vale Quilters continually come up with wonderful and outlandish ideas for their quilts. Inez Ewers asked the group for a quilt with a food theme – with wonderful results. The whole group was very excited about this theme, and the members were keen to try their hands at something beyond the usual flowers and birds. One member was so enthusiastic about Inez's plan to give the completed quilt to her daughter, who owns a popular local restaurant, that she decided to use the restaurant logo for her design. Her block has pride of place in the centre of the quilt. Inez called the quilt 'Posh Food' and now the completed quilt, adorned with all good things to eat, hangs in the restaurant, where it is quite a talking point.

Not daunted by the task ahead of her, Emma Wood, the daughter of one of the McLaren Vale Quilters, joined in a friendship round. The blocks she made were of a very high standard for a twelve-year-old and everyone was delighted to make her a clown block when it was her turn. Emma finished her wonderful quilt herself.

Beryl Hodges, now of Isaacs in the Australian Capital Territory, was one of the members of The '88s in Western Australia. The friendship group started out as an exhibition committee. The members enjoyed each other's company so much, they decided to continue meeting, making a friendship block each month. By the time it was her turn to choose a block, Beryl was living in Sydney and a 'member by post' of the group.

SHE DECIDED THAT AS SHE COULD NO LONGER MEET WITH THE OTHER GROUP MEMBERS PERSONALLY, THEY SHOULD EACH MAKE A BLOCK OF THEIR OWN FACE AS A KEEPSAKE. THEY WERE ALL HORRIFIED! BERYL RECEIVED SEVERAL OF HER BLOCKS WITH LETTERS ADDRESSED TO 'DEAR EX-FRIEND'! BUT BERYL WAS DELIGHTED WITH THE RESULTS AND IMPRESSED WITH HER FRIENDS' INGENUITY.

One had used fine Vylene to create her glasses, another chose fleece for her fluffy white hair, while a third friend had used brown satin to show Beryl the colour she had recently dyed her hair. One block even arrived unsigned, with a note suggesting that the likeness was so apparent, Beryl should have no trouble identifying the maker. Beryl arranged her blocks to resemble a gallery of portraits,

A STITCH IN TIME, *112 cm square (43⅝ in square), 1993. Made by the Marion Quilters for Mary Jarvis.*

providing her with a constant reminder of a lovely group of friends.

Mary Jarvis, a member of the Marion Quilters of South Australia, chose clocks as the motif for her quilt. She has a particular interest in clocks, because of her family history. In 1836, some of her ancestors arrived in South Australia aboard the *Buffalo*, along with Governor Hindmarsh, who was to proclaim South Australia a colony. At that time, an uncle, back in England, promised a grandfather clock to each of the four sons in the family on the occasion of their marriage. He honoured his promise and today the location of three of the four clocks is known. Mary's ancestor received the missing fourth clock and she is still trying to track it down. So the theme was a natural choice for Mary's friendship quilt. When her group was first told about the project, they were unsure of their ability to comply. But Mary encouraged everyone by cutting out appropriate photographs of clocks and explained the techniques needed to complete the blocks. The blocks could be funny or traditional. Most of the group chose to create something humorous and Mary made these into a separate wallhanging, which her grandchildren especially enjoy. She added two more blocks, featuring traditional clocks, to her collection from her group to make a second wallhanging. She inscribed on the middle left block the inscription "'Stands the church clock at ten to three, And is there honey still for tea' Rupert Brook". The block had been made by a close friend who had been heard reciting this quote when Mary announced the theme for her friendship quilt. A photograph of the French boudoir clock was sent to Mary by a visiting quilter, who is delighted to have her clock included in Mary's quilt.

DETAIL FROM 'A STITCH IN TIME' (Top).
GALLERY OF FRIENDS (Above), *130 cm x 166 cm (50⅝ in x 64¾ in), 1990. Made by The '88s for Beryl Hodges.*

TUTORS BROUGHT PEACE TO MY HEART,
138 cm x 213 cm (53¾ in x 83 in), 1988.
Made by friends at the Australasian Quilt Symposium for Margie Furness.

Give all your friends a piece of your favourite batik fabric and ask for a friendship block of any size and any design and the result will be something like Judy McDermott's wonderful quilt. The blue batik is the recurring fabric, providing a linking thread between the very different blocks from friends from all over Australia. Along with traditional blocks were blocks designed especially for Judy's quilt, including a wonderful fish on a plate – complete with knife and fork. This was Margaret Wright's contribution, who at the time was president of The Quilters' Guild and had very little time for stitching.

Judy needed the large floor space of Ranelagh House at Robertson to put the quilt together, and found many willing helpers at the Quilters' Guild Retreat, held there in 1994.

WITH OVER FIFTY QUILTERS ENTHUSIASTICALLY OFFERING ADVICE, JUDY LAID OUT HER EXTENSIVE COLLECTION OF BLOCKS ON THE FLOOR. SOME BLOCKS WERE MADE ON THE DAY, AS NEW RECRUITS JOINED IN THE FUN. JUDY HAD A GREAT SELECTION OF UNUSUAL BATIKS TO CHOOSE FROM WHEN ASSEMBLING THE SASHING STRIPS, CREATING HER UNIQUE FRIENDSHIP QUILT.

Friendship quilts have always been reminders of people and places and, often, of exciting and difficult times shared. Margie Furness of Parkdale, Victoria, recalls the unexpected gift of her friendship quilt from the Australian and overseas tutors at the Australasian Quilt Symposium in 1988:

'I was looking for something different to add a touch of merriment to the fashion parade. I was to be the compere and decided that the Australian tutors could dress up and pretend to be football players. Two teams were formed: the Crazy Quilters and the Olpha Cutters. Week One of the symposium saw the Australian tutors (Lessa Seigal, Noreen Dunn, Shirley Gibson, Elizabeth Kennedy, June Lyons, Lorraine Moran, Megan Terry, Judy Turner, Wendy Wright, and Ruth Walter), dressed in white, burst through a huge banner to the tune of 'Up There

LOAVES AND FISHES, *266 cm x 272 cm (103¾ in x 106 in), 1995. Made by friends for Judy McDermott.*

Cazaly'. Week Two and it was the overseas tutors turn. Valerie Cuthert from New Zealand performed a traditional Maori dance, Carol McClean from Canada performed 'The Snow Bird', while the Americans stole the show with Doreen Speckman, dressed as the Statue of Liberty, with Carol Bryer Fallert and Catherine Anthony holding her train.

Knowing the difficulties I was working under, after a car accident on the day of the parade, the tutors decided to make me a friendship quilt. The Olpha Cutters made the yellow and black blocks, while the Crazy Quilters used all the colours they knew I loved to make blocks typical of their work. Especially dear to me are Noreen Dunn's Bow Tie blocks. The quilt is covered by loops of rouleau, made by Wendy Wright – a reminder of her patience while she tried to teach me the technique. The quilt was really made in friendship and I am honoured that all these terrific tutors found time in their busy schedules to make this humorous, bright, happy quilt top especially for me. It is, and always will be, one of my greatest treasures.'

VICTORIAN LADIES, *173 cm x 206 cm (67½ in x 80⅞ in), 1994. Made by Fibres and Fabrics for Brigit Nicol.*

Brigit Nicol of Nambour, Queensland, makes dolls as well as quilts. As a bridal machinist, she particularly enjoys dressing her dolls and enjoys working with lovely fabrics. When she came across the pattern for her quilt, designed by Jean Teal (copyright 1991), she knew it was something she just had to make. It became her choice for her friendship quilt. She gave out the background fabric, the outline of the dress and asked for it to be made in old-fashioned colours. The dress could be changed in any way – so sleeves were changed and bustles added. Some of the members of her friendship group, Fibres and Fabrics, researched the period costumes at the library, so additions of chains at the hem, used to raise the skirt, were added. Brigit was delighted with the finished blocks, very aware that, had she made all the blocks herself, she would never have had such variety. The blocks were decorated with laces, brooches, necklaces, stand-away collars, feathers, frills and lovely bead work. When they were assembled, Brigit added the striking border and the quilt now has pride of place on her lounge-room wall, hanging from a shelf which houses her beautifully dressed dolls.

THE ROAD TO THE HOUSE OF A FRIEND IS NEVER LONG,
160 cm x 170 cm (62½ in x 66¼ in), 1993.
Made by Fibres and Fabrics for Derryn Johnson.

DETAIL FROM 'THE ROAD TO THE HOUSE OF A FRIEND IS NEVER LONG'.

FIBRES AND FABRICS

Fibres and Fabrics from Townsville, Queensland, make fabulous friendship quilts. As with many friendship groups, they have developed a style unique to their group. Because of the diversity of backgrounds of their members, including spinning and weaving, papermaking, knitting and basketmaking, their work is not restricted by the traditional skills of patchwork. Very few restraints are imposed on the makers, resulting in a varied selection of blocks, featuring a broad spectrum of techniques including all areas of creative embroidery, machine-embroidery, surface embellishments, knitting and, of course, all areas of patchwork and quilting. The themes chosen are often unusual, inspiring even greater creativity.

This group of women with such divergent interests meet happily together as Fibres and Fabrics and they have found that the bond of handwork has no age boundaries, with octogenarians interacting happily with young mothers.

For her friendship quilt, Derryn Johnson, the president of the group, asked for an Australian house in any size and using any style of making. She gave each member of the group a linen square and some fabrics in green and the 'bricky' red she liked. Each participant was asked to put something from her own house on her block.

THE KNITTER IN THE GROUP ADDED KNITTED TREES AND SPIDER WEBS TO HER VERANDAH, WHILE BIRTE, WHO IS DANISH, ADDED THE DANISH FLAG. COLLEEN WHO HAS TWINS, CROSS STITCHED TWO BABIES PLAYING IN THE FRONT YARD, WHILE ANOTHER BLOCK HAS AN ACTUAL PHOTO OF THE MAKER'S DOG AT THE WINDOW.

Derryn wanted to add something of herself to the quilt as well, so she cross stitched some of her favourite sayings about friendship and homes in the spaces between the irregularly sized blocks:

'You can't pluck a rose
All fragrant with dew
Without part of the fragrance
Remaining with you.'
and
'A house is made of brick and stone
A home is made of love alone.'

FRIENDLY FUNGI (Above left), *93 cm x 130 cm (36¼ in x 50⅝ in), 1994. Made by Fibres and Fabrics for Mavis Webster.*

VICTORIAN CRAZY PATCH (Above right), *61 cm x 102 cm (23¾ in x 39¾ in), 1994. Made by Fibres and Fabrics for Margaret Wretham.*

MAUREEN NORMAN, 'FOR SOME UNKNOWN REASON', DECIDED TO ISSUE EVERYONE WITH A RANDOMLY SHAPED PIECE OF BLACK CHINTZ FOR HER QUILT, INVITING THEM TO EMBROIDER A SPIDER'S WEB ON IT. SHE THEN INCORPORATED THE TWENTY-ONE SPIDERS' WEBS INTO HER UNIQUE INTERPRETATION OF A RAINFOREST.

The rainforest surrounding Townsville is often the source of inspiration for Maureen's work and an expression of her concern for its preservation. She says she enjoyed the challenge of the construction, attaching spiders' webs, leaves, ferns and tree trunks to the black background. Maureen employed her considerable skills in dyeing and printing fabric to create the amazing three-dimensional scene, which even includes a painted stream running through it. Machine- and hand-embroidery, couched fibres and fabrics of various textures, such as velvet and leather, were used to complete the extraordinary picture depicted on the quilt.

FRIENDLY SPIDERS, *115 cm x 198 cm (44⅞ in x 77¼ in), 1995. Made by Fibres and Fabrics for Maureen Norman.*

The 'Friendly Fungi' friendship quilt was the brainchild of Mavis Webster, another member of the amazingly creative Fibres and Fabrics group. She distributed background squares and an evening fabric that was to be incorporated into each block and asked that the group make her blocks depicting mushrooms. The returned blocks featured crazy patchwork, cross stitch, painting, appliqué, knitting and embroidery. Mavis then had the problem of how to put these wonderful blocks together. Eventually, she set them on the back of a piece of textured furnishing fabric, and reverse-appliquéd them into place. Lots of fraying gave an undergrowth effect to the whole creation. The sides were finished with covered piping cord, while the bottom was left irregular and frayed. With the blocks she received, Maureen was able to complete two quilts; on one of these, she has included a panel on which she stitched the names of everyone who contributed. 'I will treasure the completed work always,' says Mavis.

Margaret Wretham wanted a Victorian 'Crazy Patch' friendship quilt, because of her love of old fabrics, laces and buttons. Her 'not-so-old' friends lovingly sewed the blocks, using treasured pieces from a grandmother's sewing box, or lace from a great aunt's wedding gown. Margaret assembled the blocks as a merging collage, symbolising the bonds of friendship. The quilt is part of a triptych, but each section can be hung separately.

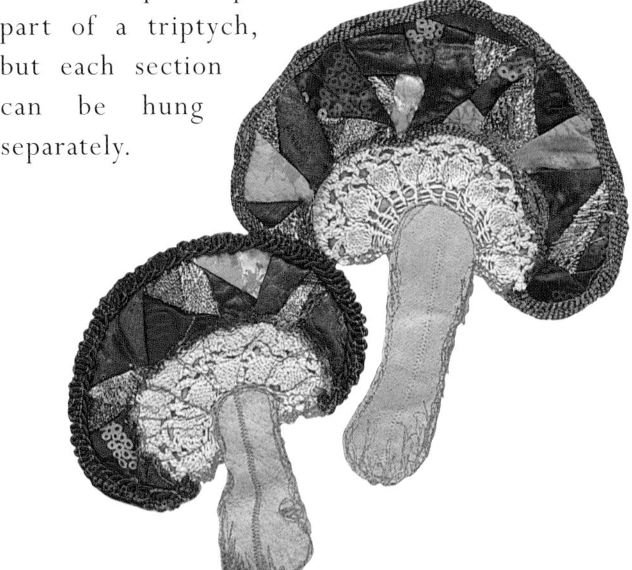

'Tumbling Friends' is another of the unique quilts created by Fibres and Fabrics. Birte Muller decided to use the traditional Baby Blocks pattern and gave each of her friends a pastel-coloured background fabric and a dark fabric for the darkest side of each cube. The creativity and ingenuity of the returned blocks delighted Birte and she was able to arrange them so the light appears to come from the top left-hand corner as instructed. Birte decided on the octagon shape, which accommodated the seventeen blocks beautifully.

Exotic fabrics, lavish embroidery and embellishments are not usually associated with the Amish, but Barbara Murphy wanted to see how their simple designs and colours blended with the exotic. She supplied the vibrant silk fabrics and some of the beads, ribbons and other embellishments in two colour schemes. Barbara now has 'Byond Amish - Blue' and 'Byond Amish – Red'.

SPECIAL FRIENDSHIP QUILTS

While the majority of friendship quilts are made by friendship groups, where every member eventually receives a collection of blocks to their specifications, occasionally a person or an event inspires the making of a friendship quilt.

During the McLaren Vale Quilters' Retreat, held annually at Douglas Scrub, all the participants make a block to the specifications given. In 1994, the request was for an animal block. Meanwhile, the group had collectively worked a gateway, lettering and people at the zoo, as a background for this unusual quilt. Some of the blocks brought to the retreat were chosen for the quilt and, as at every retreat, Dorothy Fennell secreted herself away

to join the blocks, while the rest of the group enjoyed Saturday night activities. On Sunday morning, all the participant's names went into a draw for the quilt and, to everybody's delight, Dorothy won – a fitting reward for this generous quilter.

TUMBLING FRIENDS (Top left),
135 cm (52⅝ in) octagon, 1993.
Made by Fibres and Fabrics for Birte Muller.

DOUGLAS SCRUB RETREAT ZOO QUILT (Left),
147 cm x 196 cm (57⅜ in x 76½ in), 1992.
Made by the McLaren Vale Quilters for Dorothy Fennell.

BYOND AMISH – BLUE (Above),
102 cm square (39¾ in square), 1993.
Made by Fibres and Fabrics for Barbara Murphy.

DETAIL (Right) **FROM 'BYOND AMISH – BLUE'.**

FOR ISOLDE II (Top), **FOR ISOLDE I** (Above Right), **FOR ISOLDE III** (Above), *1994.*

Made by the Epping Quilters in memory of Zolda Glockerman

The Epping Quilters have made three wonderful quilts in memory of Zolda Glockerman, a member of the group who died in 1994. Each quilter was asked to make a flower block, using the cream homespun provided as the background. Everyone wanted to participate and thirty-nine finished blocks were received and assembled into three magnificent quilts. Without any planned colour coordination, the blocks naturally fell into three groups, with purple, red and green predominating. One quilt was used as the Epping Quilters' raffle quilt for their open day in October 1995, with the proceeds going to a local charity. A second quilt was handed over to The NSW Cancer Council to use as a fundraiser and the third quilt was given to Zolda's husband, who passed it on to Lifeline, a favourite charity of Zolda's. The generous efforts of the Epping Quilters are reminiscent of the way friendship quilts were made in the past for deserving causes.

Quilters Down Under in Beenleigh,Queensland, make the most of any opportunity to acknowledge one of their members with a special friendship quilt. When May Cook turned eighty in 1993, the

MY EIGHTIETH BIRTHDAY QUILT, *118cm x 174 cm (46 in X 67⅞ in), 1993. Made by Quilters Down Under for May Cook.*

group members decided to give her a friendship quilt to celebrate the occasion. They chose a simple block, Square in a Square, and used the white centre of each block to record their birthday wishes. While most were content to record a simple message, there is also a scene of a cottage and a cow, a spider's web with the inscription, 'Friendship is hanging around together' and a delightfully decorated M in the centre. There is even a poem, written especially for the occasion:

'To marvellous May
Who just makes our day
We all miss her terribly
When she's away.'

MEGAN'S FRIENDSHIP QUILT (Top),
120 cm x 225 cm (46¾ in x 87¾ in), 1994.
Made by the Colours of Australia exhibition subcommittee for Megan Fisher.

COLOURS OF AUSTRALIA *blocks made by the exhibition subcommittee for Larraine Scouler* (Above).

Often, friendship quilts are made as a lasting acknowledgment of the high regard in which someone is held. The Colours of Australia subcommittee of The Quilters' Guild of New South Wales made just such a friendship quilt for Megan Fisher. Megan had worked very hard as the publicity officer for the committee and had become ill. The instructions were simple: make a Judy in Arabia block, including the deep aqua star points provided. The block could be changed in any way and there was a prize for the most innovative change. (Note the heart block – no sign of a 'Judy in Arabia' block there!) The simple blocks were then set on an angle to add interest. The quilt was presented to Megan, already bound and with minimal machine-quilting – a demonstration of how just a few large blocks can create an interesting quilt.

This subcommittee was responsible for the organisation and implementation of the Colours of Australia touring exhibition and book, recording the collection of forty quilts that will tour Australia until 1999. As a fitting recognition of the work of the chairperson, Larraine Scouler, each of the quilters whose work was selected for the Colours of Australia exhibition, was asked to submit a block reminiscent of their quilt in the collection. Larraine now has a unique collection of blocks from some of Australia's leading quiltmakers which will form a wonderful friendship quilt.

The committee of The Quilters' Guild of New South Wales has generously made friendship blocks for each of their retiring presidents over the last five years. One of those quilts is featured on the front cover of this book. Narelle Grieve, the president from 1990 to 1991 received a quilt top made of small blocks. Each participant had made two blocks that reflected her activity on the committee, a charming reminder for Narelle. The blocks gave some indication of what each person's function was on committee or reminded Narelle of particular incidents during her presidency. Isobel Lancashire embroidered a map on her block, as she always seemed to be navigator for Narelle – often without immediate success in finding their destination.

Quilters often get together to make quilts for charity. Narelle Grieve initiated the making of an

unusual quilt for the NSW Cancer Council for their Posh Auction in 1992. Her quilt was reminiscent of the signature quilts made by the Red Cross to raise money during times of war.

In those days, each participant would pay a small fee for the privilege of signing the quilt. Narelle decided to invite well known Australian sporting personalities to each sign a Snowball block which she then incorporated into a quilt. Of course, she didn't charge them for their generosity! Many hands helped in the making of the quilt. Val Donalson wrote the names on the blocks, so that the signatures could be recognised; Lee Cleland helped assemble the quilt top and Shirley Gibson assisted with the hand-quilting. The quilt was finished in time for the Posh Auction and the successful bid was from The Australian College of Physical Education. The quilt now hangs in the offices of the College at the Sports Centre in Homebush, Sydney, having raised a considerable amount for cancer research.

COMMITTEE MEMORIES (Top), *110 cm x 130 cm (42⅞ in x 50⅝ in), 1991. Made by the committee and friends from the Quilters' Guild for Narelle Grieve.*

SIGNATURE QUILT (Above), *270 cm square (105⅜ in square), 1992. Made by various people for charity.*

DETAIL OF 'COMMITTEE MEMORIES' (Top left).

PATRITOTIC GAMES (Above), *165 cm x 210 cm (64⅜ in x 81⅞ in), 1988. Made by the Patriotic Quilt Group for Pamela Tawnton.*

66 STARS LATER (Above right), *166 cm x 206 cm (64¾ in x 80⅜ in), 1992. Made by the Australian and overseas friends for Isobel Lancashire.*

FRIENDS OVERSEAS

More and more quilters from all over the world are joining forces and making friendship quilts for each other. Those involved say it is an enormous thrill to put a quilt together, knowing that the blocks were made by quilters across the ocean. Isobel Lancashire from Epping, New South Wales, made her quilt '66 Stars Later' with blocks from friends from South Africa and the United States. Several quilters were involved in this project, organised by Dale Ritson and Frances Thurmer in Australia. Each person was asked to make four 30 cm (12 in) star blocks – any star at all. Preferred colours were nominated – Isobel wanted burgundy and pink. She made four blue and cream stars for Marie de Whitt in South Africa and four pink and mauve blocks for Lyn Weigel in the United States. When Isobel received her blocks, she set them on point, alternating her stars with deep blue blocks to create a strong image. The finishing touch was to add sixty-six tiny stars for the delightful border.

Pamela Tawnton of Weston in the Australian Capital Territory joined the Patriotic Quilt Group in Annapolis, Maryland, by default. The group's aim was to make each member a patriotic quilt in the true red, white and blue tradition. Her long-time friend, Jean Pope, offered her a place in the newly formed group and, before Pamela could decide whether to join or not, Jean had said yes on her behalf! Somewhat reluctantly, and wondering how an Aussie would fit in, Pamela agreed to participate. For her quilt, she chose a lovely rust colour for everyone to incorporate into their blocks. 'Patriotic Games' was the result. Her friend Jean added the finishing touch with the Australian flag in the centre. Pamela made the koala block for her own quilt and one each for the eighteen other participants in the group.

THE EXCITEMENT OF BEING INVOLVED WITH QUILTERS FROM OVERSEAS IS SO ATTRACTIVE TO DALE RITSON AND FRANCES THURMER FROM EPPING, NEW SOUTH WALES, THAT THEY HAVE BEEN ACTIVELY INVOLVED IN COUNTLESS PROJECTS SINCE 1986.

Frances' first round of friendship quilts involved two pen friends from the United States – Ruth from Illinois and Kate from Texas – and Stephanie from Berowra, New South Wales. Frances, a passionate letter writer, had made contact with the American quilters through the *Quilters' Newsletter*. For their

CHECKS AND PLAIDS (Above),
170 cm x 180 cm (66¼ in x 70¼ in), 1990-92.
Made by Dale Ritson from cut-around blocks.

FLIGHT OF FANCY (Above right),
92 cm x 120 cm (35⅞ in x 46⅞ in), 1992.
Made by Frances Thurmer from cut-around blocks.

INVESTMENTS VEST (Below right), *1995.*
Made by Frances Thurmer.

friendship quilts, each participant chose a block pattern and provided templates for each other. Using these, each person cut out four blocks for everyone else. No sewing was required. Everyone agreed to this simple process and loved the idea of receiving parcels from overseas, with blocks cut out ready to sew.

Dale, always keen to be involved in whatever was happening, joined the group for the next round of blocks, when the process was streamlined. A parcel was circulated continually between the members of the group, within a predetermined time-frame. In the parcel was the selected pattern, instructions and templates for each person. When one of the group received the parcel, she cut out a block for everyone else, using the templates provided, and put it into the appropriate envelope. At the same time, she would remove all the blocks cut out for her and decide whether to continue circulating her current block (if she needed more for her proposed quilt) or change her pattern, instructions and template for a new block design. The parcel was then sent on to the next person. This parcel circulated for several years with these girls becoming firm friends, sharing their family and quilting life through letters and photographs.

The 'cut-arounds' have finally stopped, because Ruth decided she had too many UFOs (unfinished objects) and Kate's husband retired, so she had to go 'cold turkey' and withdraw from the group.

The enthusiasm for working on friendship projects with overseas quilters has not waned for Dale and Frances. They can't wait for their weekly meeting to sew, catch up on the news from overseas friends, and open the latest letter or package.

Even though the original group has disbanded, other friendship groups have been established over the years in Australia, America, South Africa, England and Germany.

At present, they are involved in a round-robin friendship quilt with participants in America, and four charm robin groups with Australian quilters. They also make large progressive quilts with two American quilters, Betty and Tammy, and small progressive quilts with Beth and Alma, from New Jersey in the United States. In addition, Frances is involved in an 'investments' round robin, where each quilter makes part of a vest for the others. As well as their enthusiastic interaction with quilters from overseas, Dale and Frances are involved in making traditional friendship quilts with their group of quilting friends at the Eastwood Patchwork Quilters.

The recent publication of *Round Robin Quilts* by Pat Maixner and Margaret and Donna Ingram-Slusser, has inspired a new type of friendship quilt. Moving away from the traditional set of square blocks, they advocated a completely new look for friendship quilts, using small and large blocks arranged on a grid. Each quilter decides on a theme, and makes a large 20 cm (8 in) or 30 cm (12 in) block that reflects her theme, together with some simple 10 cm (4 in) filler blocks (again reflecting the theme). A 10 cm (4 in) or 5 cm (2 in) grid, is drawn up on Pellon (a very thin wadding), and the completed blocks are pinned into place. Each person in turn adds to the quilt by completing small blocks that fit the chosen

HEAVENS ABOVE (Above), *Work in progress*

QUILT LABEL FROM 'HEAVENS ABOVE' (Right)

MINIATURE SAMPLER QUILT (Above),
47 cm x 51 cm (18⅜ in x 19⅞ in), 1993.
Made by Dale Ritson.

DALE'S BLUE RIBBON (Above right).
Made for her by her friend Rita.

theme. The blocks may be any size, as long as they will fit exactly into the grid. For Dale and Frances' group, each participant has to make at least six blocks, two of which are designed around the theme, while the other four are simple, geometric filler blocks. These blocks are then pinned to the grid in a pleasing arrangement, but, of course, the final decisions rest with the owner.

Dale and Frances heard about the round-robin quilts through the *Lucky Block Newsletter* and, of course, they wanted to be part of it. Frances chose for her theme 'Heavens Above' and Dale chose 'Give me a Home among the Gum Trees'. Frances began her quilt with an appliqué angel set in the centre of a brilliant yellow star block. The others added to her 'Heavens Above' quilt with more angels and stars, some appropriate verse and even a sunrise (pictured page 48). As each member of the group completed her blocks, she signed the label that Frances had prepared.

Each participant in the round robin is encouraged to include a quilt label with their instructions, so everyone can write their name and address on it as a permanent record of those involved. This is attached to the back of the finished quilt. Many round robin-members also include a journal in their parcel. The journal encourages the bonds of friendship, as it is passed from one quilter to the next, with each person recording their thoughts about the quilt, adding fabric samples, photos, and news of family and quilt happenings.

The *Lucky Block Newsletter*, to which Dale and Frances subscribe, is produced by Patricia Koehner in America. All those who register their interest send Patricia a block they have made, following the instructions in the newsletter. The blocks are then divided up into piles of twenty and one block is selected from each pile. Those selected win all the blocks in their pile. The newsletter, a quarterly publication, is also used to convey all sorts of information to its international readership, and it was here that Dale and Frances not only learned about round-robin friendship quilts, but they were introduced to Betty from Idaho and Tammy from Wyoming, and the Toledo sisters – two from Arizona and one from Mexico. They all decided to make a large progressive quilt for each other. Each participant was required to make a 40 cm (16 in) block, which was then passed on in turn to the others who were required to add borders of specified widths. Each person tried to reflect the original block when completing their borders. Frances drew up the entire border with a little help from her engineering husband, while Dale just began and worked towards the middle of each border, then used her creativity to ensure absolute accuracy. No one in the group saw their quilt top until it was completed a year later.

STAR SAMPLER,
150 cm x 226 cm (58½ in x 88⅛ in), 1992.
Blocks made by Rita for Dale Ritson.

Dale recently made a wonderful miniature sampler quilt. As she made each tiny block, she made a second identical block and sent it to Rita. Dale assembled them, using little strips of Liberty prints. To add the finishing touch to her quilt, Rita attached some of the charms Dale had been enclosing in her letters. Rita's quilt won a blue ribbon at her local quilt show, so she made a replica blue ribbon which she sent to Dale. It has pride of place, next to her quilt, displayed proudly on her bedroom wall, together with a picture of Rita's quilt.

Dale is also involved in a block swap with Edel in Germany. Dale is making Fan blocks and Edel is making Basket blocks. Dale makes two of each block and sends one to Edel, who does the same. As the result of a similar block exchange with Rita in America, Dale has many finished quilts.

Rather than being discouraged by the constant workload with deadlines to be met, Dale and Frances continue to be excited by their involvement in international friendship quilts and the people who make them.

At the same time, Dale and Frances worked on a small progressive friendship quilt with Beth and Alma from New Jersey. The small progressive quilt is similar to the large one, but has a starting block of 20 cm (8 in) and only three borders. Each step took about six weeks with completed quilt tops returned to their owners in eight months.

As well as being active members of Eastwood Patchwork Quilters, which involves them in the making of many friendship blocks including the challenge friendship blocks, Dale prepares cut-outs for Christmas for all her quilting pen friends around the world.

Just prior to Christmas 1994, she received a parcel from Rita in the United States. In it were twenty-five small butterfly blocks, ready for assembly. On the parcel was 'Don't Open Before Christmas' but Dale confesses to having the quilt completely finished before Christmas Day!

SHARED SCRAPS,
155 cm x 230 cm (60½ in x 89⅝ in), 1995.
Made by Dale Ritson from 530 shared scraps.

OVER TO YOU

This section of the book gives instructions for making some wonderful friendship quilts. Each one is an actual friendship quilt, made by friendship groups from all over Australia. You are certain to feel inspired to form a friendship group, after reading about the exciting quilts that others are making.

Forming a friendship group is a very easy process – all you need is a group of willing friends. Remember that the McLaren Vale Quilters, whose wonderful quilts are featured throughout this book, were originally a group of spinners who decided to make each other a quilt. Only a few of their group knew anything about quiltmaking, so your willing friends don't have to be experienced quilters or even quilters at all. All they need is a willingness to learn and someone to teach them.

Try to keep your group to around twelve members. This way everyone gets their blocks quickly enough so they don't lose their enthusiasm. Of course, this is not a hard-and-fast rule.

Once you are all enthused, there are some decisions to be made and guidelines to be established so that your group can run smoothly. First, decide how many you want in your friendship group and whether your group is to remain fixed. It is important to establish this in the beginning and that this number remains constant throughout the project. It can be a problem if late starters join the group, as they won't have made blocks for every member of the group.

Second, determine what restrictions you want on the type of block to be made. Some restrictions could be:

- No appliqué.
- No more than twenty pieces for a pieced block.
- No smaller than 15 cm (6 in).
- Only cotton fabrics are to be used.
- Blocks must be complex (challenge friendship group).
- No embroidery on the block.

You may decide that you are happy to have no restrictions on the blocks and accept any proposed design. Some groups use the more difficult and complex blocks as a learning tool for their less experienced members. Just keep in mind that the blocks are meant to be friendship blocks. You don't want to end up alienating your friends because they have had to spend hours and hours struggling to make your block.

Third, decide what time is to be allowed for the making of each block. Most groups allow one month for completion, while some groups expect the completed block to be handed in sooner. Castle Hill Quilters agreed to complete their blocks in three weeks to allow for every member in their group to have their set of blocks completed during the bicentennial celebrations of 1988. This was a fast and frantic pace that they were able to maintain because they had a fixed goal.

Fourth, decide what each person in your group will receive in their friendship block kit. You must include the pattern and instructions on how best to construct the block. Other inclusions could be:

- Prepared templates.
- Fabric for the background or for inclusion somewhere in the block.
- All the fabrics, either cut ready for machine-piecing, or marked for hand-piecing.
- Diagram of the quilt to be made from blocks.
- Colour preferences and mood preferences, for example pretty pinks, lavenders and lemon for a budding ballerina's bedroom.

At Eastwood Patchwork Quilters, we decided that each kit would contain the templates with a pattern sheet as well, showing block design and instructions on how to make the block. No block had more than thirty pieces and there was no appliqué allowed. If the block only had ten pieces or less, you were allowed to ask for two blocks to be made.

Fifth, to determine who will go first, put everyone's name in a hat, and draw them out in turn, recording the order and the date they are to hand out their instructions. Each member of the group should receive a copy of the draw, so they know when it is their turn. As well as circulating everyone's names, it is often helpful to include addresses and phone numbers for any tardy block makers.

Finally, you might want your friendship group to be an opportunity for instruction from a teacher, so she would choose each block to feature a specific technique. At the end of each time period, the completed blocks could be put in a bag and the lucky winner drawn out.

Whatever rules your group decides on, make sure everyone is happy with the decision and understands them. Now you are ready to begin.

GETTING STARTED

The following instructions are presented in such a way that, once you have chosen the design you like the most, you can simply photocopy the instructions and hand them out to your friends. The instruction include a block diagram, a piecing diagram and full-sized templates. As all the quilts here are friendship quilts, no fabric requirements are included. Most of the fabric for the blocks will come from your friends and you may decide to change the setting of the blocks to suit the number of blocks you receive, or add borders, as you prefer.

Each block can be made by hand or machine. To facilitate this, each template given has two lines. The outer, solid line is the cutting line, and the inner broken line is the sewing line. The lines are 7.5 mm apart (a little over ¼ in). Please use the inner dotted line when preparing templates for hand-piecing and the outer solid line when preparing templates for machine-piecing.

Details of the processes involved in piecing, appliqué and construction of a quilt are not included in this book. It is assumed that someone in your friendship group has this basic knowledge or that you have other resources you can call on. There are, in fact, many fine books available for you to refer to that are entirely devoted to basic quiltmaking. *Creative Traditional Quiltmaking* by Karen Fail (J. B. Fairfax Press, 1995) will provide you with all the information you require.

LABOUR OF LOVE, *205 cm x 255 cm (80 in x 99½ in), 1994. Made by the Eastwood Patchwork Quilters for Leigh Swain.*

DENISE'S SMOOTHING IRON QUILT

In true friendship quilt tradition, this quilt provides a generous space for messages and loving words.

Block used: Smoothing Iron

Block size: 17 cm (7 in) triangle

Quilt size: 64 blocks are needed for a 130 cm x 194 cm (51 in x 76½ in) quilt

MATERIALS SUPPLIED

Cream homespun for the centre of the block

INSTRUCTIONS

- Please provide two other fabrics to complete the block. Choose fabrics that have an old-fashioned look.
- Cut the following pieces:

1 A from cream homespun
3 B from one of your fabrics
3 C from the other fabric

Block Diagram

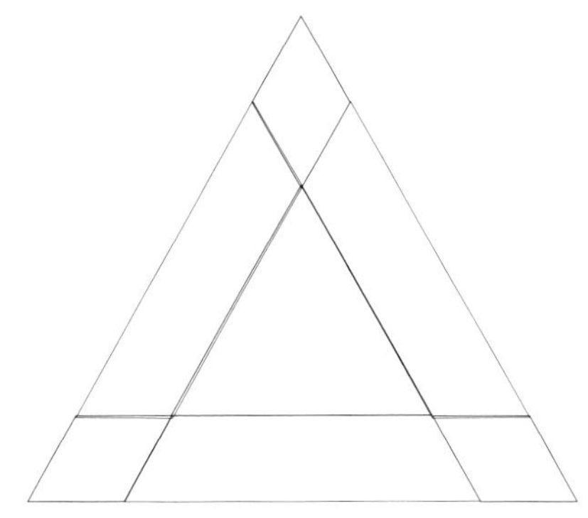

Piecing Diagram

- Complete the block, following the piecing diagram.
- Note that the triangle in the centre of the block can be pointing down or up. In your block, place the triangle pointing .... Sign your name on the cream triangle, using a permanent pen, or embroidery or paint, as was done in Victorian times. You may wish to use one of the ornate designs to be seen in various publications. If so, photocopy the design and, while it is still fresh (less than two hours old), iron it into place on the fabric. The final result will be lighter than the photocopy and reversed.

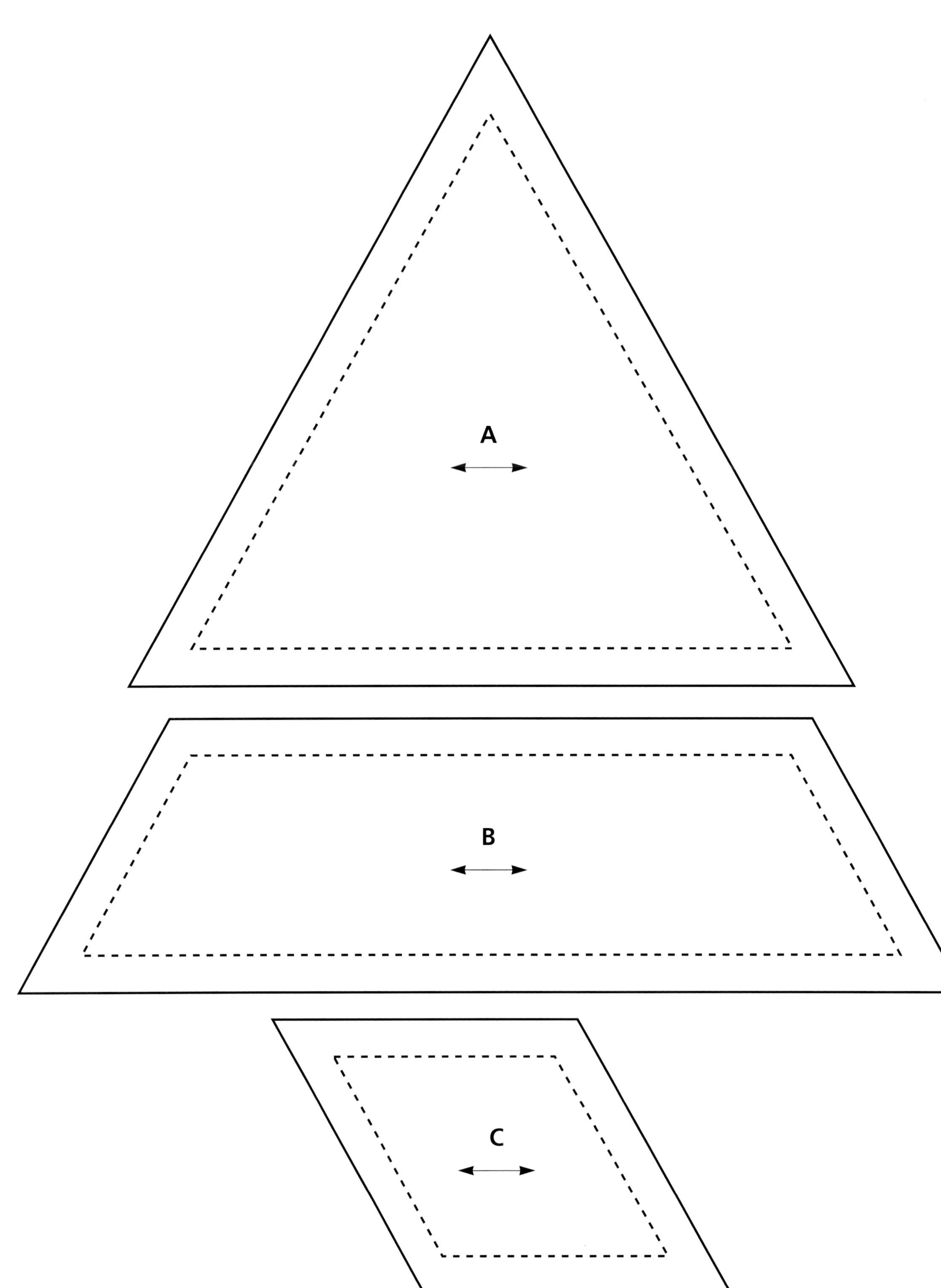
A
B
C

DENISE'S SMOOTHING IRON

For this quilt, I shared fabrics that I had purchased from a Reverse Garbage shop with my friends and ended up with a wonderful mix of zany blocks. My mother, my five sisters and their daughters, as well as my patchwork friends, made blocks for me. At work, I even held lunchtime workshops in piecing and everyone decorated their blocks at home. I treasure my quilt and all the wonderful messages on it.

THESE BLOCKS WERE MADE BY FAMILY, QUILTING FRIENDS AND WORK FRIENDS FROM CASTLE HILL, NEW SOUTH WALES, 1995, FOR DENISE EAST.

FOREVER FRIENDS

Combining the simple Log Cabin blocks with hearts, a lasting symbol of friendship, makes a wonderful quilt to be treasured.

Block used: Log Cabin Heart, adapted from '*A Celebration of Hearts*', by Jean Wells

Block size: 20 cm (8 in)

Quilt size: 64 blocks are needed for a 198 cm x 267 cm (78 in x 105 in) quilt

MATERIALS SUPPLIED

Aqua square background for appliqué

INSTRUCTIONS

- Prepare a heart in the colour of your choice and appliqué it to the aqua square, using your preferred method of appliqué. Embroider your name on the heart.
- Choose shades of the same colour for the logs, varying the colour from light to dark. Cut 4 cm (1½ in) wide strips from each fabric for the logs. These measurements include a 6 mm (¼ in) seam allowance. You will need four light, four medium and two dark strips of your chosen colour. Attach them to only two sides of the aqua square following the diagram for order of placement. Trim each strip to the required length after it is stitched.

Block Diagram

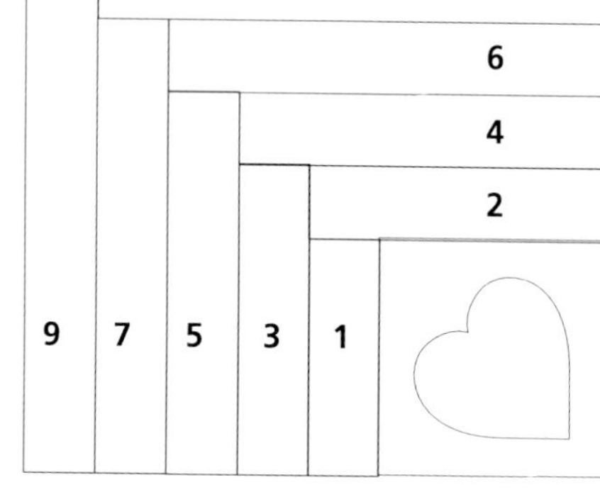

Piecing Diagram

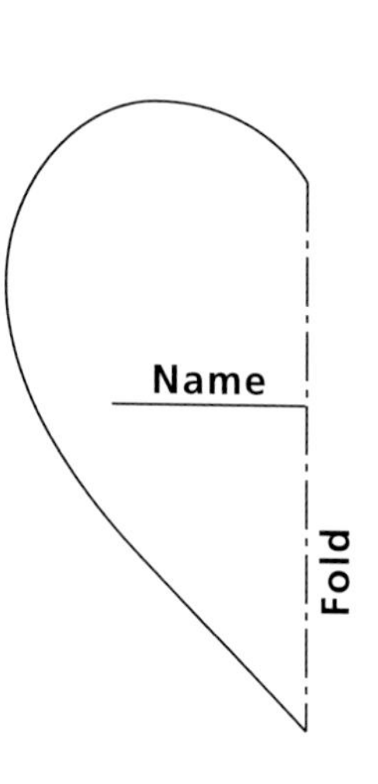

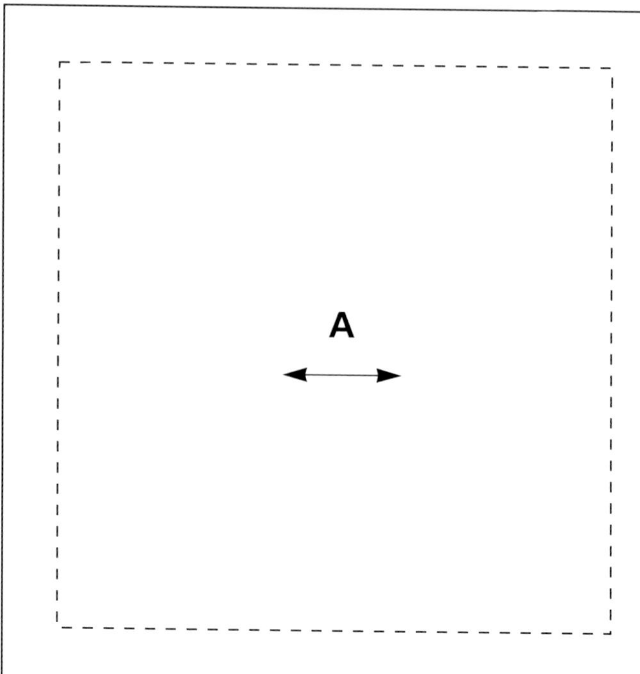

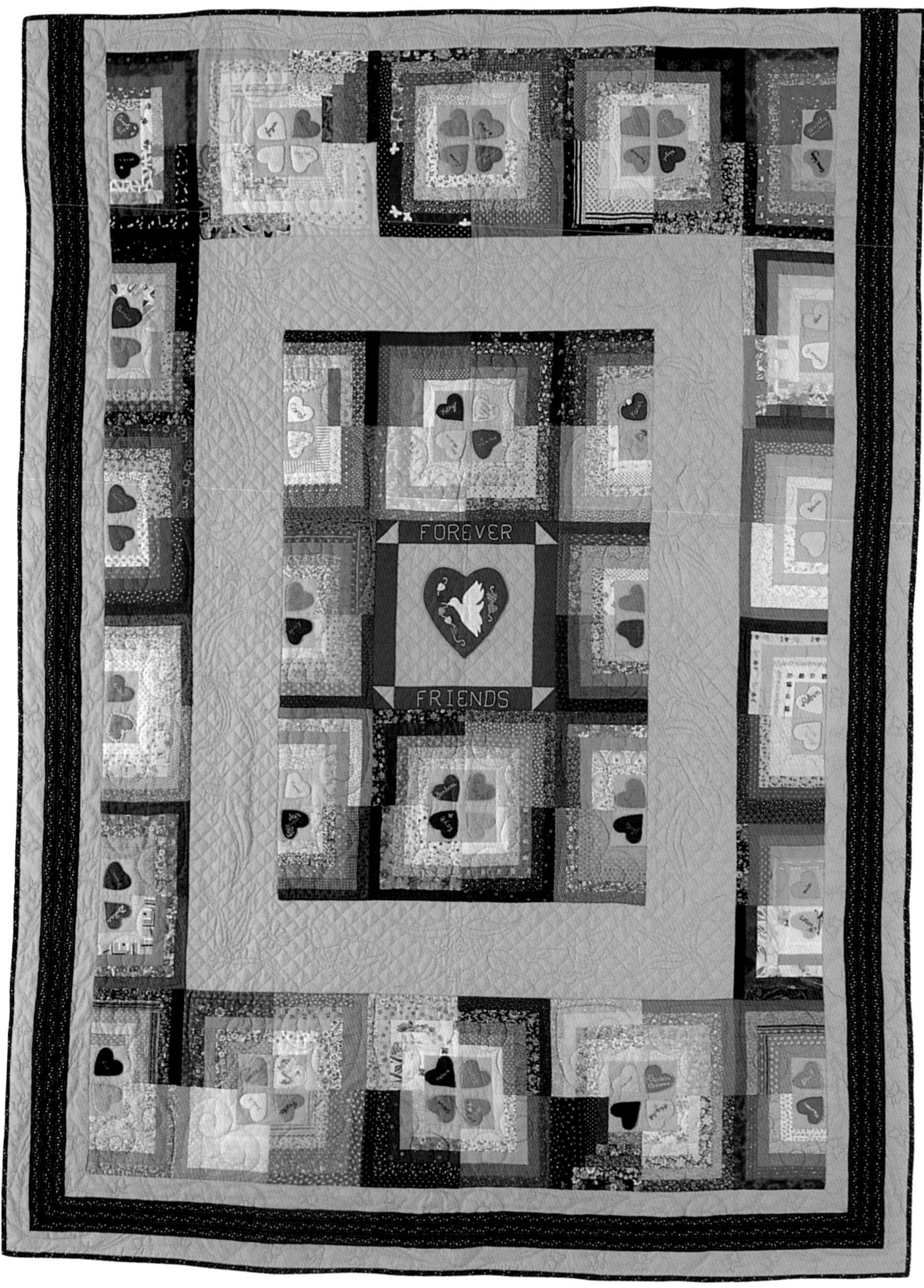

FOREVER FRIENDS

While delighting in the heart motif as a symbol of friendship, I was keen to put an 'Aussie' stamp on my quilt. So gum leaves and gumnuts, cockatoos and butterflies feature in the crosshatched aqua borders. As a recently arrived American, I just love my quilt. What a welcome to Tasmania and what an assurance of friendship to a newcomer.

THESE BLOCKS WERE MADE BY THE TASMANIAN QUILTERS GUILD, THE LAUNCESTON PATCHWORKERS AND QUILTERS AND THE TUESDAY QUILTERS, ALL OF TASMANIA, 1990, FOR TINY KENNEDY.

AUTUMN BASKETS

Tiny baskets create a treasure to remember friends by. Temporarily gluing paper templates to the fabric makes handling the tiny pieces easier.

Block used: Flower Basket

Block size: 8 cm (3 in)

Quilt size: 9 pieced blocks are needed for a 45 cm (18 in) square quilt

MATERIALS SUPPLIED

White fabric for the background

INSTRUCTIONS

▣ Copy or draft the entire block onto grid paper, using a sharp pencil. Cut out carefully on the lines and paste the paper templates onto the fabric, taking note of the grain lines. Cut out the fabric, leaving seam allowances as you would normally do when you mark fabric. You can prepare plastic templates and mark the fabric if you prefer. If you choose the latter method, make sure you sew along the inside of the pencil line, so that you preserve the size of the block.

▣ Cut the following pieces:

6 A from an autumn-toned fabric (you may make a scrappy basket or use only one fabric)

2 B from a different autumn-toned fabric

8 A, 2 C and 1 D from the white fabric

▣ Using the paper templates as a guide and following the piecing diagram, hand-piece the block.

Block Diagram

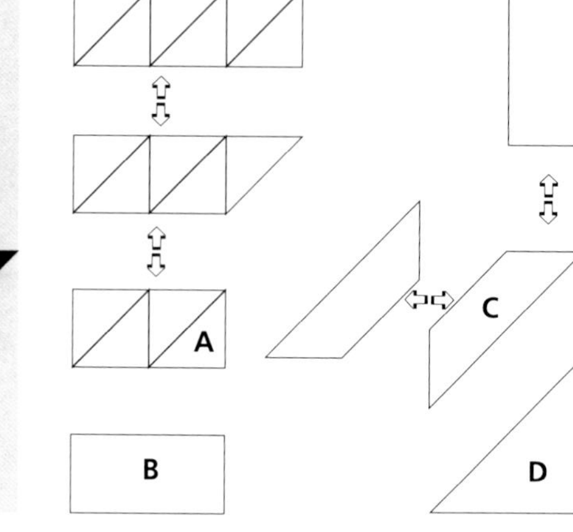

Piecing Diagram

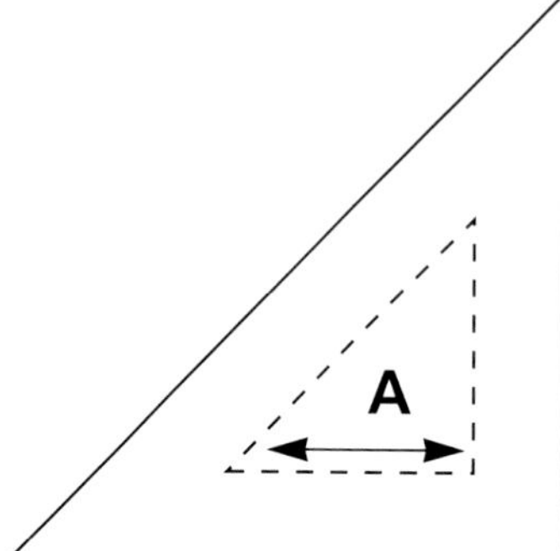

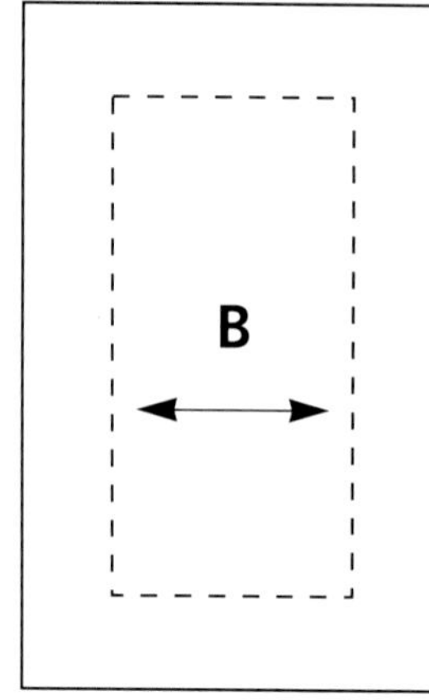

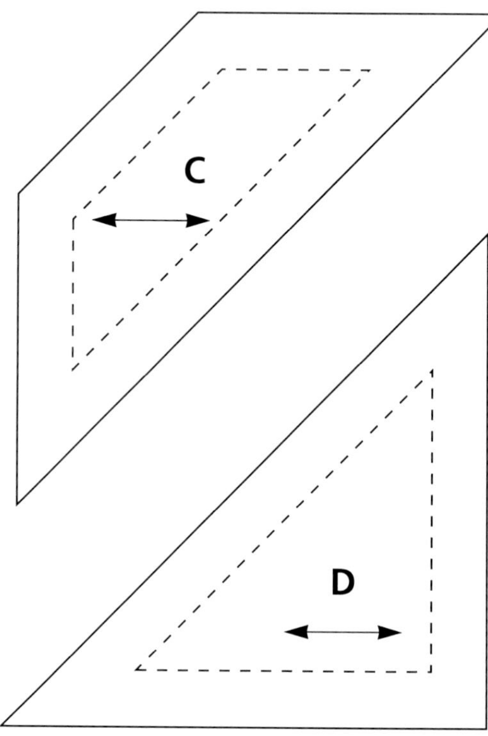

AUTUMN BASKETS

Our group had only just started, and we wanted to make friendship quilts that we would be sure to finish, so we decided to make miniatures. Often the templates were so small that marking the fabric increased the size of the block. So we resorted to using paper templates. I was delighted with my tiny baskets and pleased I had chosen a spotted fabric for the alternating blocks. That meant I could simply follow the dots to complete the crosshatching. Long after the quilt was finished, I took it off the wall and quilted the white inner border, a finishing touch I had not thought necessary when my little quilt was first deemed 'finished'.

THESE BLOCKS WERE MADE BY GOOD INTENT QUILTERS, RYDE, NEW SOUTH WALES, 1989, FOR LOIS COOK.

FRIENDLY FANS

Create fanciful bow ties with the traditional Grandmother's Fan block set in this unusual way.

Block used: Grandmother's Fan

Block size: 20 cm (8 in)

Quilt size: 49 blocks are needed for a 160 cm x 190 cm (63 in x 75 in) quilt

MATERIALS SUPPLIED

Square of cream homespun with a 20 cm (8 in) square marked on it

Blue fabric for the fan centre

INSTRUCTIONS

- Prepare templates A and B, following the general instructions on page 52.
- Using template A, cut 6 from 6 different blue print fabrics – a little pink or mauve is quite acceptable.
- Cut 1 B from the blue fabric supplied.

Block Diagram

- Join the fan sections together. For a smooth curve, run a small running stitch, just inside the seam allowance of each curved segment. Place the template on the stitching line on the wrong side of each fabric section and draw up the thread. Press the segment firmly, giving a nice smooth curved edge.
- Appliqué the fan into position with the edges matching the background edges.
- Embroider your initials on the completed block.

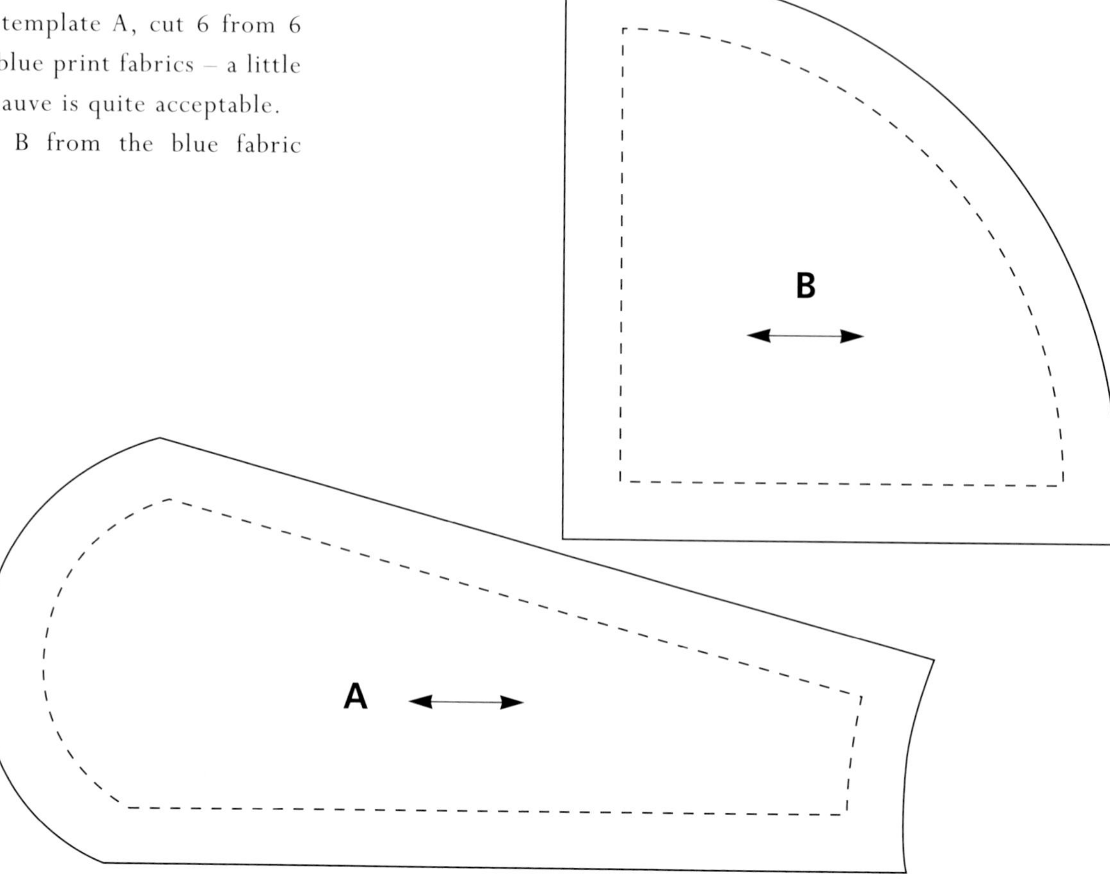

FRIENDLY FANS

Soft dusky pinks and blues are my very favourite colours, and I really like traditional patterns. When I saw a photograph of a quilt with the Grandmother's Fan block set in such a pretty way, I decided to make this my next friendship block. Everyone embroidered their initials on their block, so I can read them and remember who helped me make my delightful quilt.

THESE BLOCKS WERE MADE BY THE EPPING QUILTERS, NEW SOUTH WALES, 1995, FOR ELAINE GALLEN.

CHATELAINE FRIENDSHIP

The charm of this quilt is in its simplicity. A simple sixteen-patch block divided diagonally into light and dark creates a visual delight.

Block used: Shaded Four-patch

Block size: 30 cm (12 in)

Quilt size: 16 blocks are needed for a 180 cm (71 in) square quilt

MATERIALS SUPPLIED

Some floral fabrics to set the mood

INSTRUCTIONS

- Please supplement the fabrics provided with florals from your own collection. Keep dark fabrics on one side of the block and light fabrics on the other. Remember that the dark fabrics only need to be darker than the light fabrics in the block.
- Cut the following pieces:

6 A from the light fabrics
6 A from the dark fabrics
4 B from the light fabrics
4 B from the dark fabrics

- When piecing the block, join the light and dark triangles first to form four squares, then complete the block, following the piecing diagram.
- Embroider your initials on your block, when it is finished.

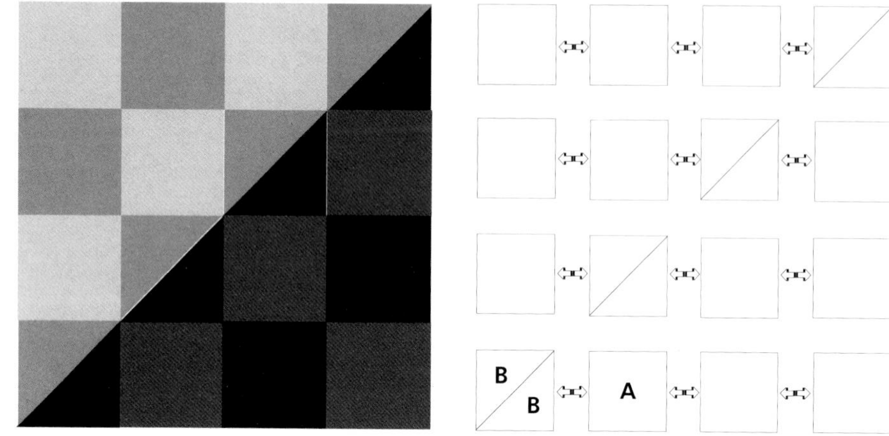

Block Diagram

Piecing Diagram

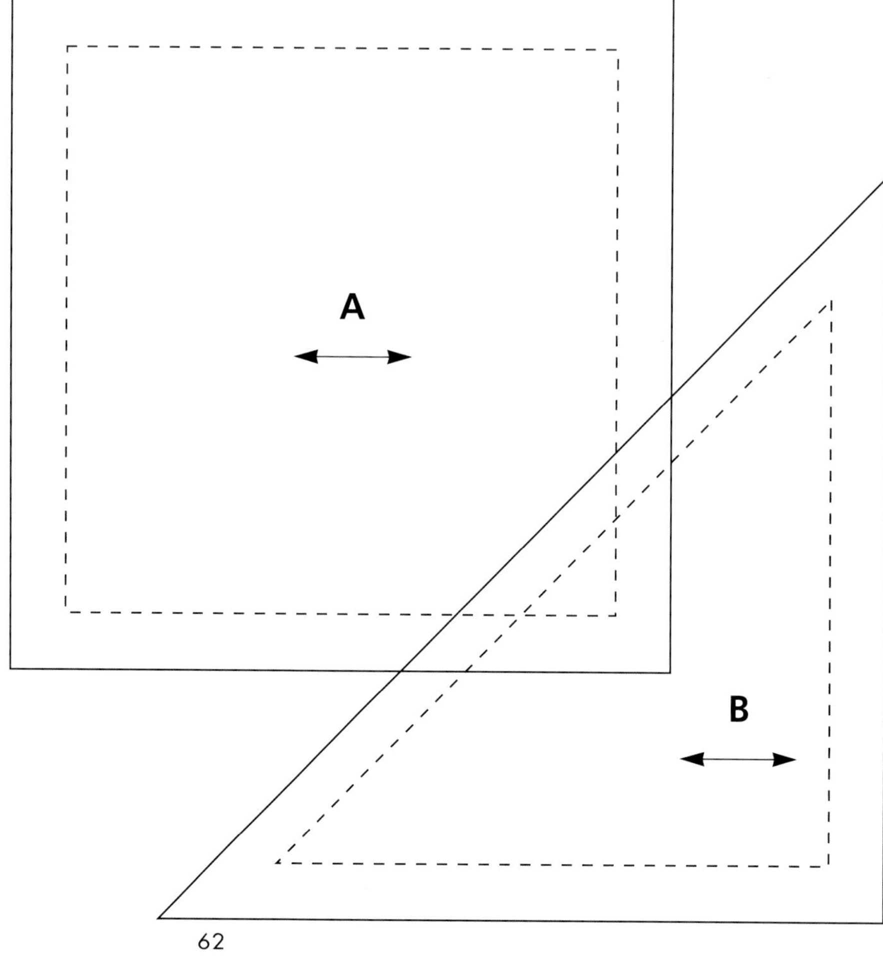

CHATELAINE FRIENDSHIP

When these blocks were made in 1989, nobody had heard of colour-wash, and the variety of florals was limited. I was keen to keep my friendship block simple, while creating a dynamic quilt. I achieved this by extending the pieced blocks into the border and accentuated the catherine wheel effect by quilting with straight lines.

THESE BLOCKS WERE MADE BY THE CHATELAINE QUILTERS, SYDNEY, NEW SOUTH WALES, 1989 FOR CAROLYN SULLIVAN.

DAISIES

With such a variety of fabrics used in the petals, and each alternate block cut from a different fabric, this quilt achieves a wonderful scrappy thirties look.

Block used: Daisy

Block size: 20 cm (8 in)

Quilt size: 48 daisy blocks and 35 plain blocks are needed for a 170 cm x 225 cm (67 in x 88 in) quilt

MATERIALS SUPPLIED

22 cm (9 in) cream square for the background

INSTRUCTIONS

- The finished quilt should look like a scrappy, old quilt with each petal in a different fabric.
- Trace the templates onto template plastic. A coin with a 3 cm (1¼ in) diameter is a useful template for the centre circle.
- Mark 1 petal on the right side of each of 8 different fabrics.

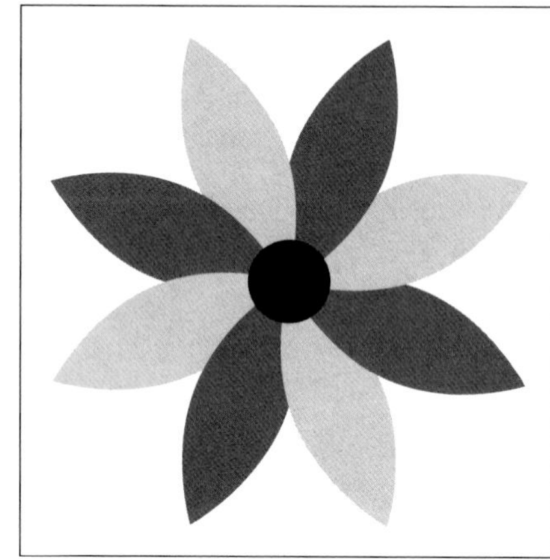

Block Diagram

- Fold the background square in half horizontally and vertically and along each diagonal to locate the centre and provide guidelines for placement of the petals. Press the folds.
- Baste under the seam allowance on each petal. Put a pin through the tips of the petals, then through the centre of the background, then stick the pin through into the ironing board cover to anchor it. Carefully spread out the petals and line them up on the pressed lines on the background fabric. Pin each petal into place as you go. Each petal slightly overlaps the preceding one. At the end, tuck the edge of the last petal under the first one.
- Centre the coin on the petals and lightly mark its outline with a pencil. Trim each petal to within 6 mm (¼ in) of the pencil line to reduce bulk.
- Appliqué the petals, using a thread to match each petal.
- Using a solid fabric of your choice and the round template, mark the circle on the right side of the fabric. Appliqué the circle in the centre of the petals.

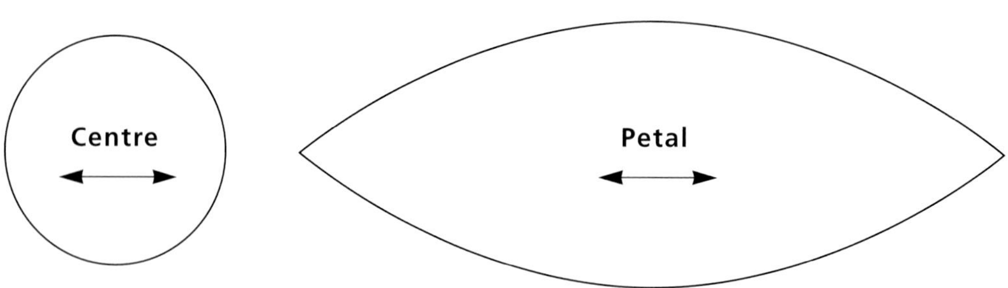

DAISIES

Having made many friendship quilts, my group decided that it was time to make blocks with a challenge. So I chose this overlapping appliqué flower. I belong to several groups who all wanted to join in, so I have blocks from my friends in all my groups. Some of the blocks were cut out by three quilting friends in the United States and I pieced them. It was a great way to get an interesting variety in the fabrics used.

THESE BLOCKS WERE MADE BY THE EASTWOOD PATCHWORK QUILTERS, THE GOOD INTENT QUILTERS AND THE HILLSIDE QUILTERS, ALL OF NEW SOUTH WALES, 1991, FOR LEIGH SWAIN. (SOME BLOCKS WERE CUT-AROUND BLOCKS WHICH LEIGH STITCHED TOGETHER). THIS QUILT WAS INSPIRED BY 'LENA' QUILT ART ENGAGEMENT CALENDAR 1990.

BEARLY THERE FRIENDSHIP QUILT

Using a traditional block and the same fabric for the sashing strips, the setting triangles and the borders allow the blocks in this delightful quilt to float on a sea of green.

Block used: Farmer's Daughter

Block size: 25 cm (10 in)

Quilt size: 23 blocks are needed for a 156 cm x 241 cm (61 in x 95 in) quilt

MATERIALS SUPPLIED

Cream homespun for the background

INSTRUCTIONS

- You will need two, three or four scraps of various pink and green fabrics.
- Prepare templates A, B and C. (See page 52 for how to prepare templates.)
- Cut the following pieces:

4 A and 4 B from the cream homespun

4 C and 4 C(r) from one of the pink and green fabrics or 4 C from one fabric and 4 C(r) from another fabric

1 A for the centre square

4 A for the surrounding squares from a different pink or green fabric

- Assemble the block following the piecing diagram.

Block Diagram

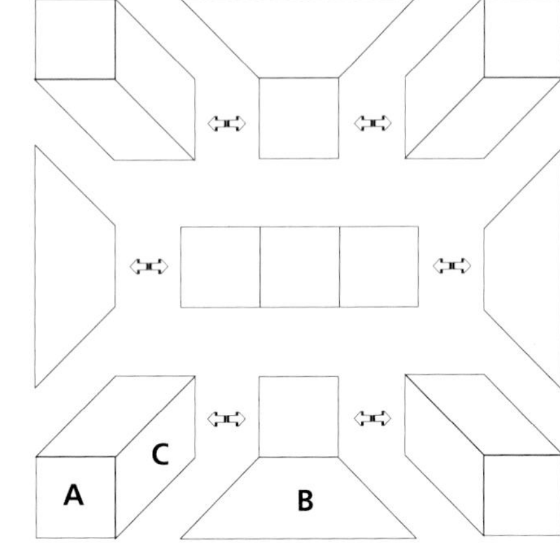

Piecing Diagram

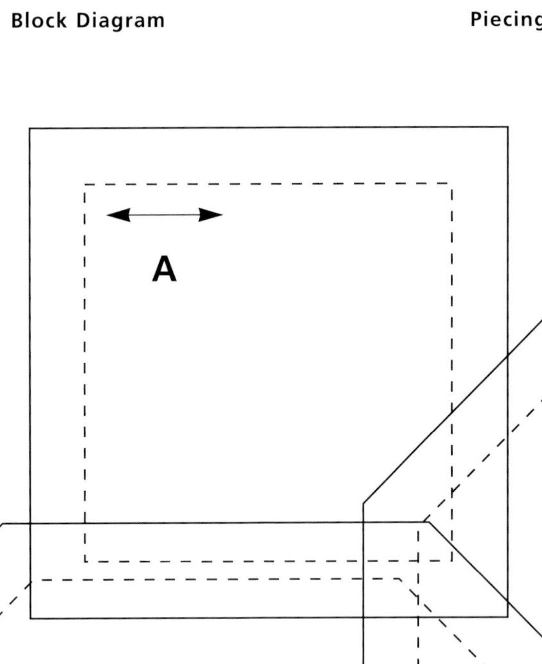

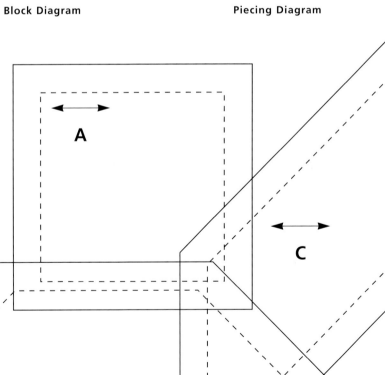

BEARLY THERE FRIENDSHIP QUILT

To herald our adventures into friendship blocks, the group chose this design and the colour scheme. We seemed very casual about the whole process and I wasn't particularly keen to win the finished blocks as I had planned a Bear's Paw block for my quilt. Having recovered from the shock of winning, by the time I had put the quilt together completely by hand and finished the hand-quilted grid, I really loved it. I find myself thinking as I look at each block, 'Ah yes! That's so like Sue or Judy'. It's wonderful to have such a permanent reminder of each of my quilting friends.

THESE BLOCKS WERE MADE BY THE BEARLY THERE QUILTERS, HORNSBY, NEW SOUTH WALES, 1993, FOR JENNY EVANS.

MY LITTLE GIRLS

Sunbonnet Sue again captures everyone's heart as she chases butterflies and catches balloons in this traditional block.

Block used: Sunbonnet Sue

Block size: 30 cm (12 in)

Quilt size: 15 blocks are needed for this 110 cm x 180 cm (43 in x 71 in) quilt

MATERIALS SUPPLIED

35 cm (14 in) white square
Double-sided fusible webbing

INSTRUCTIONS

- Choose any colours you like, except green!
- Trace the pattern pieces onto the paper side of the fusible webbing. Do not remove the paper at this stage. Note, when extra is needed for the underlap it is denoted by the dotted lines. Cut out each pattern piece leaving 1 cm (3/8 in) outside the pencil line. Iron these pieces onto the wrong side of your selected fabric, using a dry iron on the wool setting. Turn the fabric over and firmly press on the webbed area. Cut out each piece on the pencil lines. Peel off the paper backing.

Block Diagram

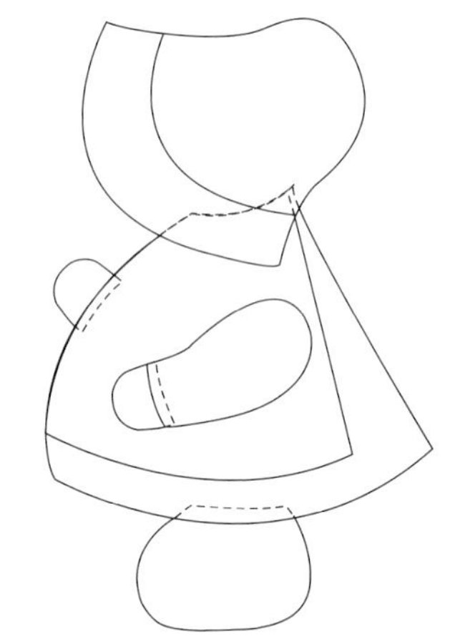

Piecing Diagram

- Assemble the motif on the background fabric, making sure that your Sunbonnet Sue is in the centre of the square. Be sure to place the underlaps correctly. Use the numbers on the templates to guide you. Your Sunbonnet Sue is now facing right. Press the pieces lightly into place, one at a time. When you are happy with the assembly, put a light pressing cloth over the lot and press again, very firmly. Turn over and press again on the wrong side.
- Hand-stitch in herringbone or buttonhole stitch around the edges of all the pieces. Add embellishments such as flowers, insects and the like, then embroider your name in the bottom left corner in a colour to match your Sunbonnet Sue.

MY LITTLE GIRLS

I had just discovered appliqué in 1990 and fell in love with Sunbonnet Sue. A friendship quilt was a great way to get the blocks quickly, as I was not convinced that I would finish the fifteen blocks needed. Each block was quilted before assembly, with a sun in one corner, clouds in another and the outline of hills, grass and a crazy path along the bottom. The quilted blocks were then joined using the 'quilt as you go' method. Lastly the broderie anglaise frill was added. I was delighted!

THESE BLOCKS WERE MADE BY THE MARION QUILTERS, SOUTH AUSTRALIA, 1992, FOR MOYA MARSHALL.

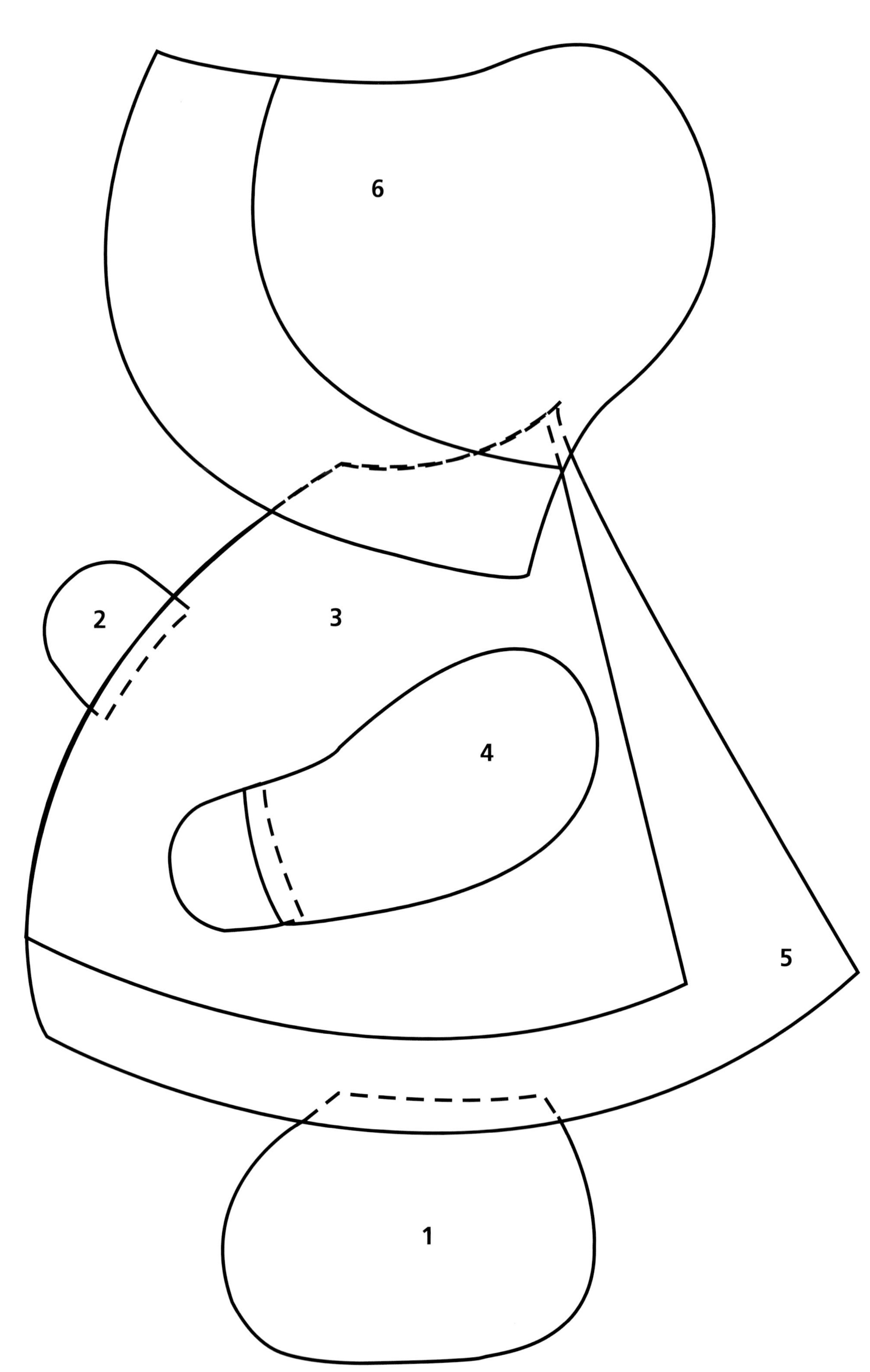
6
2
3
4
5
1

SCOTTIES

Scottie dogs with smart plaid jackets line up ready to march straight into your heart. A dog biscuit provides the template for the unique quilting pattern.

Block used: Scottie

Block size: 25 cm (10 in)

Quilt size: 15 blocks are needed for a 135 cm x 205 cm (53 in x 81 in) quilt

MATERIALS SUPPLIED

Cream fabric for the background

Black fabric for the body

INSTRUCTIONS

- Use a plaid or checked fabric for the Scottie's coat.

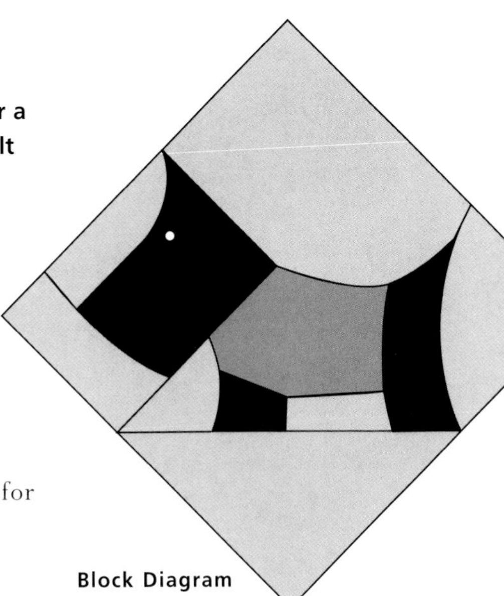

Block Diagram

- With your finished block, include enough of your plaid or checked fabric for four 9.5 cm ($3\frac{3}{4}$ in) squares.
- Cut the following pieces:
  1 E, 1 F, 1 G, 1 H, 1 I, 1 J and 1 K from the background fabric
  1 A, 1 B, 1 C from the black fabric
  1 D from the plaid or checked fabric
- Piece the block, following the piecing diagram.

Piecing Diagram

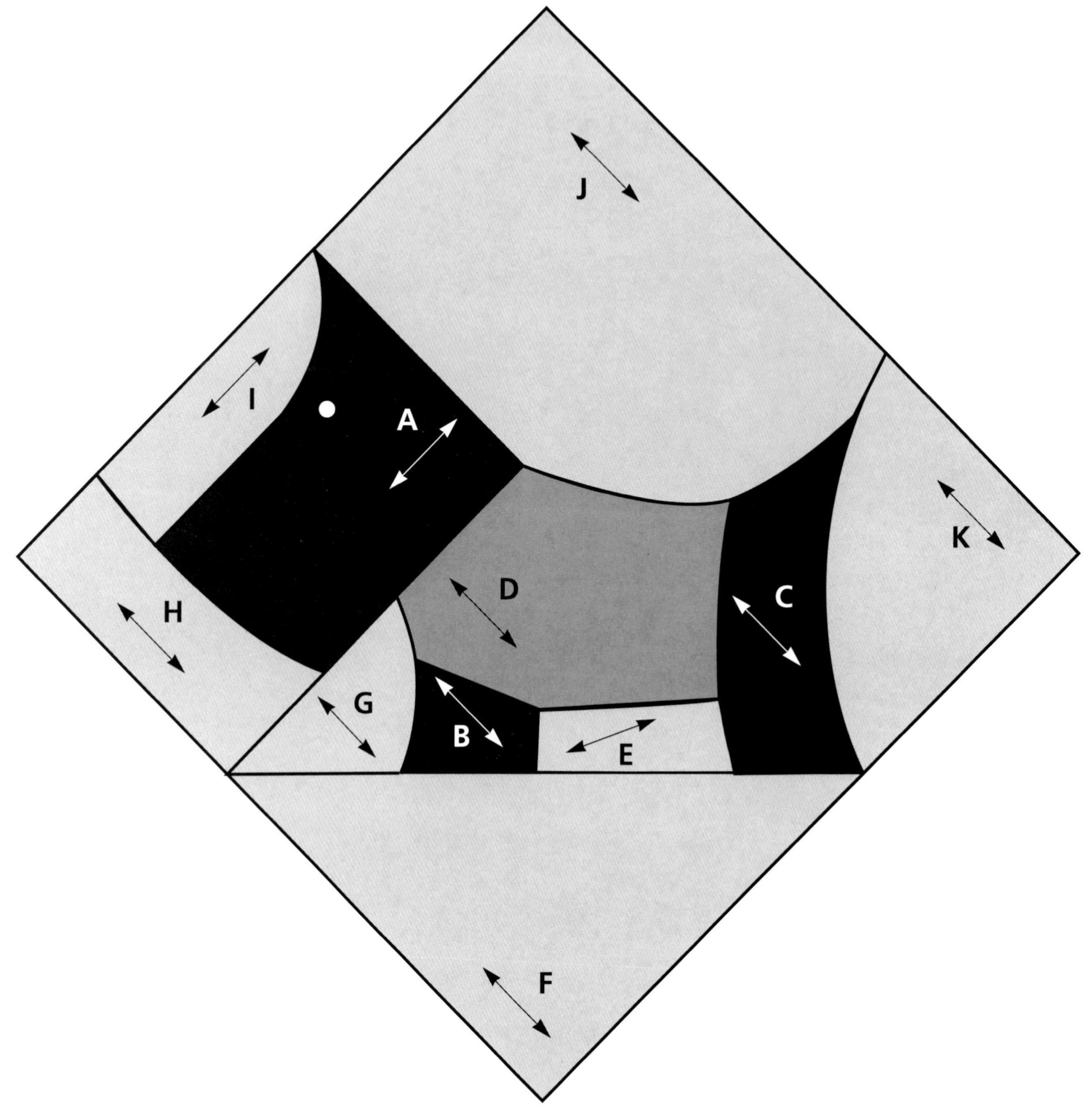

Templates and Piecing Diagram

Note: Templates are given at 50% of actual size. Enlarge to 100%

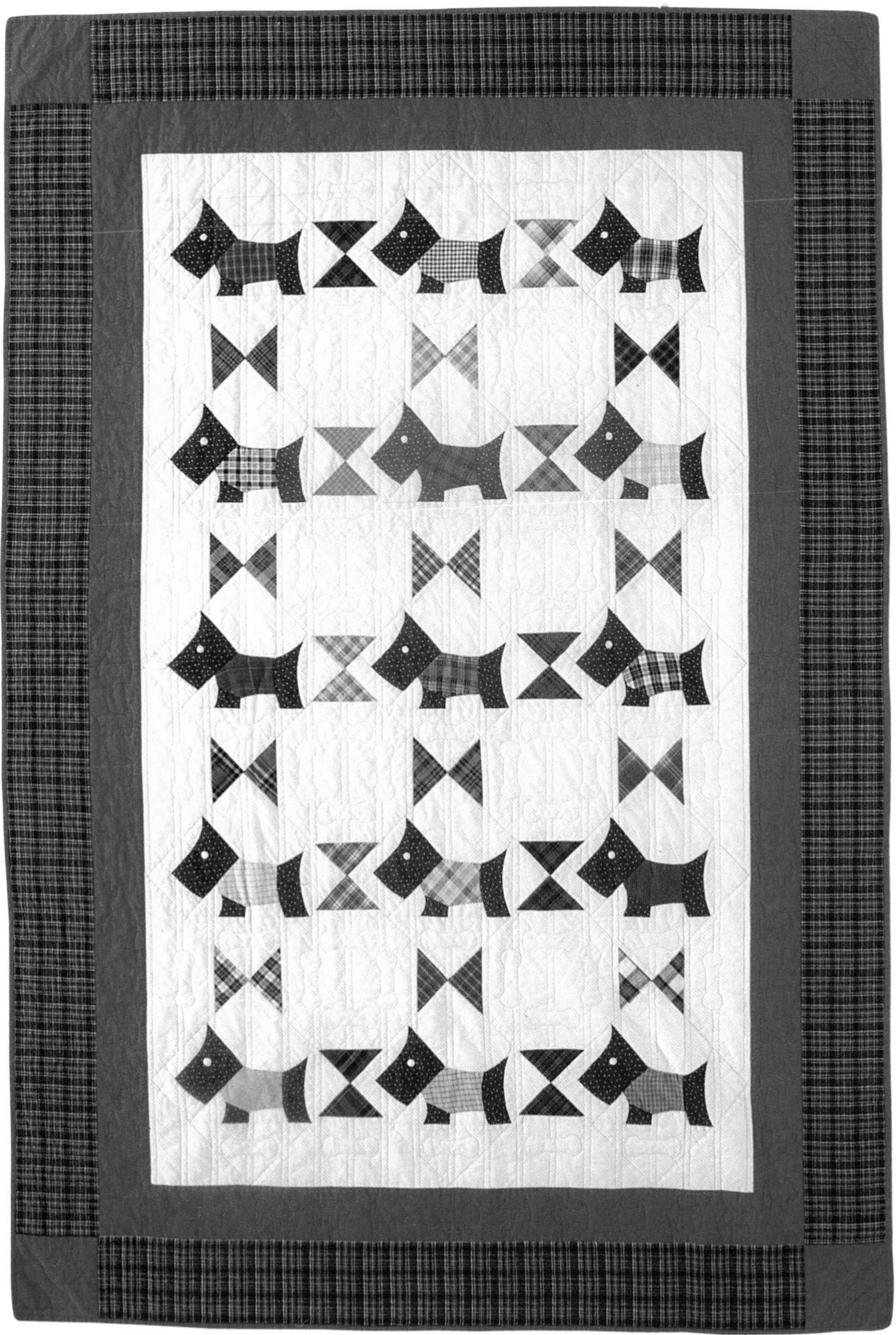

SCOTTIES

I thought this quilt would be a lot of fun to make and would be enjoyed by my family. My friends chose plaids that reflected their preferences in fabrics, so I can tell who made which block. One of the Eastwood quilters, Lea Lane, suggested using the Snowball block in the alternate rows, an idea which worked well. For the quilting, I traced around a dog biscuit to make the template. Finally, I quilted miniature Scotties in the red border.

THESE BLOCKS WERE MADE BY EASTWOOD PATCHWORK QUILTERS, NEW SOUTH WALES, 1991, FOR EVELYN FINNAN.

BUTTERFLY REQUIEM

This pretty version of the traditional Dresden Plate block creates a quilt with an old-world charm that looks like it was made many years ago.

Block used: Modified Dresden Plate

Block size: 30 cm (12 in)

Quilt size: 25 blocks are needed for a 185 cm (73 in) square quilt

MATERIALS SUPPLIED

32 cm (12½ in) square of cream homespun

Cream homespun for the centre circle

One of my favourite fabrics to use, if you want to

INSTRUCTIONS

- Prepare templates for the petal (A) and the leaf (B).
- Cut the following pieces:

16 A from scraps, but every petal does not have to be different

16 B from colours that coordinate with the petals, but the leaves do not have to be green

1 B from the homespun

1 B from thin cardboard

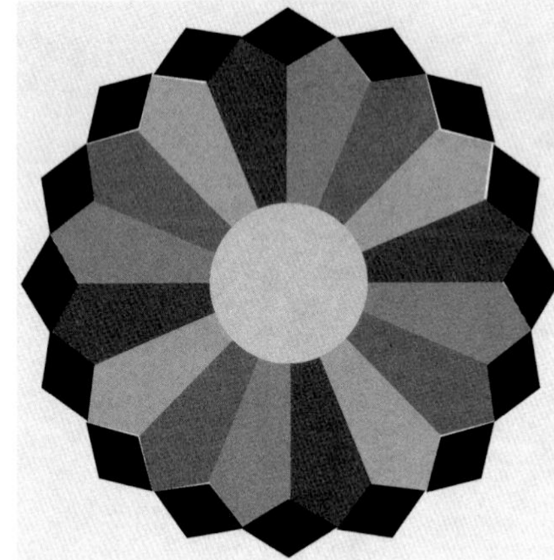

Block Diagram

- Sew two of the petals together and place a leaf diamond between them. Repeat this eight times, then join the pairs to form a circle.
- Baste the homespun circle around the cardboard template C. Press. Place the circle in the centre of the ring of petals and appliqué it in place. Remove the basting and take the cardboard out from the back, through the centre hole.
- Appliqué the completed plate to the background block, making sure it is centred.
- Please embroider your name in the centre circle and appliqué or embroider a butterfly there as well.

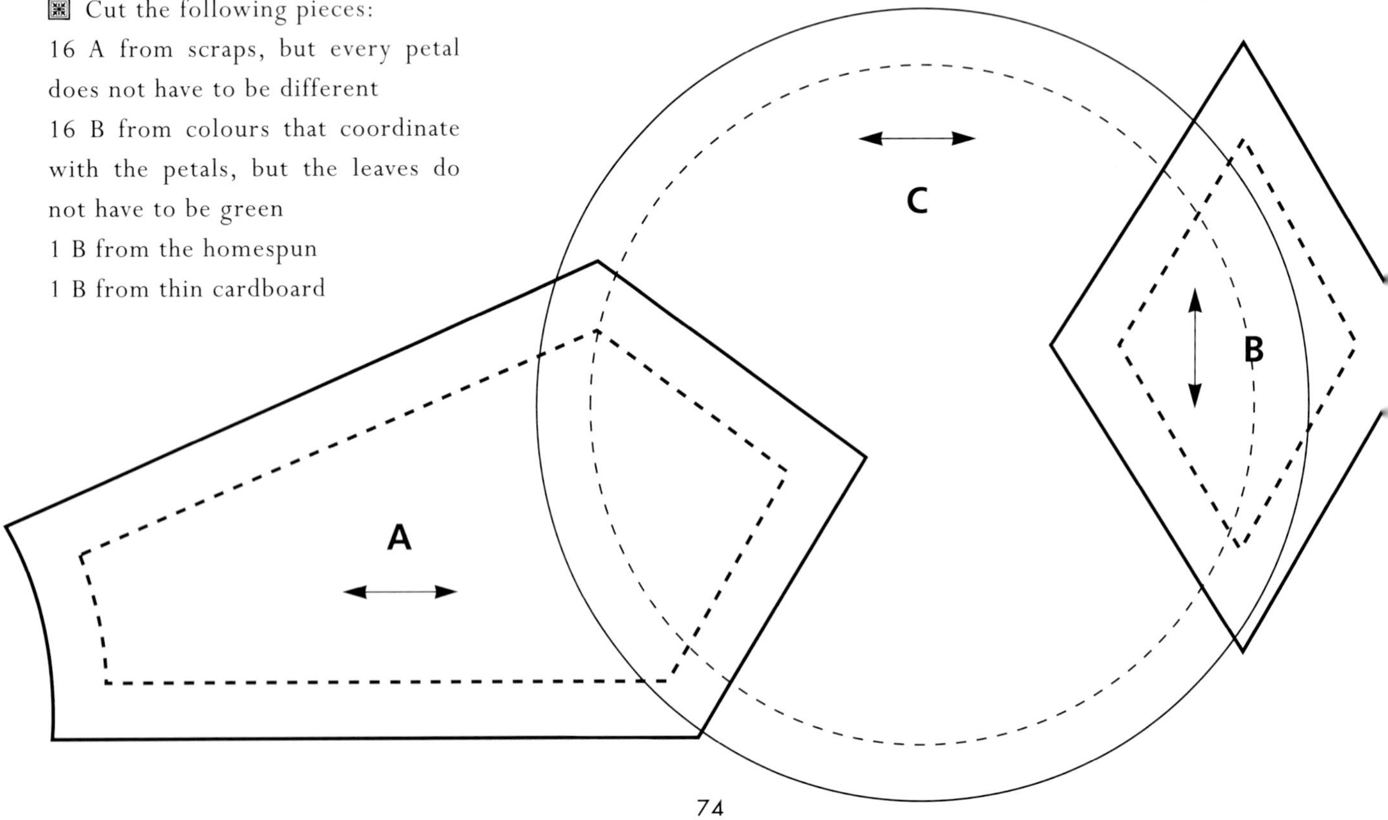

C

A

B

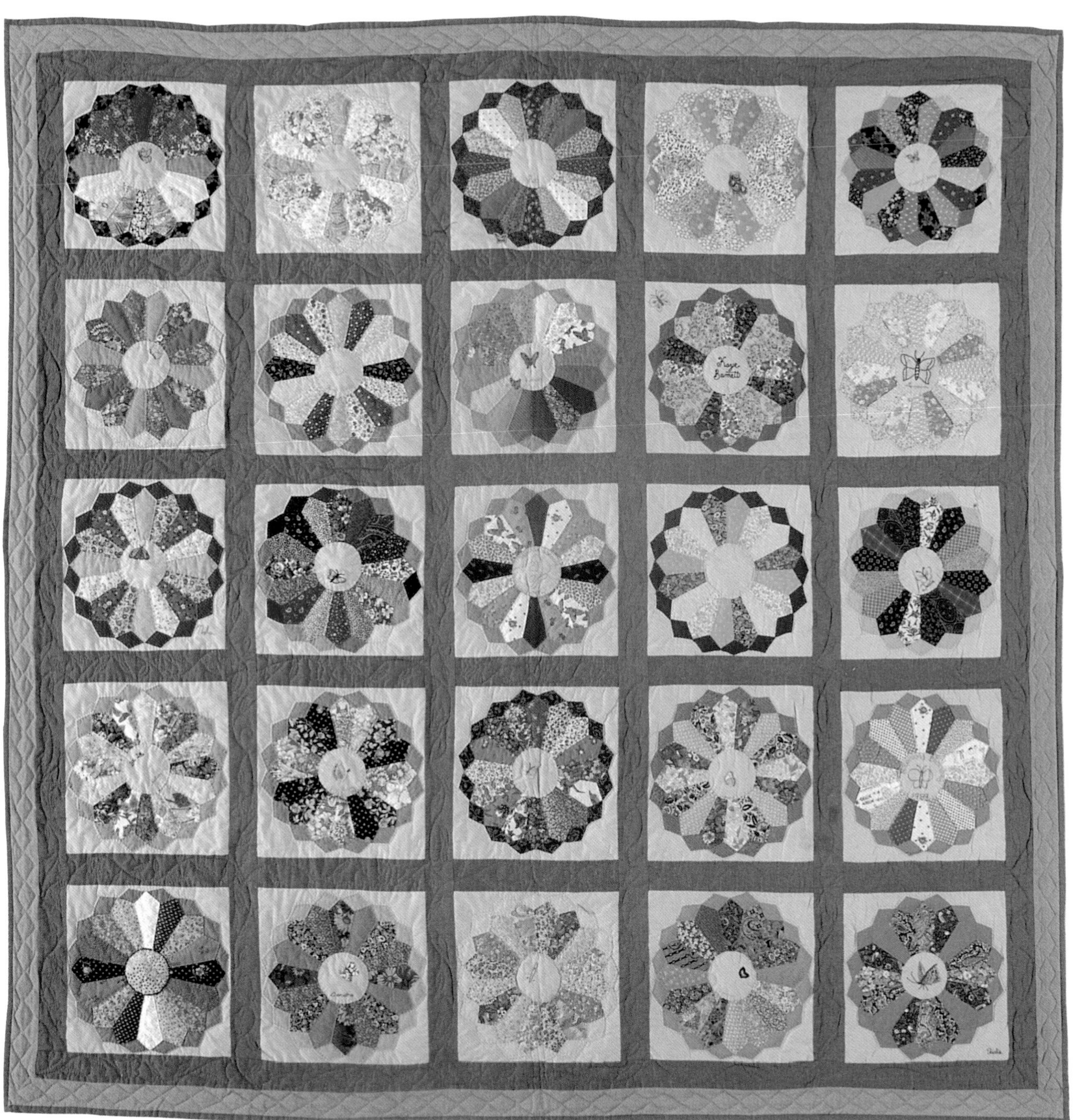

BUTTERFLY REQUIEM

The things that give me pleasure are Botticelli paintings, potpourri, old roses, and the colours of the sea at dusk, so my friendship quilt needed to look as though it had been made in a bygone time, faded by years of loving wear and care. I chose greyed shades of green, pink and blue, with lots of lovely scrappy fabrics from my friends. They added butterflies to their completed blocks, and I quilted butterflies on the ivy leaf quilting design. The butterflies are symbolic of the spirits of my three friends, who died while the quilt was being made, but who are now flying free.

THESE BLOCKS WERE MADE BY QUILT CONNECTIONS, WESTERN AUSTRALIA, 1990, FOR CYNTHIA BAKER.

ROSE STAR

The colours of a bygone era are used in this quilt to give it an old-world look, as the rose stars float on a background of cream, their black points providing sharp contrast to the otherwise gentle colours.

Block used: Rose Star (an Alice Brooks pattern)

Block size: 48 cm (19 in) diameter hexagon

Quilt size: 22 complete blocks and 6 half blocks are needed for a 2.5 m (2¾ yd) square quilt

MATERIALS SUPPLIED

Cream fabric for the background

INSTRUCTIONS

- Only one kite-shaped template is required for this block.
- Please choose fabric with an old-world look.

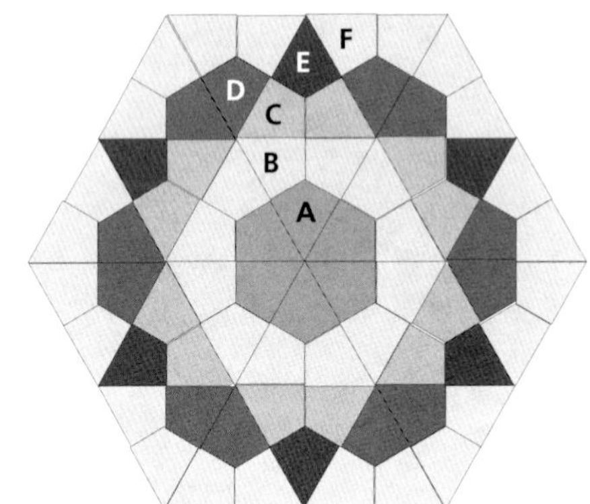

Block Diagram

- Make the points on the rose out of a black fabric with a small print.
- For each block, cut the following pieces, using the template provided:

6 from a light fabric for A
12 from a dark fabric for B
12 from a medium fabric for C
12 from a medium dark fabric for D
6 from the black fabric with a small print for E
24 from the cream fabric, supplied, for F

- Piece the block, following the block diagram.

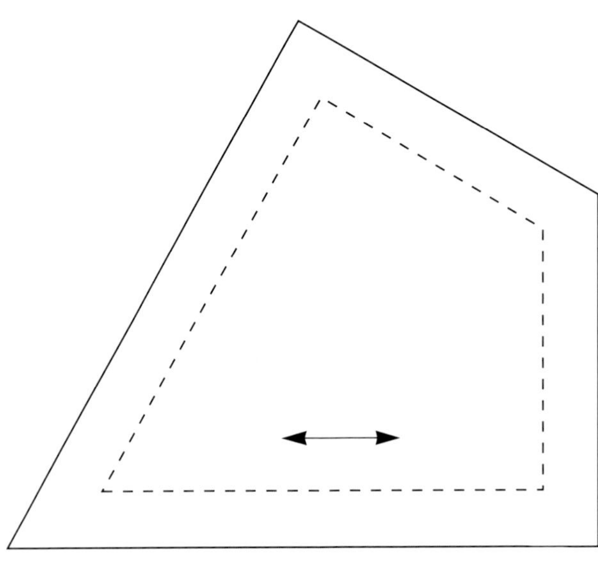

ROSE STAR

As the block, Rose Star, used only one template I thought it would provide a challenge for the group. They loved the look of the finished quilt and it now covers my mother's old eiderdown, remade to fit the quilt.

THESE BLOCKS WERE MADE BY THE GOOD INTENT QUILTERS, EPPING, NEW SOUTH WALES, 1994, FOR ELIZABETH BOSWELL.

TIME FOR TEA

Piecing in strips and a controlled palette make this quilt a real winner and a delightful reminder of shared moments over a cup of tea.

Blocks used: Teapot, Cup and Saucer

Block size: 30 cm (12 in)

Quilt size: 16 blocks are needed for a 168 cm (66 in) square quilt

MATERIALS SUPPLIED

Cup and saucer print fabric
Pale blue print fabric
Medium blue fabric
White fabric for the background

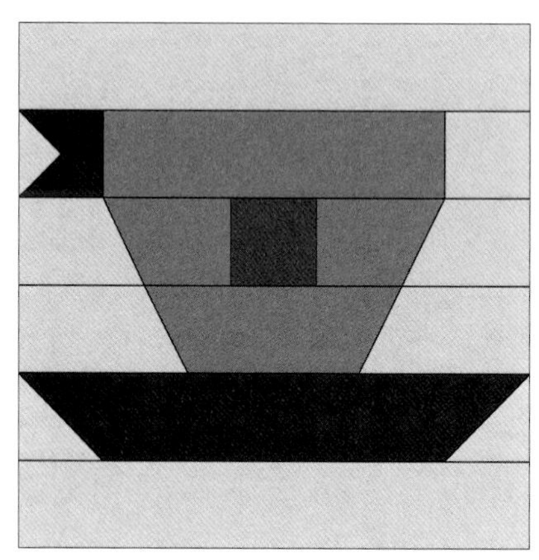

Block Diagram

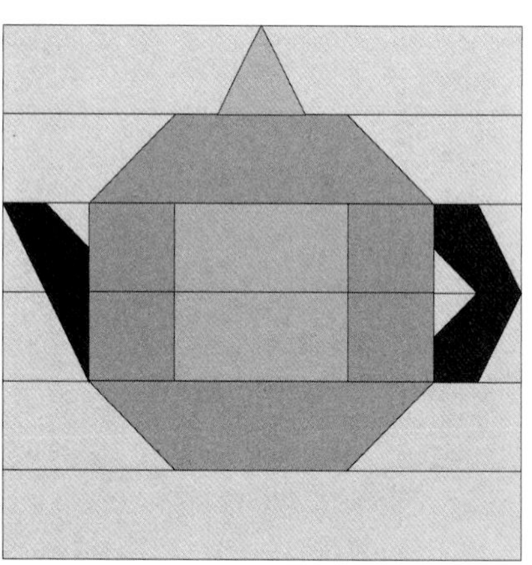

Block Diagram

INSTRUCTIONS

- Decide if you want to make a teapot or a cup and saucer. Following the appropriate piecing diagram, prepare a 30 cm (12 in) square. Divide it into 5 cm (2 in) squares. Draft your chosen design and prepare your templates.
- The blocks are easily pieced into five horizontal strips which are then joined together.
- You may add touches of one other fabric of your choice.
- If you wish, you may appliqué a curved handle on the cup, rather than piece it.

TIME FOR TEA

Inspired by a painting of teapots and cups, I decided to design two blocks for my friends in Burnie to make. Somehow I thought that these blocks really summed up the group well – twelve ladies all gathered round, with cups in hand, sharing tips as they work towards completing their first quilt.

THESE BLOCKS WERE MADE BY THE BURNIE QUILTERS, TASMANIA, 1995, FOR LINDA CARTER.

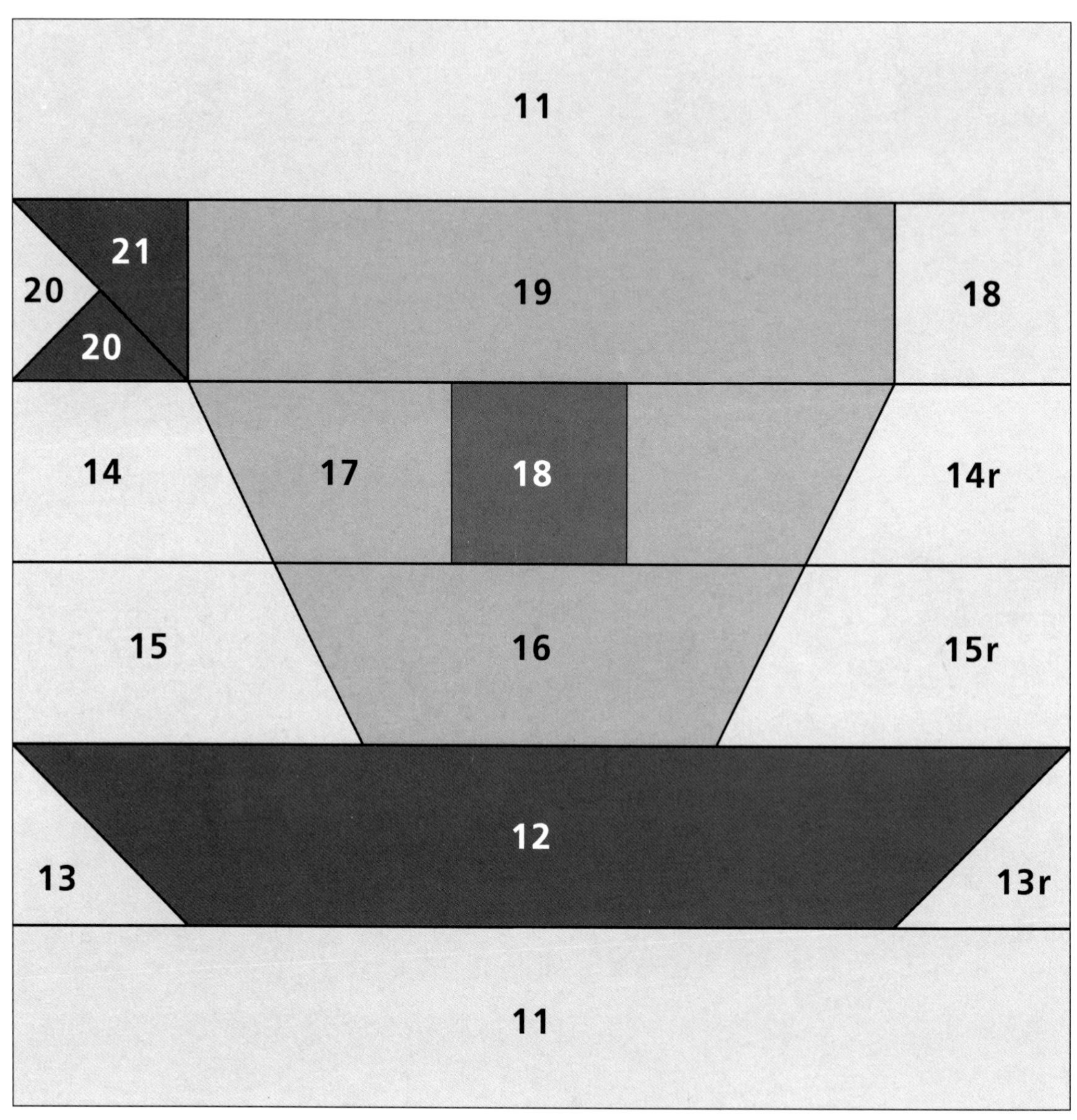

Templates and Piecing Diagram

Note: Templates are given at 50% of actual size. Enlarge to 100%

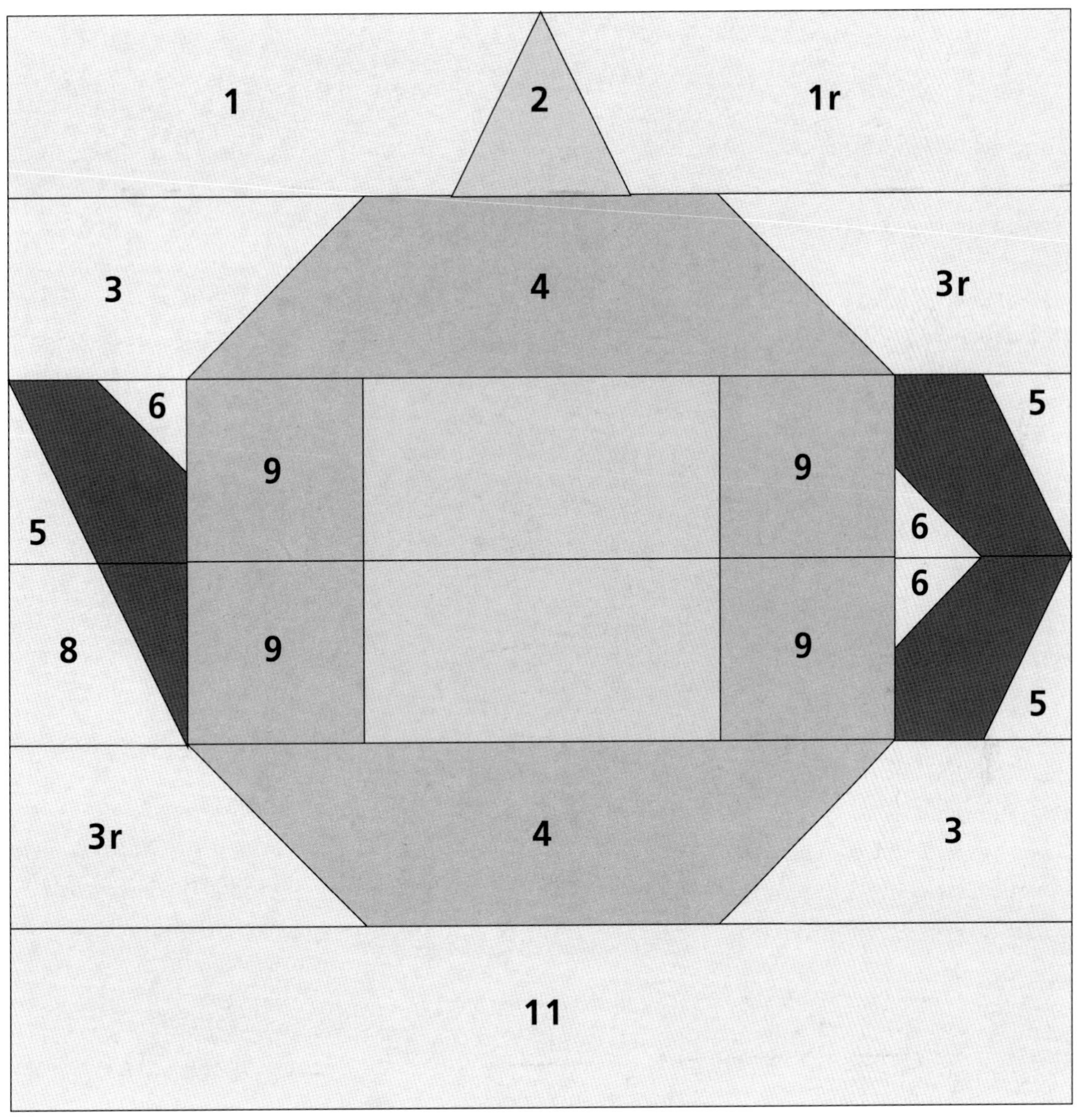

Templates and Piecing Diagram

Note: Templates are given at 50% of actual size. Enlarge to 100%

FAR EAST, FAR OUT, FAR OFF

Original blocks can create fun friendship quilts, especially when an outlandish batik fabric is the starting point. A wild border adds the finishing touch.

Block used: original design 'Larraine's Star'

Block size: variable from 7 cm (2¾ in) to 45 cm (17¾ in)

Quilt size: the number of blocks needed varies, depending on the sizes chosen. This quilt is 140 cm x 180 cm (55 in x 71 in).

MATERIALS SUPPLIED

Piece of multicoloured batik

INSTRUCTIONS

- You only need to use one fabric, other than the one supplied, to complete the block. An unusual or bizarre fabric would be great!
- The block can be any size, except 30 cm (12 in). No templates are provided. One quarter of a 24 cm (9½ in) block is given as a guide.

Block Diagram

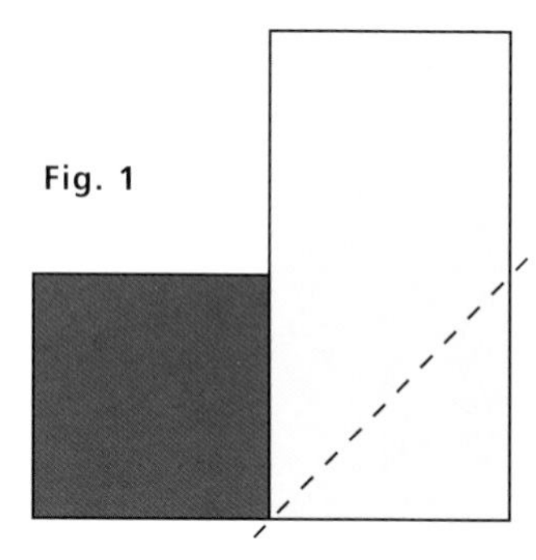

Fig. 1

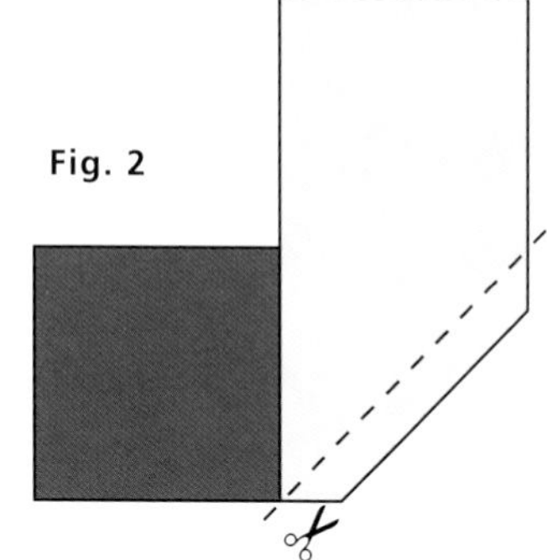

Fig. 2

Fig. 3

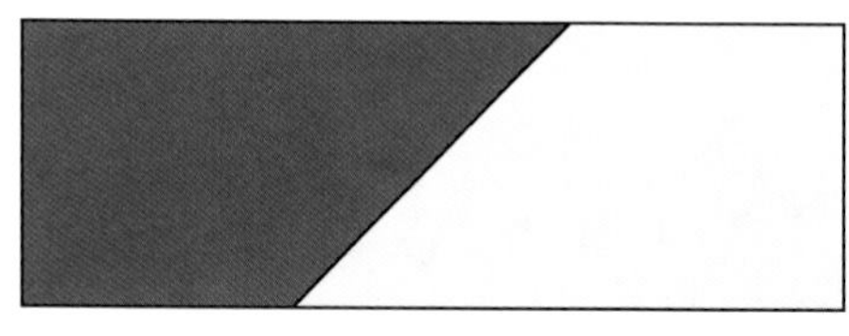

Decide what size you want your block to be and draft the appropriate templates. Note that only one-quarter of the block needs to be drafted.

- Quick machine- and hand-piecing construction tip: a block can be made from four long and four short rectangles, cut from each of the two fabrics. Add your preferred seam allowance to the finished sizes as calculated below:

Finished length of long rectangle A = finished size of block divided by 2

Finished length of short rectangle B = finished size of block divided by three

Finished width of both rectangles = finished size of block divided by six

- Piece the block, following figures 1-3 and the block diagram.

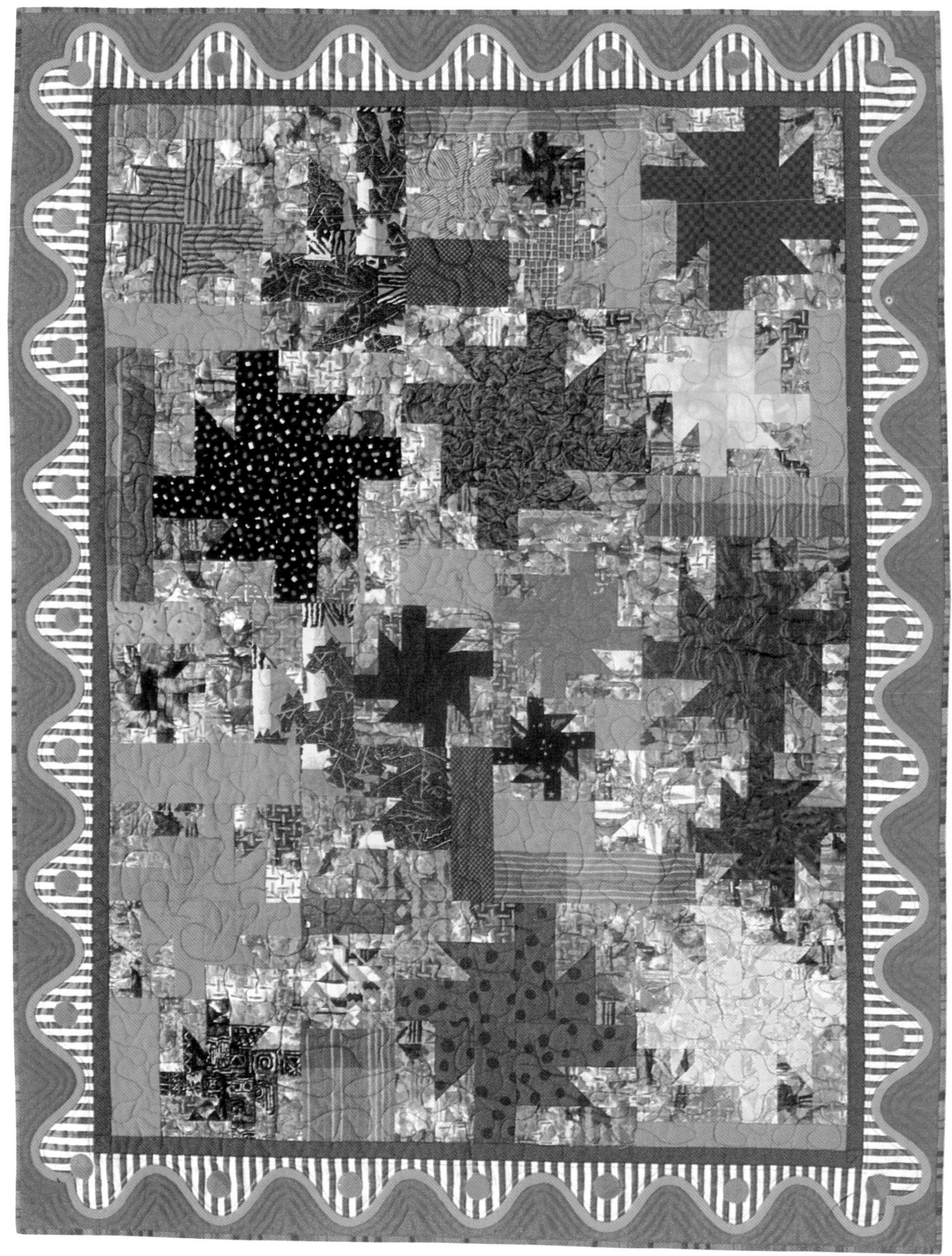

FAR EAST, FAR OUT, FAR OFF

Our group has attempted many difficult blocks as we endeavour to challenge ourselves and improve our skills. I decided to set an easy task to show that patchwork does not have to have a lot of pieces to be interesting. The only 45 cm (17¾ in) block was pieced, using a dinosaur print with fluorescent orange glasses. It dominated the layout – no matter where it was placed. In the end, I split it into several parts. My asymmetrical layout meant I had gaps to fill. For these, I used special fabrics collected from friends over the years. Even the backing fabric, a bright blue and red print, was donated by a friend who confessed to not sleeping at night while it was in her home.

THESE BLOCKS WERE MADE BY THE RUSTY PINS QUILTERS, GLENBROOK, NEW SOUTH WALES, 1994, FOR LARRAINE SCOULER.

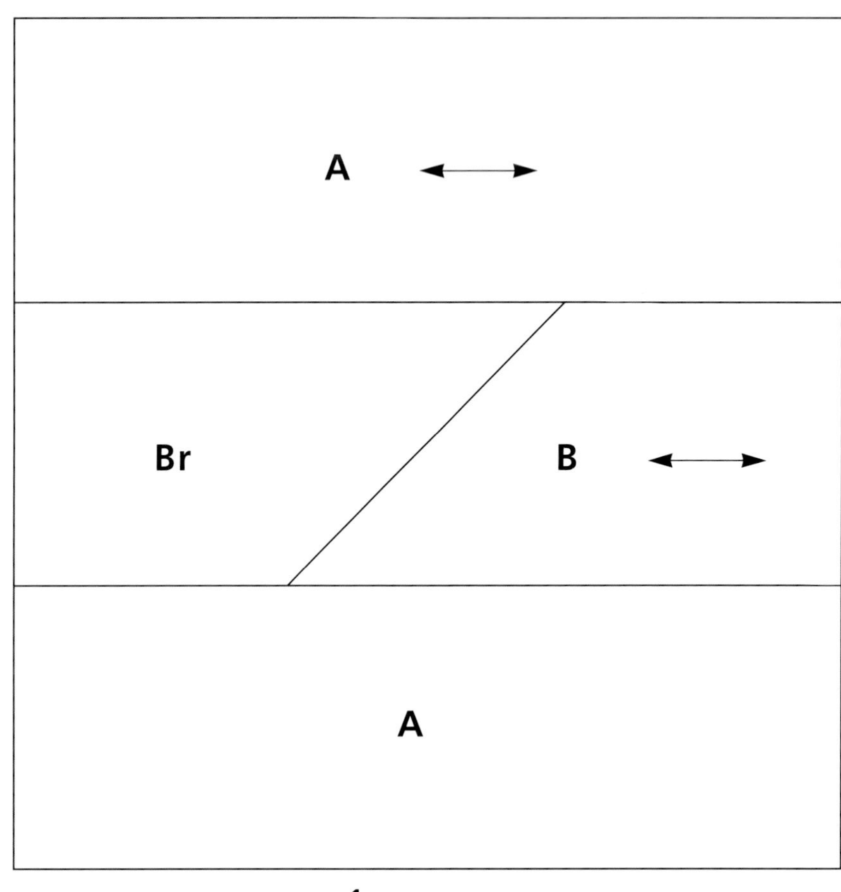

1/4 Block

AMISH SPARKLE

Using the traditional Amish colours against a black background in a mosaic tile pattern creates a stunning graphic effect in this quilt.

The mosaic tile pattern is adapted from *Mosaic Tile Designs* by Susan Johnson (Dover Publications. New York)

Block used: Mosaic Tile

Block size: 30 cm (12 in)

Quilt size: 24 blocks are needed for a 162 cm x 220 cm (64 in x 87 in) quilt

MATERIALS SUPPLIED

Black fabric for the background

INSTRUCTIONS

- The quilt is to be in Amish colours, from the cool side of the colour wheel, that is dark red through to green. Only plain fabrics are to be used to maintain the Amish theme. You will need to select fabric in three colours.

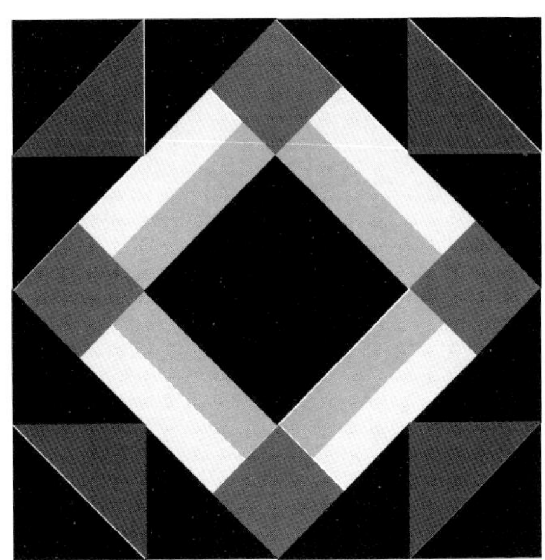

Block Diagram

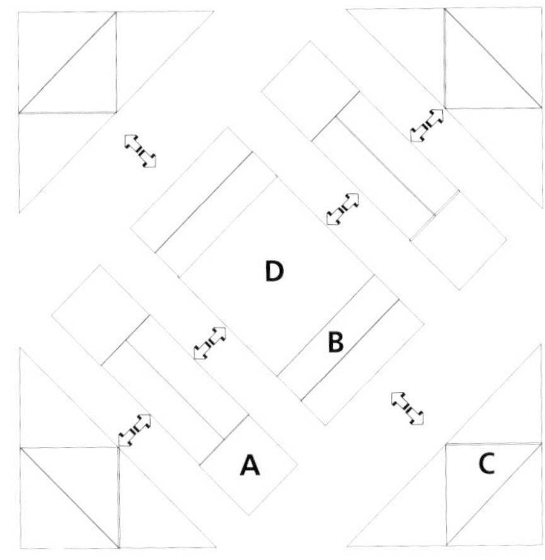

Piecing Diagram

- Prepare templates A, B, C and D. (See page 52 for how to prepare your own templates.)
- Cut the following pieces:

4 A from fabric one
4 B from fabric two
4 B from fabric three
4 C from fabric three
12 C from the black fabric
4 C from fabric two
1 D from the black fabric

- Assemble the block, following the piecing diagram.

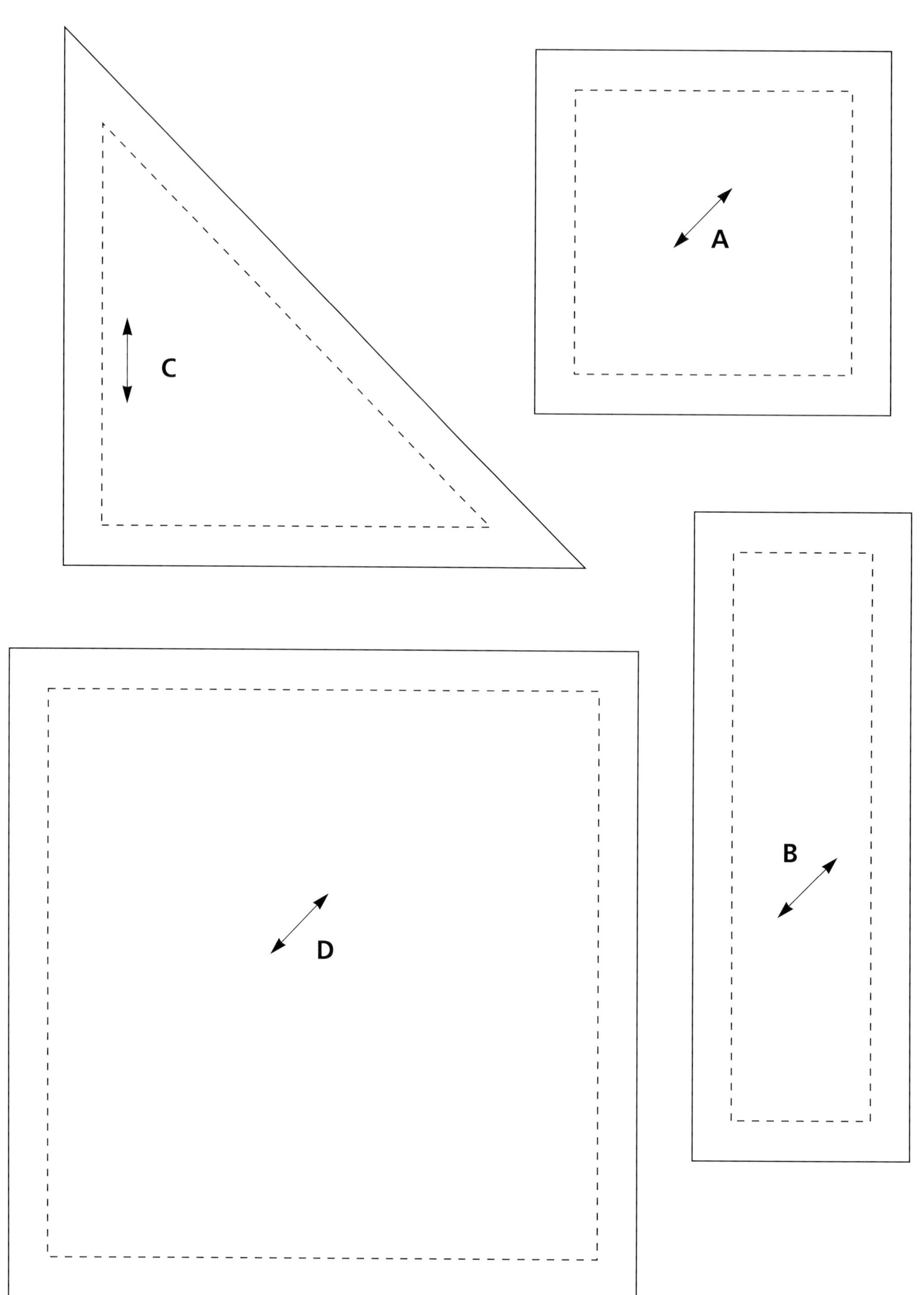
C
A
B
D

AMISH SPARKLE

My block with twenty-eight pieces just fell within the limit of thirty pieces for a friendship block, set by my group. I really love the Amish designs and thought this tile pattern, although not traditionally Amish, would work well using an Amish palette. I couldn't have been more pleased with my blocks and find them a lasting memory of friends in my quilting group at that time.

THESE BLOCKS WERE MADE BY EASTWOOD PATCHWORK QUILTERS, NEW SOUTH WALES, 1993, FOR JULIE WOODS.

JUDY IN ARABIA

Create a twinkling sea of stars with this simple block. A soft palette adds to the appeal of this quilt. Construction of the top is easy too, with blocks sewn directly together.

Block used: Judy in Arabia

Block size: 21 cm (9 in)

Quilt size: 48 blocks are needed for a 165 cm x 215 cm (65 in x 86 in) quilt

MATERIALS SUPPLIED

No special fabrics needed. Please use your own scraps.

INSTRUCTIONS

- You will need four or five fabrics, mostly the aqua and mauve palette and including at least two light, two medium and one dark fabric.

Block Diagram

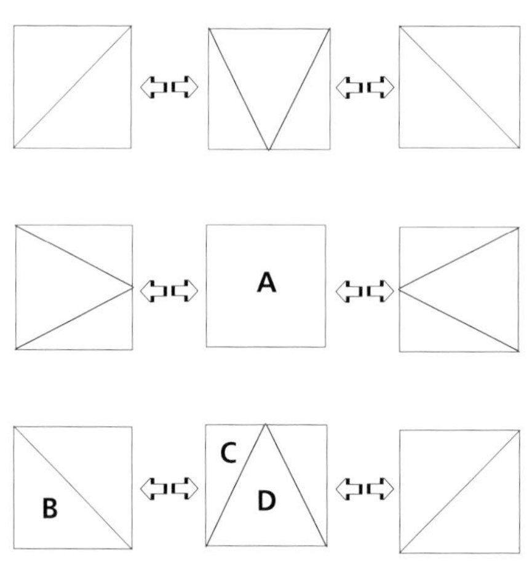

Piecing Diagram

- Prepare templates A, B and C. Cut the following pieces:

1 A
4 B from a light fabric for the corners
4 B from a medium or dark fabric to complete the corner squares
8 C for the star points (make sure they provide a good contrast with the corners)
4 D from medium to dark fabrics for between the star points

- Assemble the block, following the piecing diagram.

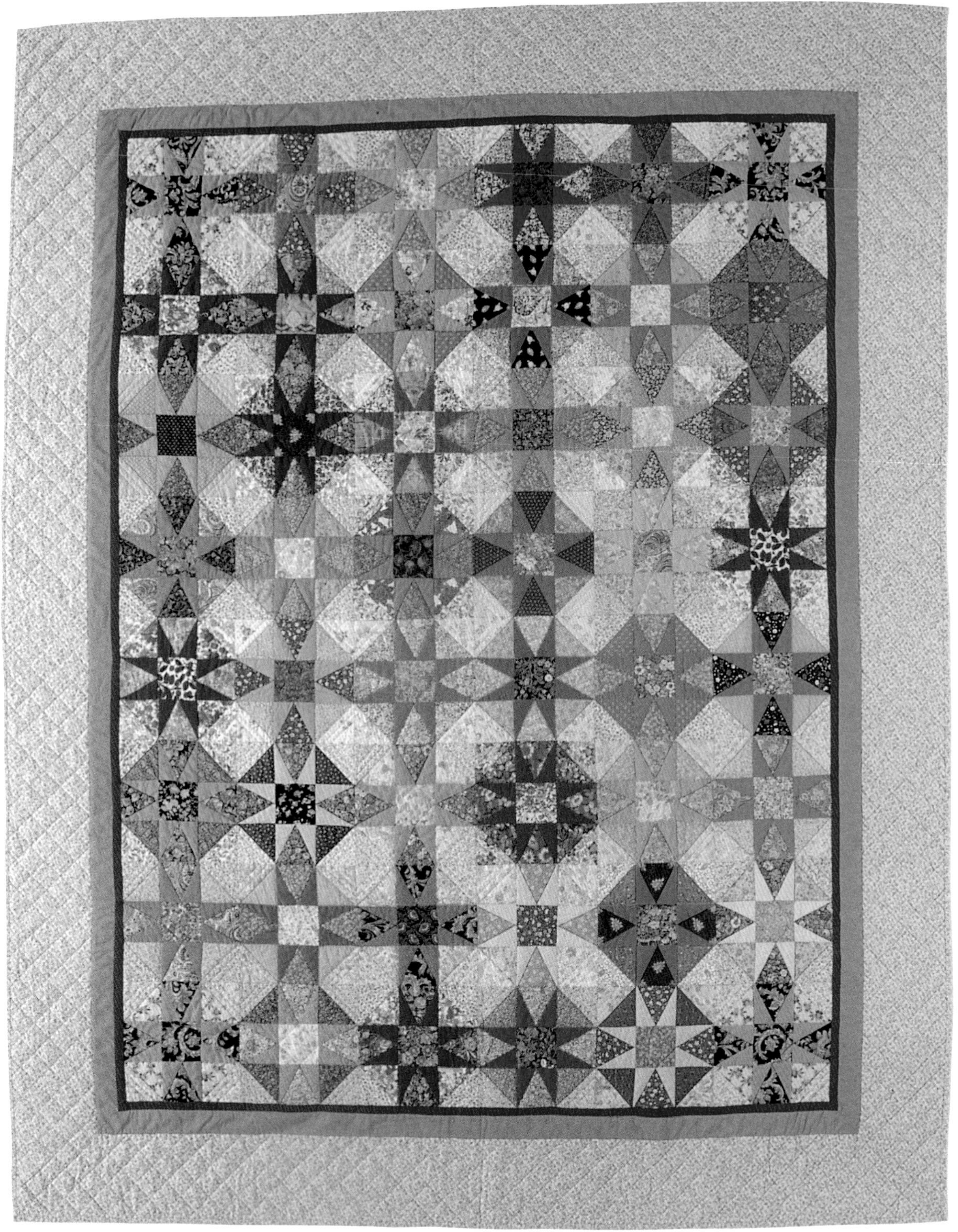

JUDY IN ARABIA

Aqua is my favourite colour. I have decorated my house with an aqua and apricot theme, and especially love the soft aqua walls in my bedroom and my window seat covered in aqua velvet. When I found sample bags of off-cuts of Liberty fabric at the local school fete, I bought the lot – including a wonderful collection of aqua pieces which I decided to include in my friendship quilt. The irregular pieces provided quite a challenge in cutting out, but eventually I had enough blocks prepared to give two to each of my friends at Northbridge Quilters. I still needed to make several blocks myself to complete my wonderful quilt, which has provided the finishing touch to my bedroom.

THESE BLOCKS WERE MADE BY THE NORTHBRIDGE QUILTERS, NEW SOUTH WALES, 1990, FOR SUSAN KELLY.

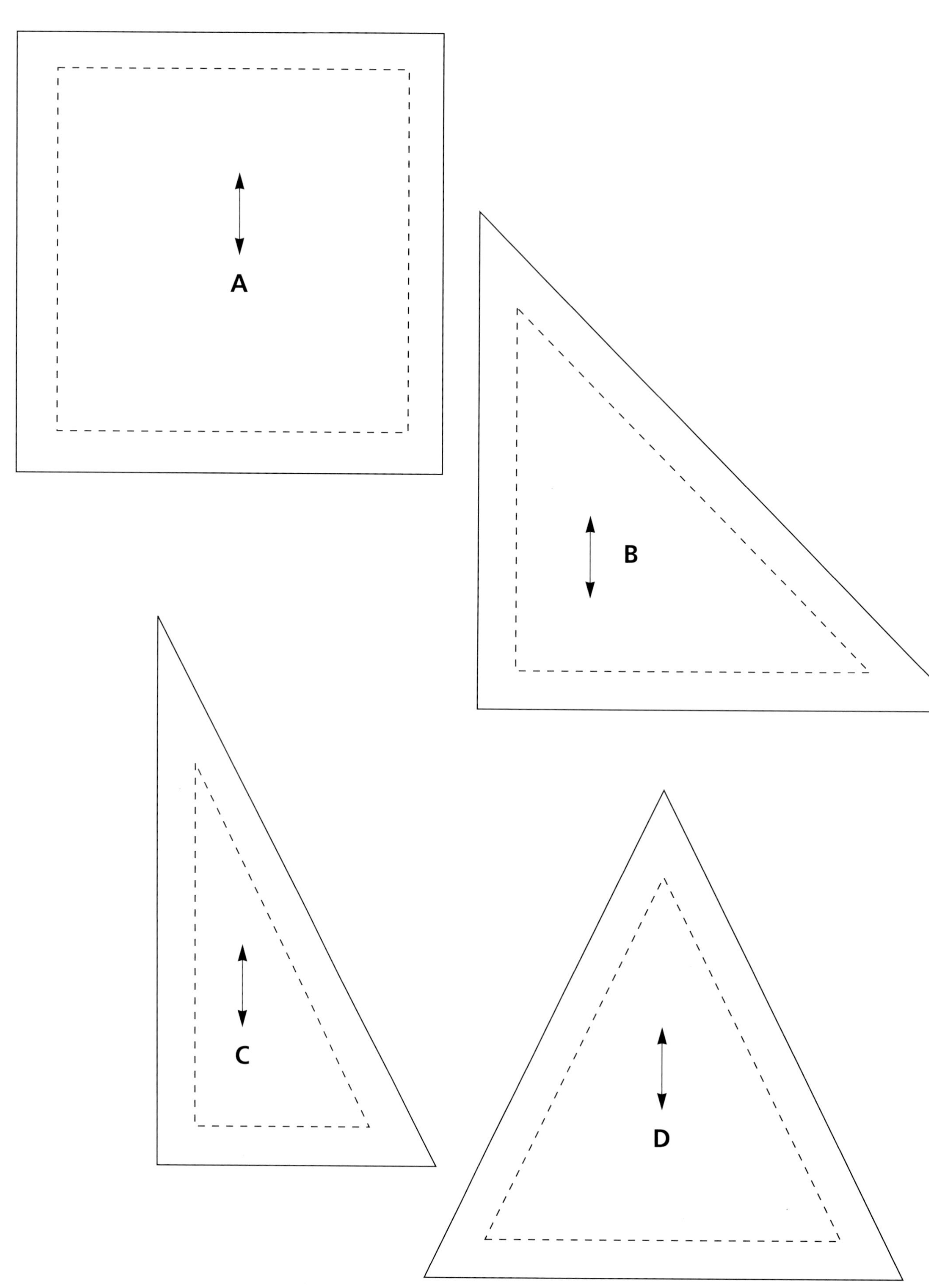
A
B
C
D

MY FRIENDSHIP QUILT

This quilt is made from one version of the Friendship or Album block. The Friendship block retains the characteristic common to all these blocks – the white space in the centre for friends to sign.

Block used: Friendship or Album

Block size: 25 cm (10 in)

Quilt size: 30 blocks are needed for a 150 cm x 180 cm (59 in x 71 in) quilt

MATERIALS SUPPLIED

White fabric for the centre square
Dark fabric for the outer rectangles

INSTRUCTIONS

- Use any fabrics that coordinate with the dark fabric that is supplied. The triangles can be calico or cream print fabrics.

Block Diagram

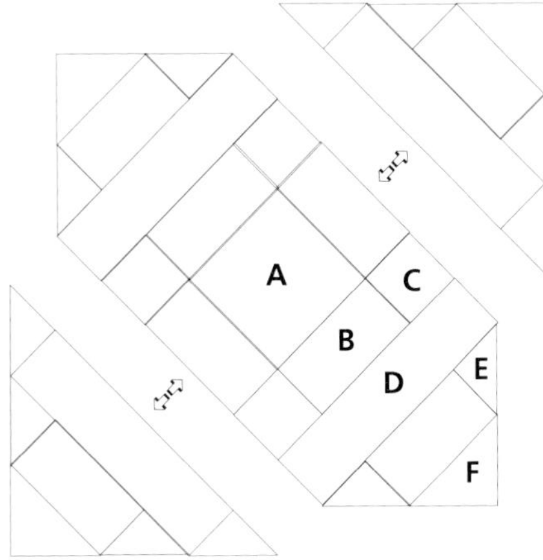

Piecing Diagram

- When you are marking the fabric, take care to follow the grain.
- Cut the following pieces:

1 A from the white fabric provided
4 B from the dark fabric provided
4 C from a light fabric provided
4 D from a medium value fabric
12 E from a cream fabric
4 F from a cream fabric.

- Feel free to embroider a motif, as well as your name, in the white central square.
- Assemble the block, following the piecing diagram.

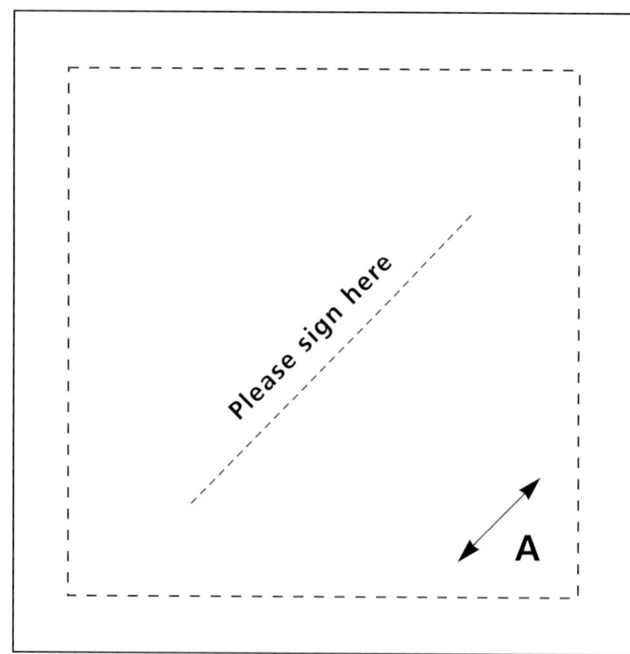
Please sign here
A

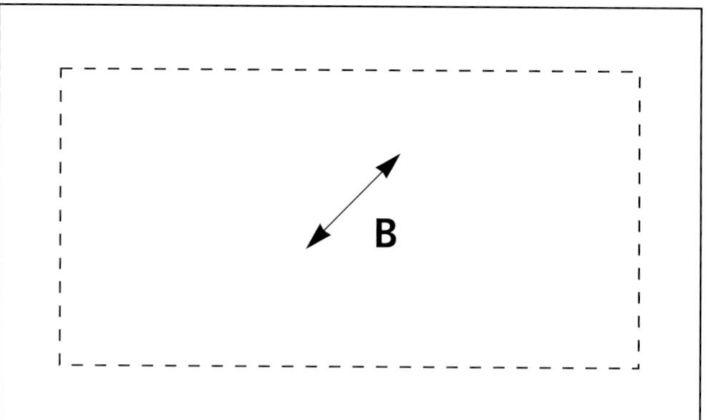
B

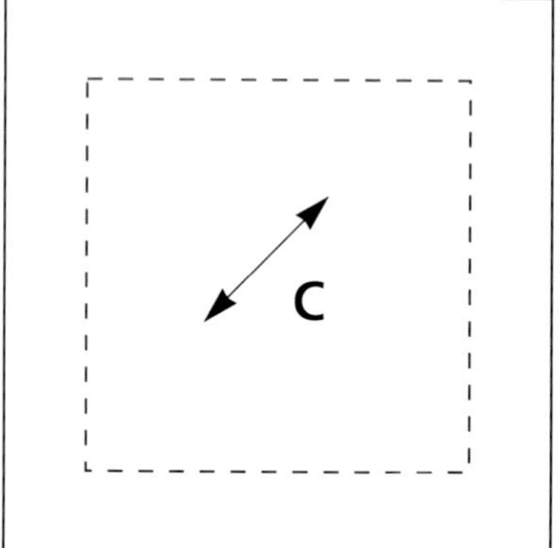
C

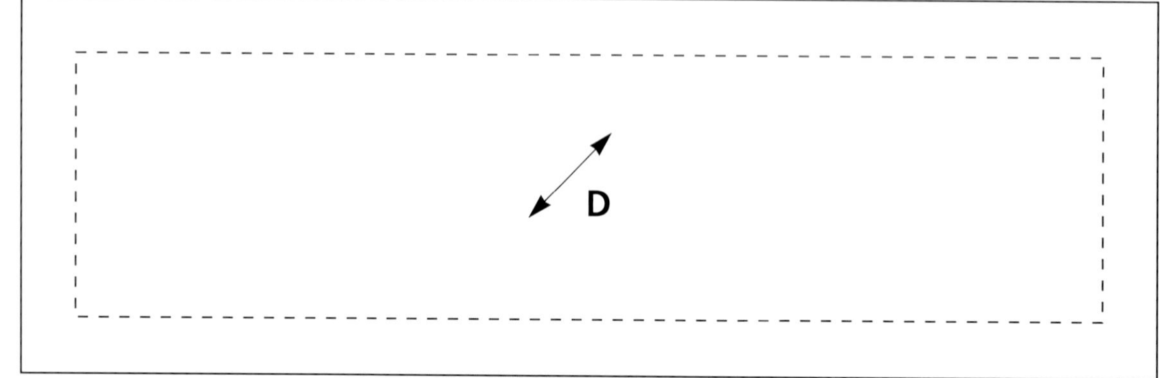
D

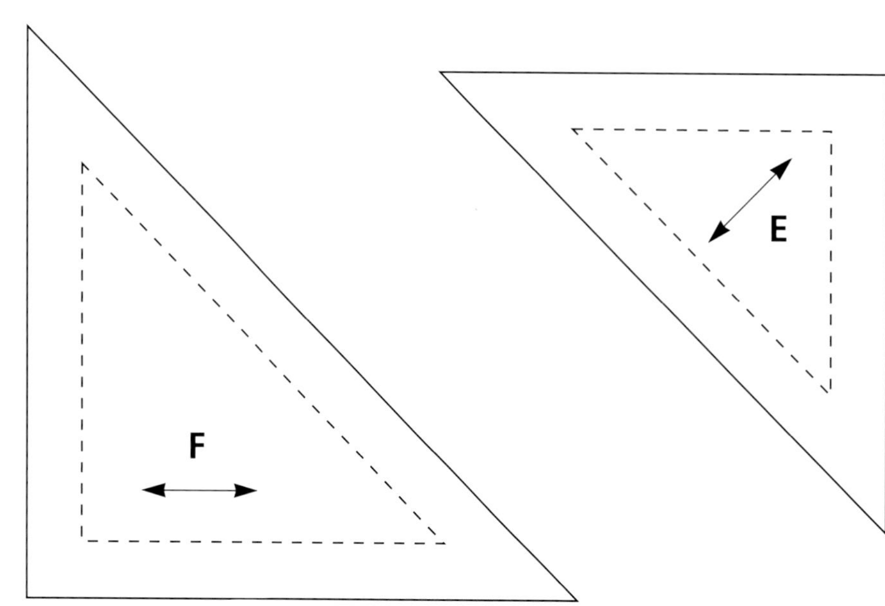
F
E

MY FRIENDSHIP QUILT

Because I really wanted to re-create the look of the old friendship quilts, I was keen to employ one of the patterns traditionally used to record names and messages by friends and family. When the blocks were finished, complete with names and other interesting motifs, I counted ninety-three different fabrics!

THESE BLOCKS WERE MADE BY FRIENDS IN LAUNCESTON, TASMANIA, 1993, FOR LILLIAN ATKINSON.

CHARMING STARS

Every star is different in this delightful scrap quilt, each reflecting the personality of the maker. The stars are signed, using pen or thread, with an occasional signature done by machine.

Block used: Five-pointed Star

Block size: small blocks, 16 cm x 28 cm (6½ in x 11 in), large blocks, 22 cm x 28 cm (8⅝ in x 11 in)

Quilt size: 49 blocks are needed for a 172 cm x 220 cm (68 in x 87 in) quilt

MATERIALS SUPPLIED

Select all the fabrics from your own collection.

Block Diagram

Block Diagram

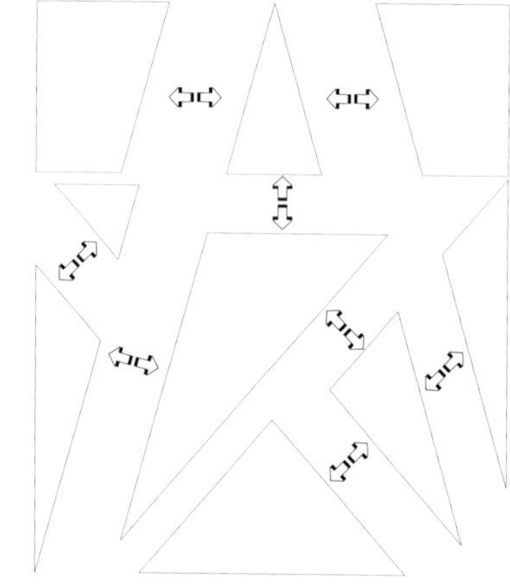

Piecing Diagram

INSTRUCTIONS

- You may choose to make a small or a large star block.
- Use the templates provided or draw up your own full-sized block if you wish.
- When marking the back of the fabric, make sure you reverse the irregular-shaped templates A, C, D, E, G, I, L, K, M, O.
- Use fabrics from your own collection, so the block reflects your particular style of quiltmaking.
- Assemble the block, following the piecing diagram.
- Sign your block.

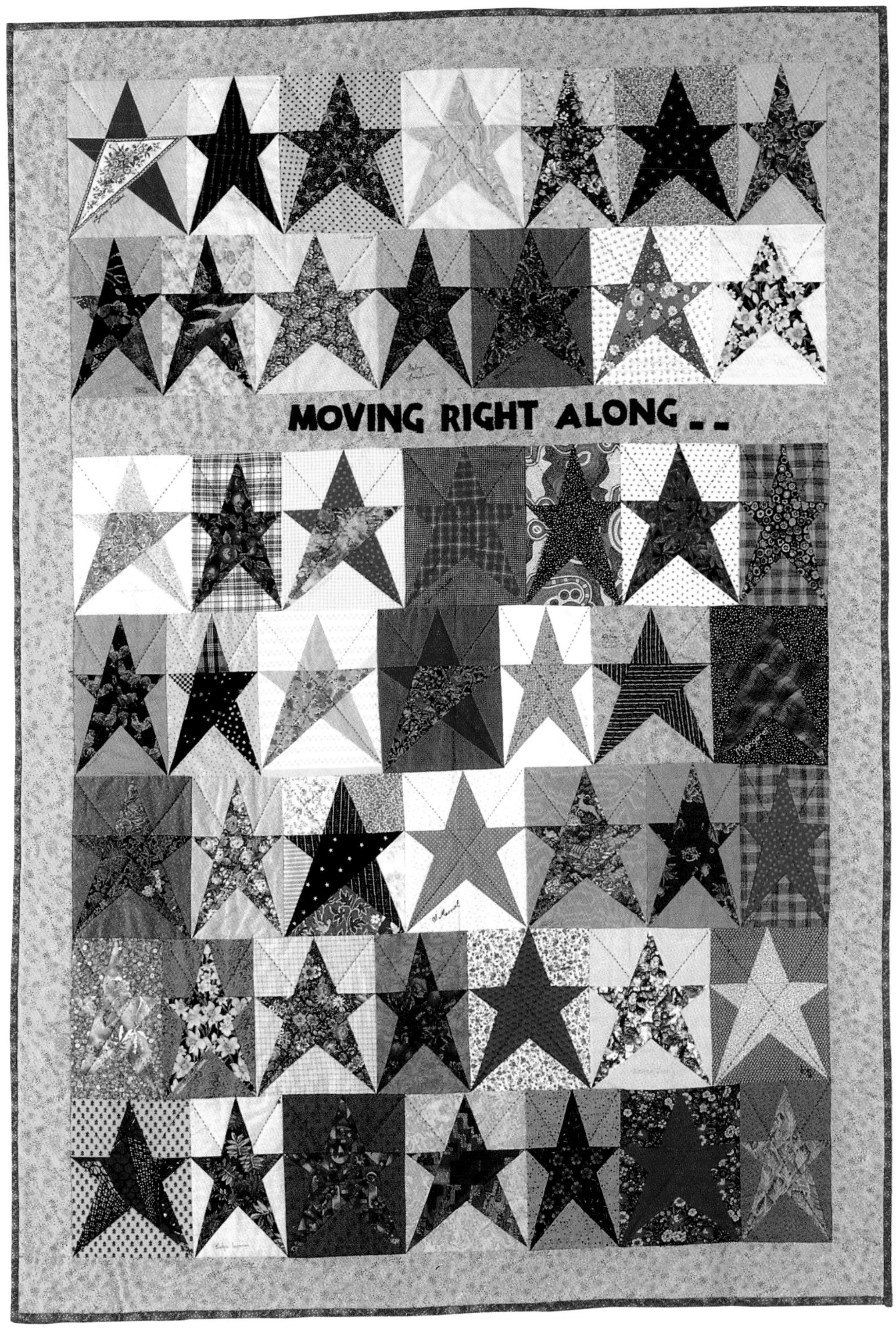

CHARMING STARS

During my presidency, the tenth anniversary Quilt Show of the Quilters' Guild was held in Sydney in 1992. It was called 'The Decade of Stars' and featured not only members' work, but a retrospective of ten years of Quilt Show favourites. The theme category that year was 'Stars'. The following year's Quilt Show was 'The Charm of Quilts' with the theme category being 'Charm and Scrap Quilts'. Both of these themes are in my friendship quilt, a wonderful reminder of the people I worked with during that time. 'Moving right along' was apparently a favourite saying of mine during committee meetings, but I don't ever remember saying it!

THESE BLOCKS WERE MADE BY THE COMMITTEE AND FRIENDS OF THE QUILTERS' GUILD OF AUSTRALIA FOR KAREN FAIL ON HER RETIREMENT AS PRESIDENT OF THE GUILD 1992-93.

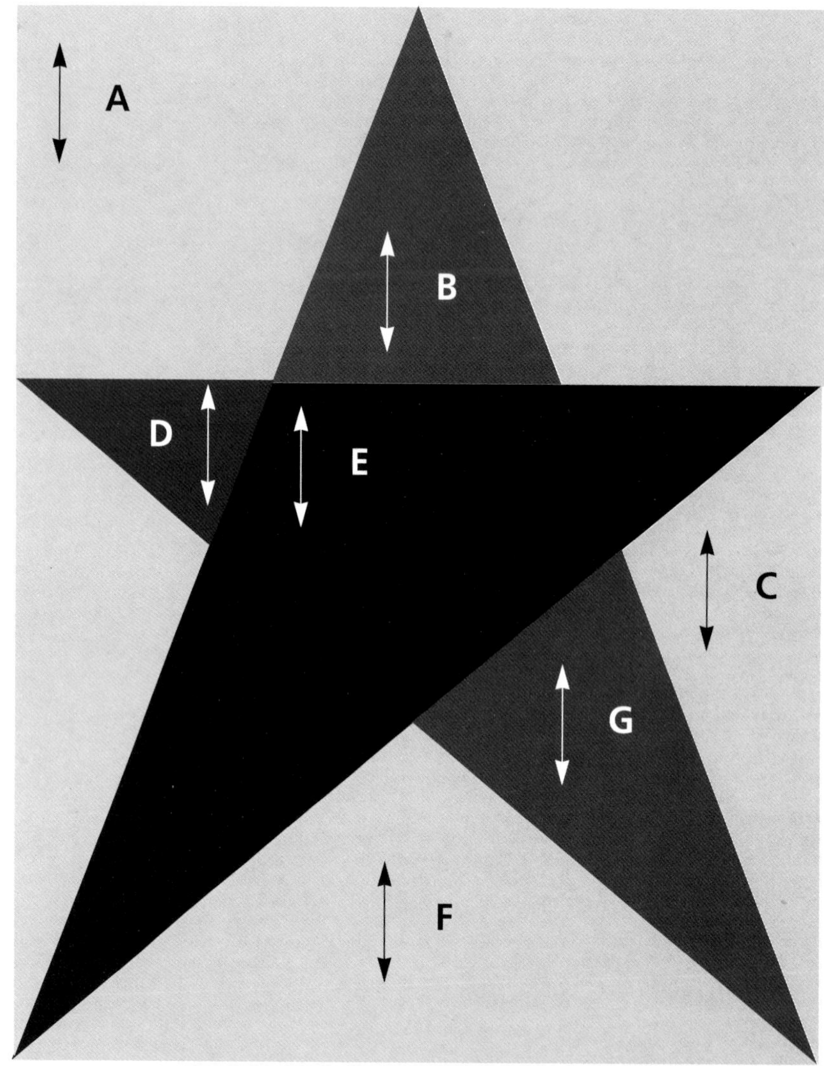

Templates and Piecing Diagram

Note: Templates are given at 50% of actual size. Enlarge to 100%

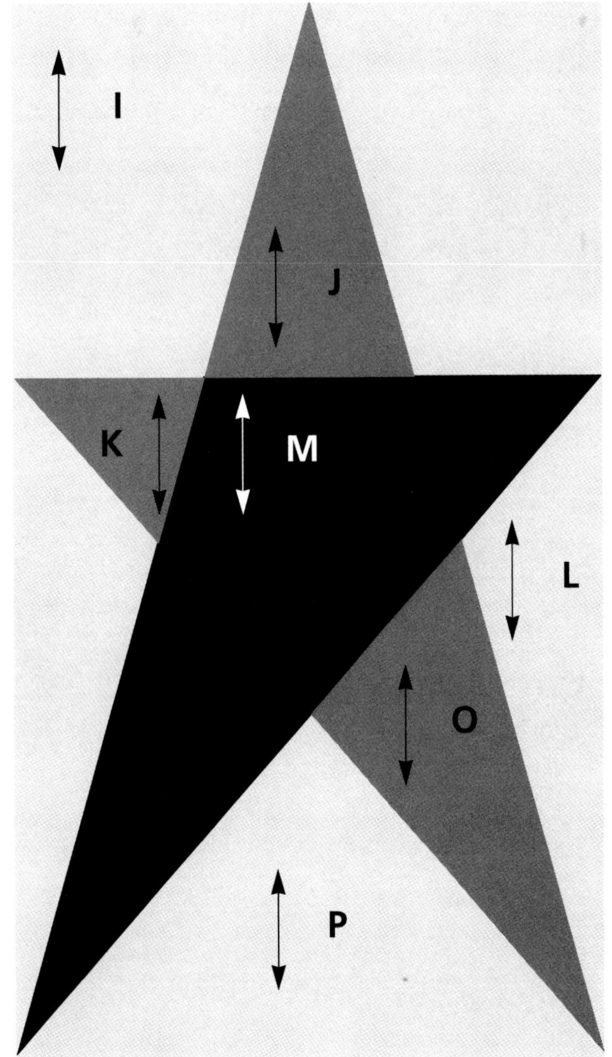

Templates and Piecing Diagram

Note: Templates are given at 50% of actual size. Enlarge to 100%

BOLD AND BEAUTIFUL

A light, bright small quilt, using the same central fabric in each block. No fabrics with white or cream backgrounds are used.

Blocks used: Sawtooth Star and Puss in the Corner

Block size: 16 cm (6 in) (These block sizes are not simple conversions of one another. For ease of drafting, choose to work in either imperial or metric and choose the appropriate block size for your choice. Metric templates are provided.)

Quilt size: 13 Sawtooth Star and 12 Puss in the Corner blocks are needed for a 100 cm (37½ in) square quilt

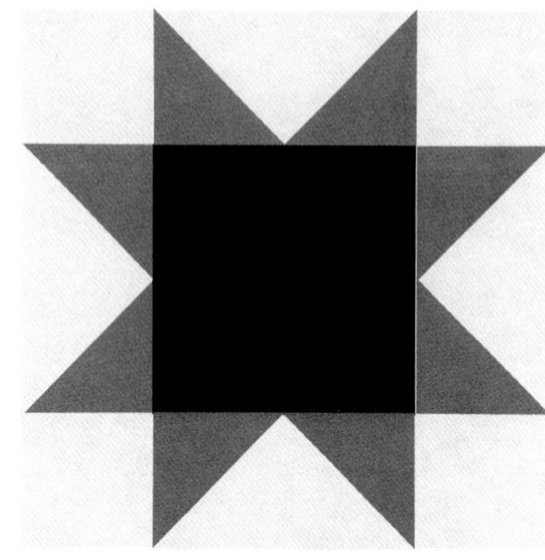

Block Diagram - Sawtooth Star

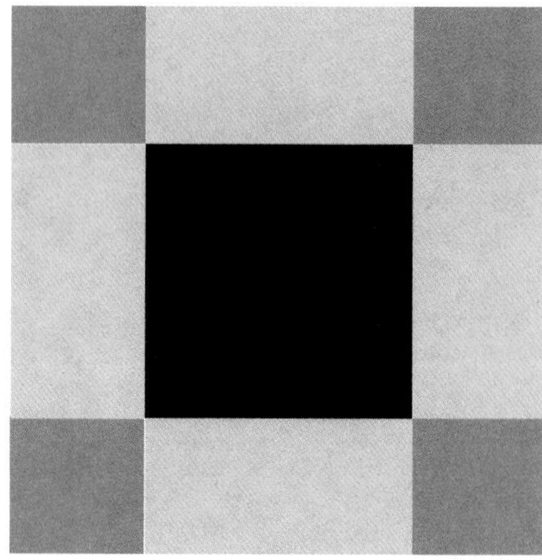

Block Diagram - Puss in the Corner

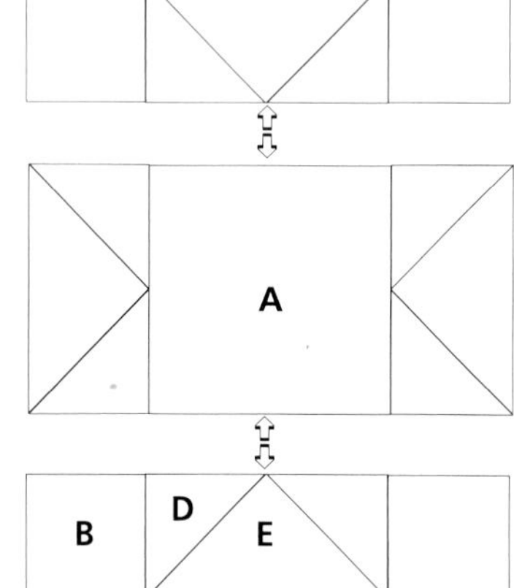

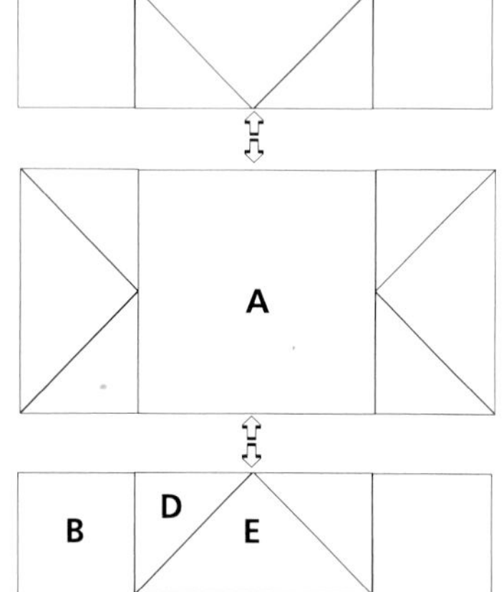

Piecing Diagram

A

B

C

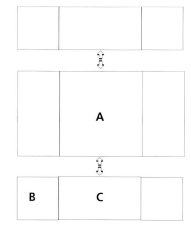

Piecing Diagram

MATERIALS SUPPLIED

Multicoloured fabric for two centre squares

Templates A, B, C, D, and E

INSTRUCTIONS

- Make one Sawtooth Star and one Puss in the Corner block.
- For the Sawtooth Star block, cut the following:

1 A from the fabric provided
4 B from a medium value fabric
4 E from a medium value fabric
8 D from a dark fabric.

- For the Puss in the Corner block, cut the following:

1 A from the fabric provided
4 B from a dark fabric
4 C from a fabric of your choice

- No white or cream background fabrics are to be used.
- Use medium and large print fabrics with abandon.
- Strong plains work well also.
- Assemble the blocks, following the piecing diagram.

BOLD AND BEAUTIFUL

I struggle to use blue in my quiltmaking, so I thought I could get my friends to help. The fabric I chose for the centre square had quite a lot of blue in it (according to me), but my friends saw mainly the many warm hues. In a last-ditch attempt to have a blue quilt, I had to add the cornflower blue border myself!]

THESE BLOCKS WERE MADE BY EASTWOOD PATCHWORK QUILTERS, NEW SOUTH WALES, 1992, FOR KAREN FAIL.

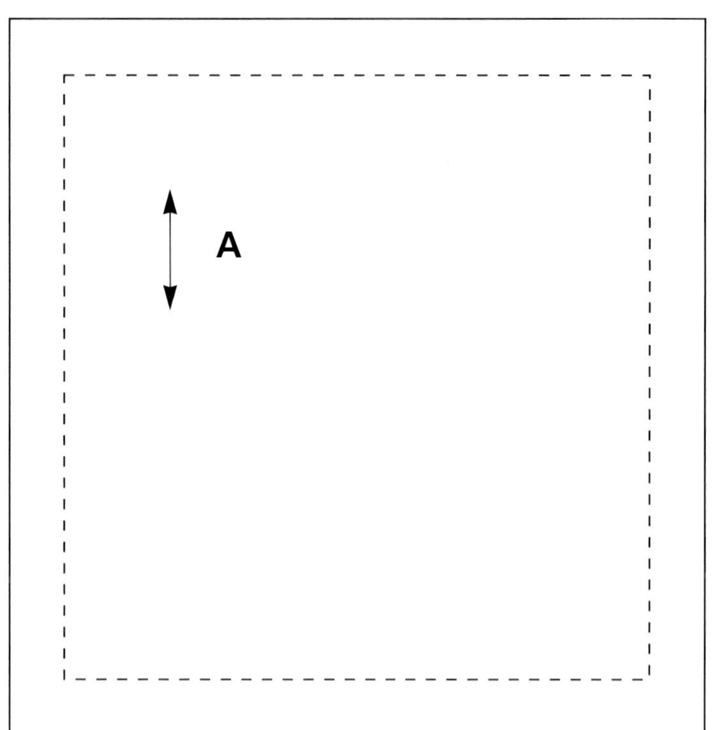
A

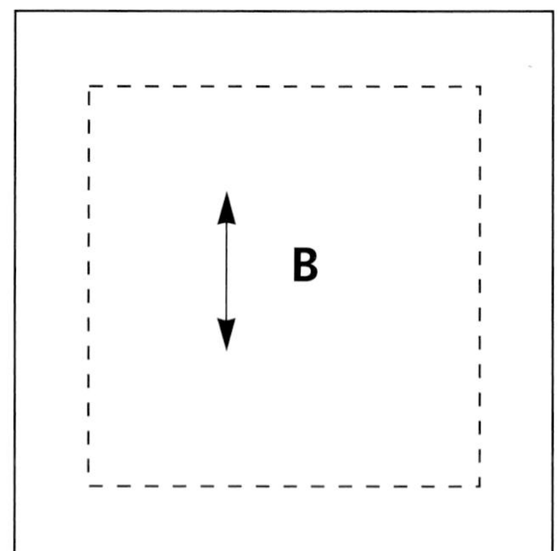
B

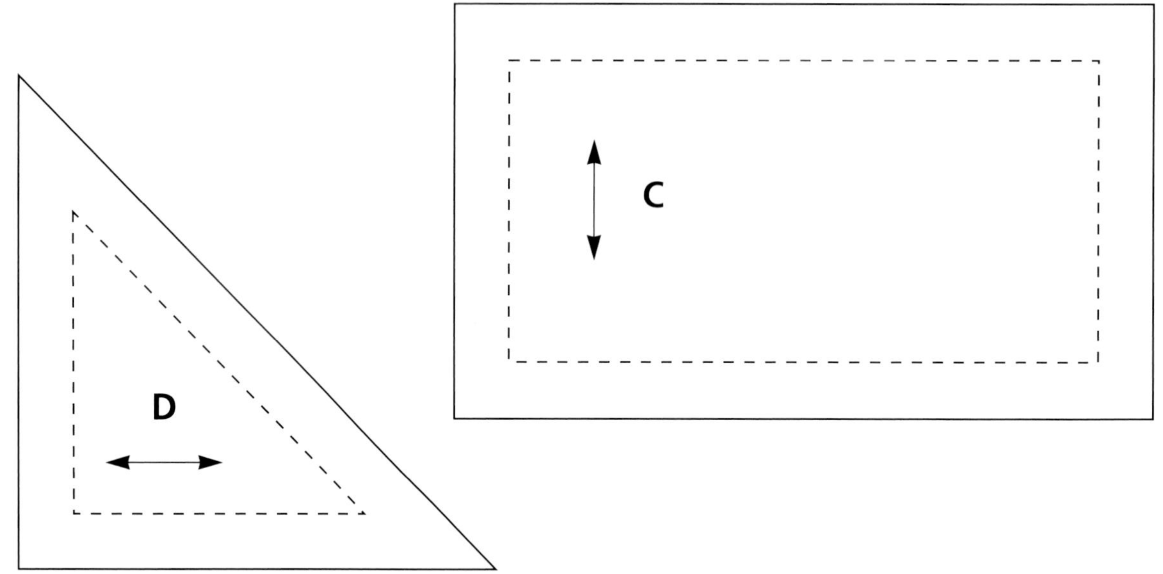
C
D

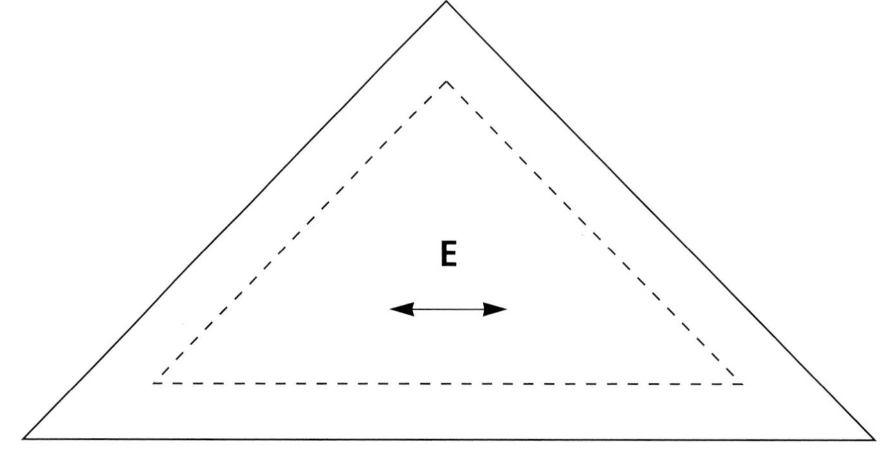
E

JEWEL OF THE ORIENT

Two simple blocks and a striking colour scheme create a stunning quilt that glows like a Persian carpet.

Block used: Original design, adapted from 'New England Sojourn' *Quiltmaker* Spring/Summer 1985

Block size: 24 cm (9 in)

Quilt size: 31 Block 1, 20 red Block 2 and 12 blue Block 2 are needed for a 183 cm x 230 cm (72 in x 90½ in) quilt.

MATERIALS SUPPLIED

Four fabrics are provided so that the quilt will resemble an oriental carpet: turquoise fabric, red or blue fabric, a small khaki print and a large khaki print

INSTRUCTIONS

- Please make one Block 1 and two Block 2.
- For Block 1, cut the following pieces:

1 C from the turquoise fabric
4 D from the red or blue fabric
4 E from the the large khaki print

- For Block 2, cut the following pieces:

4 A from the red or blue fabric
4 B from the small khaki print fabric

- Each of the blocks requires simple hand-piecing. Refer to the piecing diagrams.
- Use a grey thread to blend in with all the colours.

Block Diagram

Block Diagram

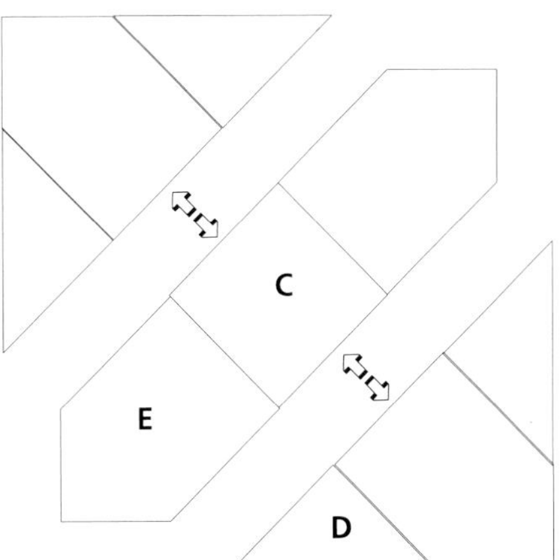

Piecing Diagram

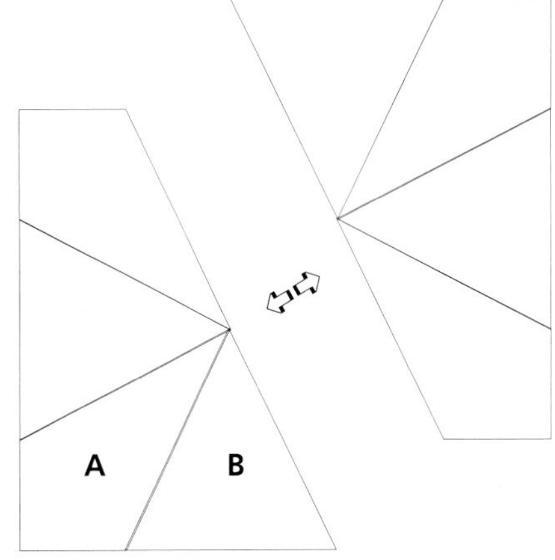

Piecing Diagram

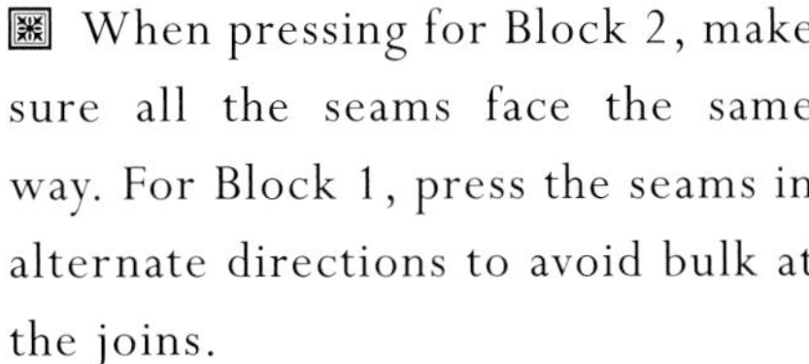

- When pressing for Block 2, make sure all the seams face the same way. For Block 1, press the seams in alternate directions to avoid bulk at the joins.
- Embroider your name or add a little 'something' especially from you in the turquoise centre.

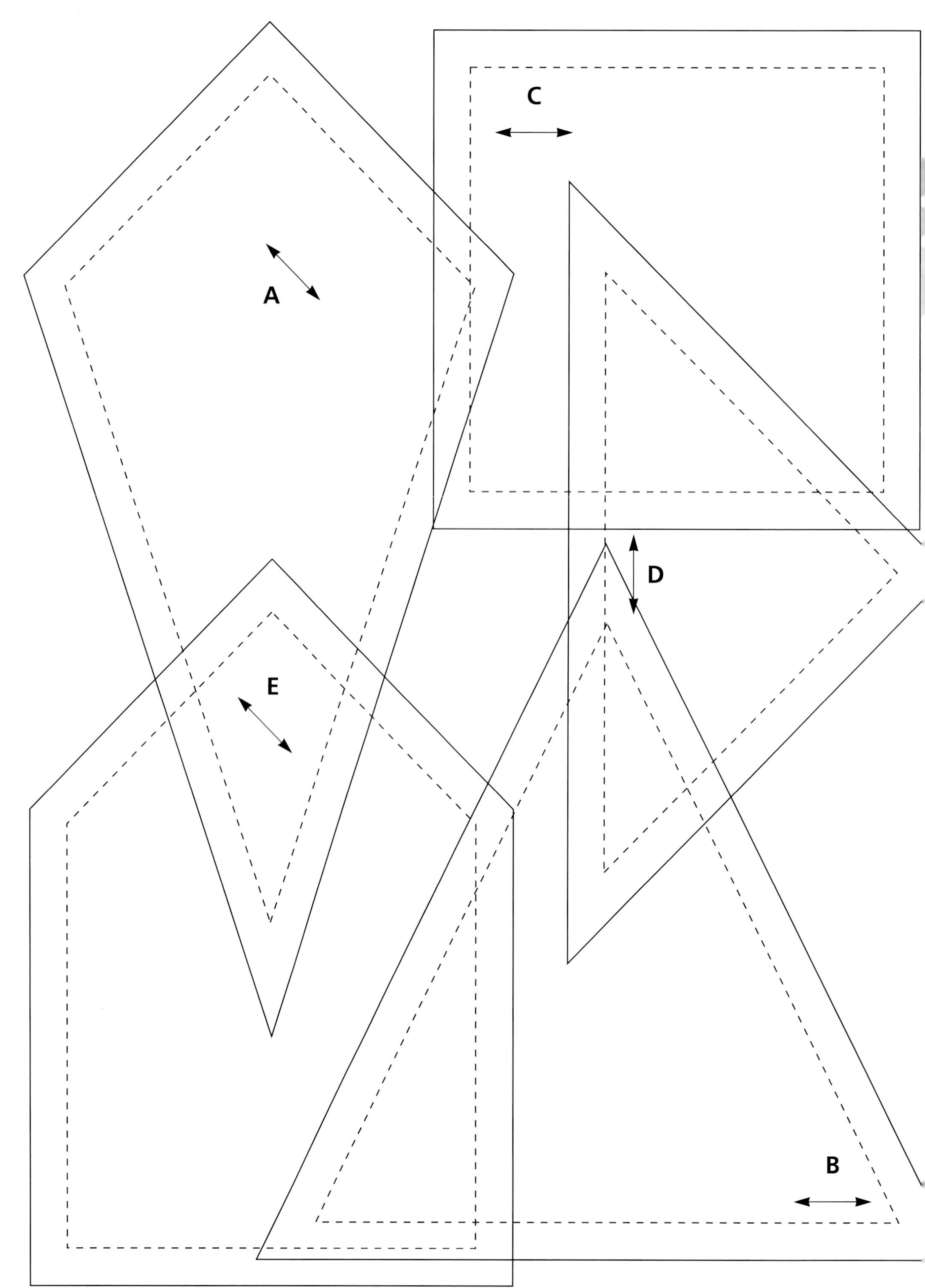
C
A
D
E
B

JEWEL OF THE ORIENT

I really love traditional quilts and enjoy experimenting with colour within a traditional framework. I also like simplicity of design. This quilt met all my criteria. With so many willing helpers, the quilt blocks were finished in no time, even though I thought the bright turquoise would mean I would have to supply sunglasses along with the instructions. I am delighted with the finished quilt, which reminds me of a Persian carpet with its glowing colours. The names of my friends in the turquoise squares add the finishing touch.

THESE BLOCKS WERE MADE BY CASTLE HILL FRIENDSHIP QUILTERS, NEW SOUTH WALES, 1990, FOR LYN SHAYLER.

Friendship quilts are always treasured as they are such wonderful reminders of friends and shared moments in our busy lives. In the light of this, I want to thank sincerely the many quilters all over Australia who sent their special quilts to me to be photographed and shared their stories with me. I am very thankful for their trust and their enthusiasm for this book.

As always, the idea of friendship quilts is contagious. At J.B. Fairfax Press the staff has established a friendship group with most of the office involved in making blocks for one another. Two off-site employees (both wives of staff members) participate enthusiastically and one ex-employee waits patiently for her next instructions by mail. It is so exciting to see the latest block arrive at the office. It is also exciting to watch confidence blossom in sewers and non-sewers alike as they learn the intricacies of patchwork and discover the joy of working with fabric. Their enthusiasm has been a wonderful affirmation of all that this book seeks to communicate.

Karen Fail

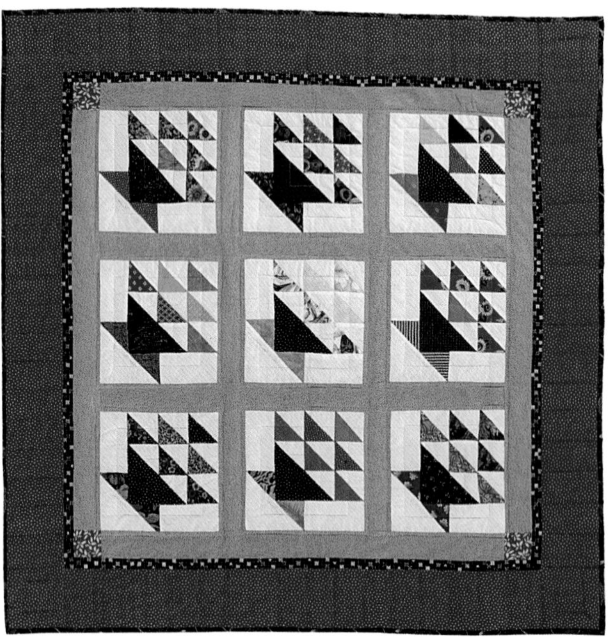

BASKETS OF FRIENDS, *150 cm square (58½ in square), 1995. The blocks for this quilt were made by the staff of J. B. Fairfax Press Limited and were won by Judy Poulos in a draw.*

Published by Creative House
An imprint of Sally Milner Publishing Pty Ltd
PO Box 2104
BOWRAL NSW 2576
AUSTRALIA

National Library of Australia Cataloguing-in-Publication data:

Fail, Karen.
Creative friendship quilts.

ISBN 1 877080 09 8.

1. Friendship quilts - Australia. 2. Quilting - Australia. 3. Patchwork. I. Fail, Karen. Between friends :quilts to share. II. Title.

746.460994

Printed in Hong Kong

EDITORIAL
Managing Editor: Judy Poulos
Photography: Andrew Payne
Illustrations: Nicki Rein
DESIGN AND PRODUCTION
Manager: Anna Maguire

Inside Front and Inside Back Cover: **WASHING DAY**, *164 cm x 204 cm (64 in x 79 ½ in), 1994. Made by the McLaren Vale Quilters for Pam Whaite.*